H50 117 428 5

AF689993

Hertfordshire

7/12

Please renew or return items by the date shown on your receipt

www.hertfordshire.gov.uk/libraries

Renewals and enquiries: 0300 123 4049

Textphone for hearing or speech impaired users: 0300 123 4041

L32b/05.17

Hertfordshire

PERCIVAL LOWELL

Percival Lowell

The Culture and Science of a Boston Brahmin

David Strauss

HARVARD UNIVERSITY PRESS
Cambridge, Massachusetts
London, England · 2001

Copyright © 2001 by the President and Fellows of Harvard College
All rights reserved
Printed in the United States of America

Frontispiece: Lowell's favorite photographic portrait (Lowell Observatory photograph)

Library of Congress Cataloging-in-Publication Data

Strauss, David, 1937–
Percival Lowell : the culture and science of a Boston Brahmin / David Strauss.
p. cm.
Includes bibliographical references and index.
ISBN 0-674-00291-1
1. Lowell, Percival, 1855–1916. 2. Astronomers—United States—Biography. I. Title.
QB36.L849 S77 2001
520′.92—dc21
[B] 00-058080

COMMUNITY INFORMATION: LIBRARIES
H50 117 428 5
BC 03/01
£30-95

For Dhera

Contents

◆ ◆ ◆

Illustrations

Acknowledgments

◆ ◆ ◆

One of the pleasures of researching Percival Lowell has been the opportunity to track him to some of his favorite places: Beacon Hill in Boston and Mount Auburn Cemetery in Cambridge, Oak Creek Canyon and Mars Hill in Arizona, Mount Ontake and the Noto Peninsula in Japan. In the course of these travels, Priscilla Macmillan and Grier and Hélène Monroe in Boston; William Lowell Putnam, Robert Millis, and Mary Lou Evans in Flagstaff; and Onishi Naoki and Uchida Katsuichi in Tokyo have been gracious hosts.

It is a pleasure to acknowledge generous archival support from the following institutions: Aichi Normal University, Arnold Arboretum, Boston Athenaeum, Boston Public Library, Dedham Historical Society, Isabella Stewart Gardner Museum, Harvard University Archives, Houghton Library at Harvard University, International Christian University (Tokyo), the Library of Congress, Lowell Observatory Archives, Massachusetts Historical Society, Mary Lea Shane Archives of the Lick Observatory, Massachusetts Institute of Technology, Peabody Museum of Salem, Regenstein Library of the University of Chicago, University of Arizona, Waseda University, Yale University, Yerkes Observatory, and Yokohama Archives of History.

The editors of *Annals of Science, Journal for the History of Astronomy,* and *Journal of American–East Asian Relations* have kindly permitted me to use material from articles that I first published in their journals.

This book would not have seen the light of day without the assistance of the Lowell family in locating materials on their illustrious ancestor; I thank especially Harriet Belin, Michael Putnam, William Lowell Putnam, Lawrence L. Reeve, A. Lawrence Rotch, and Anna Lowell Tomlinson for sharing family correspondence with me.

I also wish to acknowledge support of the following Japanese colleagues who have shared their research with me or helped in locating relevant materials: Fujita Fumiko, Ito Hisako, Miyazaki Masaaki, Onishi Naoki, Sakashita Tamaki, Yamaguchi Seiichi, and Yokoo Hiromitsu. (Here and in the text, I follow the Japanese custom of putting the family name before the given name.)

At the Lowell Observatory, four individuals have been particularly helpful in guiding me to the appropriate Lowell documents: Antoinette Beiser, Mary Lou Evans, Henry Giclas, and Martin Hecht. I am also indebted to Dorothy Schaumberg, curator of the Mary Lea Shane Archives of the Lick Observatory.

Though not on Lowell's itinerary, Kalamazoo College has been the site of most of the work on this project. I have been generously supported by two sabbatical leaves, several grants from the Faculty Development Committee, and the tireless response of the library staff to an unending series of interlibrary loan requests. A very timely faculty seminar on Charles Darwin, organized by Paul Sotherland and supported by the Center for Western European Studies, shed interesting light on the intellectual climate in which Lowell operated. In addition, my colleagues Rose Bundy, Franklin Presler, Robert Stauffer, and Paul Sotherland have offered useful advice for revising sections of the manuscript.

Two National Endowment for the Humanities summer seminars at Yale University, directed respectively by Alan Trachtenberg and Martin J. Klein, provided new perspectives for this project, while a grant from the National Science Foundation supported the completion of the manuscript.

Scholars in both the history of science and Japanese-American relations have generously read appropriate sections of the manuscript and provided me with helpful references. For their assistance, I wish to thank Richard Baum, Stephen Brush, David DeVorkin, Steven J. Dick, Clark A. Elliott, Robert Ellwood, Norriss Hetherington, Akira Iriye, Jordan D. Marché II, Albert Moyer, Ronald Numbers, Donald Osterbrock, Howard Plotkin, Robert Rosenstone, Marc Rothenberg, Michael Sokal, and Thomas Williams.

At Harvard University Press, thanks to Michael Fisher, Donna Bouvier, and Julie Hagen, whose careful reading of the manuscript much improved it.

Michael Crowe took me in hand early in this project and provided invaluable bibliographic suggestions. His hosting of the Notre Dame History of Astronomy workshops has generated a supportive environment for all scholars in the field.

Three colleagues, William Sheehan, Robert Smith, and George E. Webb, have read large sections of the manuscript and offered useful advice for revising it, while two others have suffered through endless drafts of the entire manuscript with only an occasional murmur. My Kalamazoo history colleague David Barclay, with his keen sense of narrative, has devoted himself to improving the flow of the book. John Lankford has provided guidance from the earliest stages of this project. His perceptive reading of the manuscript and enthusiasm for the project have energized me from the outset.

Dhera Strauss has been a delightful companion in tracking Lowell to his—and now our—favorite haunts. She has provided a patient sounding board for my endless speculations about him and has shared my enthusiasm for the subject. In completing this project, we both lose an excellent travel guide, whose adventurous spirit has enriched our lives over the past few years. Lowell will be a difficult act to follow.

PERCIVAL LOWELL

Introduction: Percival Lowell and the Boston Brahmins

It is not difficult to conjure a picture of Percival Lowell at the pinnacle of his life in the first decade of the twentieth century. Accounts of his family and friends and surviving photographs depict a man of medium height, slim and handsome, with an athletic build and an intense expression. His erect bearing and fastidious dress contributed to a commanding presence that was universally noted. This air of authority derived as well from a powerful impulse to make and execute decisions on important matters with great—sometimes too great—dispatch.[1]

In his "exhaustless energy" Lowell resembled President Theodore Roosevelt, whose first cousin had married Lowell's sister. Like the president, Lowell was an exceptionally versatile individual whose talents led him into a variety of fields, each of which he pursued with great vigor. One friend remarked that Lowell "grew so enthusiastic and demonstrative on the subject of Japanese wrestling that I was almost apprehensive . . . he certainly does delight in his canals of Mars . . . He has the very fountain of health and strength in a great and hopeful work." Furthermore, his ambition and his vivid imagination drove him to undertake challenging projects on a grand scale, including his famous searches for Planet X and for evidence of life on Mars. His energy and enthusiasm excited the public and ensured the devotion of his own assistants. And, again like Roosevelt, he performed the various aspects of his public role with great gusto, whether campaigning with journalists for his view of Mars or delivering captivating addresses to large audiences.[2]

However, his professional career as an astronomer was capped with disappointments. His claims for the existence of life on Mars were rejected by most professional astronomers, and he was unable to find Planet X.

For a confident, sometimes arrogant individual, these failures were clearly painful.[3]

Public disappointments no doubt help to explain the unhappiness that, by his own admission, characterized his private life. Lowell was known by friends and family alike for his affectionate nature, displayed in his fondness both for them and for animals. Despite his peripatetic lifestyle, he found time for relationships with family and friends to whom he remained loyal throughout his adult life. Their memoirs and letters, in turn, reflect admiration for his accomplishments and genuine affection for the man. All the same, Lowell was obsessively concerned with personal autonomy, acutely sensitive to the repressive features of upper-class Boston culture, and highly self-conscious. His continuing search for a distinctive identity even drove him to embrace the egotism and expressive posturings of Oscar Wilde. As a result, Lowell found it difficult to develop an emotionally satisfying private life that might have brought him the enduring happiness he sought. Because of his high-strung nature, frenetic work pace, and constant travel, he was also subject to nervous disorders, which on one occasion incapacitated him for almost four years. These "instabilities of mood and of nerve" complicated his life at important junctures and contributed to his feelings of unhappiness. Like Roosevelt, however, he accomplished a great deal during a relatively short life, which was terminated, also as in Roosevelt's case, at the age of sixty-one.[4]

Of course, many readers will already have met Percival Lowell through recent newspaper accounts of space flights to Mars or of the possible existence of microorganismic life. Such accounts invariably pay tribute to Lowell's earlier quest for evidence of intelligent life on Mars, as do a spate of recent books by academicians. And those readers who do not know Lowell the astronomer may have met Lowell the interpreter of Japanese culture, whom the writer Lafcadio Hearn regarded as the most brilliant guide to understanding that country. Still other readers have encountered Lowell the Brahmin socialite, who served as best man to Edith Wharton's husband, Teddy, and as the lively interlocutor of Bernard Berenson, William James, and Henry Adams. In each case, however, the reader glimpses only a segment of Lowell's life. And so Percival Lowell is known most often as a Brahmin, an observer of Japan, or an explorer of Mars, but rarely in all three capacities.[5]

In fact, this segmentation is an inevitable result of efforts by a highly specialized academic community to deal with Lowell's polymath and universalist career. His ability to cross the rapidly growing divide between the two cultures at the end of the nineteenth century has created problems for recent scholars. To combat this segmentation of Lowell's life and experience, I have attempted to put Lowell back together so that we can see him whole again. To do this requires an effort to restore the upper-class Boston context of which he was a part, for as we shall see, Lowell and many of his friends dedicated themselves in significant ways to addressing problems that beset Bostonians of their class as they came of age. Thus, to understand the values Lowell espoused, the career decisions he made, and the nature of his research, we need to clarify the cultural outlook he acquired during his upbringing.[6]

The Brahmins were upper-class Bostonians who had turned an already secure financial situation, based on commercial success, into substantial wealth as a result of creating the American cotton industry in the 1820s. Reinvestment of profits from the cotton mills over two generations provided these magnates with secure fortunes. In addition, intermarriage among the descendants of the founders of the industry concentrated this wealth. Percival Lowell, as the scion of two great cotton dynasties, the Lawrences and Lowells, was, in this respect, a representative Brahmin.

The social as well as the economic effect of intermarriage was also important. Along with wealth, intermarriage separated Brahmins from the larger population of Boston. Close interaction of members of the class over a number of years nurtured in them a powerful sense of common identity. In addition to their involvement in the cotton industry, Brahmins were soon investing their wealth in a variety of local cultural enterprises. Among their favorite charities was Harvard College, which they financed, managed, and populated with their sons. Accordingly, Brahmins acquired a justified reputation for their strong commitment to learning. Compared to American upper classes from New York and Philadelphia, they paid more attention to cultural affairs in general and higher education in particular. Thus, superior economic status translated over time into a strong sense of social and cultural superiority.

In the pre–Civil War period, Boston's Brahmins maintained a rough sense of parity with their counterparts in New York, Philadelphia, and elsewhere. This situation changed dramatically, however, in the period after the Civil War, when the new oil and steel industries experienced rapid growth. The owners of those businesses acquired fortunes far beyond those of the Boston capitalists and began to form a new upper class with a national rather than a local outlook. In the face of this competition, Boston was rapidly turning into a financial backwater, and the Brahmins faced the prospect of becoming an irrelevant class. This development contributed to the strong sense among Lowell's generation of a crisis in the making. Despite the comfort and security in which they had been bred, Lowell and his friends could not avoid the perception that their influence was in decline. Compared to economic rivals in New York and Chicago, they would have relatively little influence on the affairs of the nation.

Percival Lowell's life makes sense only when we view it against the background of this Brahmin crisis. The major events of his life—including his challenge to some of the old Brahmin values, his curious career search, his brief expatriation to East Asia, and the creation of the Lowell Observatory in the Arizona Territory—were all attempts to revitalize the Brahmin class by addressing the crisis on the home front. What makes Lowell's life particularly interesting, however, is that he entered the fray without any clear notion of how to accomplish this task. He tested various careers until he found one that particularly suited his talents. Even after entering astronomy, however, Lowell found himself peculiarly out of place. Only inadvertently and against the mainstream of American astronomy did he establish a place for himself and his observatory.

Early on, Lowell struggled to break out of what he felt were the narrow confines of Brahmin Boston. He clearly chafed against the expectation that individuals would sacrifice their own interests and needs for the benefit of the community. He also challenged the constraints on individual expression. This impulse to resist, however, was intended not to undermine Brahmin hegemony but to individualize and modernize the Brahmin ethos and bring it into line with the more expressive world of the emerging age of consumption. Lowell's short business career succeeded not only in shoring up his personal finances but also in clarifying for him the fact that Bostonians would not be able to compete with their New York and Chicago counterparts. Lowell discovered as well that a business career would keep him in the shadow of Boston and thus reinforce the localism that contributed strongly to his sense of entrapment.

To fit himself and his class for modern America, Lowell broke from the confines of both business and parochialism. He and his friends would prepare for the world—and the cosmos—by leaving behind business for careers in science and culture. Accordingly, these Bostonians adopted cosmopolitanism as their watchword. Lowell would measure Boston against the cultures of this world and others to achieve a broader outlook and a keener sense of Boston's place in the larger scheme of things. His ventures to East Asia, Arizona, and Mars need to be viewed in this light. Each destination provided a new perspective from which to reflect on the localism in which he had been trapped. In the process, Lowell would not only become a cosmopolitan—he playfully listed his address as "cosmos" in Isabella Stewart Gardner's guest book—but would also find a suitably liberating career in which he could become a national leader.

In fact Lowell's careers were marked by a complicated mix of success and failure. Too imbued with Brahmin values and superiority to empathize with East Asians, Martians, or native Arizonans, Lowell never really became the cosmopolitan he thought he was. His search for the canals of Mars and for Planet X failed. And he became a kind of pariah in the world of astronomy for violating the standards of the new professional establishment. Lowell entered that field as a gentleman amateur relying on a small, modestly equipped observatory, only to find himself in competition with a new breed of highly specialized professional astronomers who were in the process of creating factory observatories with costly instruments and large research staffs. Accordingly, his observatory was threatened with the same backwater status as the Brahmins of Boston's Back Bay.

Nonetheless, Lowell salvaged from this potential wreckage a good measure of success that would have satisfied lesser men. Like his friend Isabella Stewart Gardner, who built an Italianate palace in Boston and decorated it with Renaissance paintings, Lowell succeeded in pioneering the establishment and management of his own cultural institution. In the process, he showed other Brahmins the way to individualize a career. He also played a strong role in expanding the horizons of Boston's upper class by creating a home for himself on the Arizona frontier. In this way, he served not only to satisfy his own yearning to break the constraints of his class but also to spread the Brahmin passion for enlightenment to a distant frontier. And despite the failure of much of his own research, the Lowell Observatory gained a measure of respect in the world of astronomy, based largely on the success of his assistants.

To carry out this biographical project, I have deliberately rejected an exclusively chronological approach to Lowell's life in favor of treating themes and topics that comprehend various, sometimes overlapping, segments of his life. My purpose has been to clarify the impact of Lowell and the Brahmin community on each other in the course of his upbringing and young adult years. I have also sought to explain the nature of and rationale for his career shifts as well as underlying conceptual continuities within his major works dealing with Japan and Mars. Accordingly, each of the three parts of this study explores a single theme or topic in order to illuminate these issues. Although the sections are not arranged in a strictly chronological fashion, they do proceed from the most private to the most public dimension of Lowell's life and work.

Part I examines elements of Lowell's personal life in order to understand his disposition, sensibility, and values as they emerged in relations with his family, his male cohort, and intimate women friends. This section puts Lowell at center stage as a dynamic and complex individual, while also clarifying his impact on other Brahmins and vice versa. An important part of this story is Lowell's almost twenty-year search for a suitable career, from the time of his Harvard graduation to the establishment of the Lowell Observatory. To better understand this search, I examine briefly each of the careers he tried. In the process, it becomes clear that Lowell sought both increasing versatility and a way of cushioning himself against the shock of change. To accomplish these goals, he carried forward to each subsequent career skills developed in the previous one. He also vacillated between resistance to and affirmation of traditional Brahmin values by relying on Brahmin friends and family as role models, consultants, and collaborators even as he sought to maximize his independence from Boston and his family.

Part II is concerned with issues of continuity and change in a different sense. I examine in some detail Lowell's writings on East Asia and Mars, all published in the second half of his life, to establish the common intellectual approach he employed in both domains. Historians have long realized that Lowell was a Spencerian but have not clearly shown how deeply Herbert Spencer's work shaped his views of both Japan and Mars. Nor have they understood how thoroughly Lowell appropriated Spencer's concept of disciplinary hierarchy, which privileged astronomy over all other fields. This concept of fields, in turn, helps us to understand how Lowell crossed disciplinary lines so readily.

Part III deals with Lowell's astronomical career, and, thus, with the last third of his life, but eschews a detailed recounting of the much-studied debate over the Martian canals. It focuses instead on the formative effect of Lowell's Boston connections in the establishment of the observatory and the impact of an entirely different context on the debate over the canals. As it turns out, the founding of the Lowell Observatory coincided with the coming of age of modern American astronomy, including the rise of "factory" observatories, the domination of astrophysics, and the creation of new national organizations to advance these developments. My concern has been to illuminate the ways in which this environment turned Lowell, the founder of a small observatory with conventional instruments and a sensational research program, into an instant outsider. The story of the conflict clarifies the distinctive identities of mainstream observatories and the Lowell facility and helps to explain the eventual and ironic convergence, over the course of two decades, of Lowell's program and methods with those of establishment observatories.[7]

The conclusion circles back to the subject of Part I: the link between Brahminism and Lowell. It reassesses this impact in the light of Lowell's growing attachment to Flagstaff and his eagerness to embrace an elitist ideology designed to counteract intensifying ethnic and class conflict in the United States. On the Arizona frontier, the Brahmin values that Lowell had resisted in his youth took on new meaning. As he made a home in Flagstaff, Lowell belatedly embraced a latent version of his father's ethos.

It should be clear from this introduction that Lowell's polymath inclinations challenge modern scholarship but also promise significant rewards. As an important figure in the Brahmin community, a major contributor to the early cultural dialogue between America and Japan, and a significant player in the formative years of American astronomy, Lowell's life can illuminate a surprisingly large number of subjects. In the final analysis, however, his life is a story worth telling because Lowell's charm, wit, and intelligence, as well as his keen sense of adventure, have made him a worthy companion for the author, whose interest in the subject will, I hope, be conveyed to the reader.

I

The Private Lowell

Percival Lowell belongs on the short list of scientists who deserve attention for their complex and interesting private lives as well as their significant professional careers. Noteworthy in its own right, the private life takes on a special importance in view of Lowell's career shifts and his unique place in the world of American astronomy. Furthermore, we cannot clarify Lowell's complicated role in American astronomy as both maverick astronomer and institution builder without considering his struggle for autonomy within his family. Lowell grew up as at once an outsider, challenging Brahmin inhibitions, and an insider who believed in the principle of authority and the importance of strong institutions. As we shall see, these competing psychological impulses helped to shape his choice of research topics and his approach to the astronomical community.[1]

Lowell's two most important career shifts were closely correlated with crises in his private life. Following graduation from Harvard in 1876 and a year-long grand tour in Europe, Lowell settled in Boston for six years and embarked on a career investing family funds. In late 1882 he broke an engagement to a Brahmin woman, at considerable psychic and social cost, and, in the aftermath of this decision, left the investment trade. For the next eleven years, Lowell lived the relatively unstructured life of a traveler and author whose experiences in East Asia were the subject for widely acclaimed poems, books, and articles.

The second great turning point in his life, which also involved major emotional trauma and a career shift, occurred in 1894 and 1895. Lowell established his Flagstaff observatory in 1894, just as his mother, to whom he was very attached, entered the terminal phase of her struggle with Bright's disease. Two years later Lowell himself succumbed to neurasthenia, which incapacitated him for four years. Following his recovery in 1901, Lowell dedicated the remainder of his life to the work of the observatory, and he apparently found solace and companionship in his 1908 marriage to a former Boston neighbor and long-time acquaintance.

Lowell's substantial commitment to the new culture of expressive individualism offers the key to understanding his private life. He grew up at a time of transition and embodied in his own attitudes a certain ambivalence toward the institutions and values that his parents and other Brahmins of their generation had accepted without question. Lowell's enormous zest for life, personal flamboyance, and vivid imagination, found expression in his travels to distant lands and his quest for a challenging career; these experiences, in turn, brought him into conflict with traditional Brahmin values of hard work, reserve, self-restraint, and understatement. Thus Lowell partially rejected the older individualism that measured achievement by productivity in the workplace and frequently objected to the Brahmin preference for tempering individualism to accommodate communal obligations. Accordingly, Lowell, often in the company of his friends, carved out a lifestyle that departed in important ways from that of his parents. He sought personal development through recreation and travel as well as through deeper and more meaningful personal interactions. In the process, he and his friends shifted the balance from a utilitarian life toward an expressive individualism emphasizing aesthetic and emotional development.[2]

Although Lowell challenged Brahmin attitudes toward the work ethic, sexual behavior, and liberated lifestyles, he later came to endorse the Brahmin commitment to authority and the importance of stable institutions. He was also a belated convert to the work ethic, remained an ardent investor all his life, and participated enthusiastically in the Boston social clubs with his Harvard cronies. Inevitably, his vacillation between utilitarian and expressive individualism was not fully satisfying. Lowell's attempts to achieve emotional and sexual intimacy were impeded at various times by his intense involvement in astronomical work, his obsessive attention to personal autonomy, and strong inhibitions that were inculcated by his family. Nonetheless, he challenged the prevailing Brahmin lifestyle and thus paved the way for a more liberated version among upper-class Bostonians in the twentieth century.

Percival Lowell's career decisions were closely intertwined with his challenge to the ethos of his father's generation. Lowell distanced himself from his father both professionally and geographically by emphasizing cultural as opposed to business activities and pursuing them at a far remove from Boston. He chose lines of work that gave greater scope to individual as opposed to community needs. Nonetheless, he, like his father, sought to augment the family fortune as a way of advancing his cultural and scientific causes. This task, however, was rendered more difficult by the challenge of the newly rich industrialists in other parts of the country who threatened to displace Brahmins like Lowell. To meet this challenge, Lowell, his family, and friends sought to bolster their financial position by shifting investments from cotton to newer, more profitable industries. In addition, they used their financial resources to create new cultural and scientific institutions that would bring them a different kind of recognition. In this way, they made science and culture, rather than business, their primary focus.

Furthermore, Lowell and his Brahmin contemporaries broke the pattern established by previous generations who had spread their resources among a number of institutions, such as Harvard College and Massachusetts General Hospital, and then shared the governance of those institutions with other Boston leaders. Instead, they concentrated their support to create new institutions (the Lowell Observatory, the Blue Hill Observatory, the Gardner Museum) that could provide an outlet for their cultural and scientific interests as well as their aptitude for governance. As we shall see, however, individual initiatives of this sort did not preclude using family and friends to support these institutions as well.[3]

Lowell's generation also engaged in cultural and financial projects designed to replace Brahmin parochialism with a more cosmopolitan orientation. For example, they invested in western and southern railroads located at a considerable distance from Boston. At the same time, Lowell was part of the westward movement of culture, along with Frank Hamilton Cushing and Sylvester Baxter, who brought Bostonians into contact with the Zuni Indians in New Mexico, and Edward Sylvester Morse, who pioneered the exploration of Japan. This westward and southern march of money and brains enabled Bostonians to revitalize their financial and cultural position in the face of competition from Chicago and New York and find a role for themselves beyond the confines of Boston.[4]

Lowell's contribution to the regeneration of Brahmin culture notwithstanding, he was an American incarnation of the leisurely gentleman amateur. Lowell was nearly forty years old before he found a career that truly suited him. During this quest, however, he demonstrated an extraordinary diversity of talents by becoming an astute investor, a traveler who opened East Asian frontiers to the American public, and a versatile writer, speaker, and photographer. To be sure, he could not have afforded this journey without a substantial family fortune, but the fact remains that few men of his time, rich or poor, demonstrated such versatility and, in the dawning age of specialization, few would dare to undertake a similar journey. During his adventurous career search, Lowell produced a valuable body of work and learned skills that served him well in his astronomical career. At the same time, Lowell's work enhanced the prestige of his own class, to which he remained fiercely loyal.

While it would be tempting to divide Lowell's career into three phases—business, Asian travel, and planetary astronomy—in fact it would be more accurate to think of him as acquiring new aptitudes without giving up those he had already learned and practiced. According to his brother, Lowell maintained an intense interest in several activities at the same time. He remained a businessman all his life, though he shifted his focus from investing in and managing cotton companies to doing the same for an observatory. From 1883 to the end of his life, he traveled regularly and systematically, though his travel destinations shifted from East Asia to the Arizona territory, Europe, and, yes, Mars. His vigorous prose style served equally well to convey his images of Japan and Mars. And he considered his research on Japan and his observations of Mars as two parts of a single grand scientific enterprise. In this way, Lowell's astronomical career drew on the skills he had mastered on his curious vocational odyssey. In becoming an astronomer, Lowell nonetheless remained a businessman, a traveler, and a writer. And through it all he experienced the satisfaction of challenging the routines, expectations, and values of Brahmin tradition.[5]

1

◆ ◆ ◆

The Making of an Improper Bostonian

There is strong evidence to indicate that Percival Lowell experienced intensely the conflict between expressive and utilitarian individualism. Shortly after his death, Lowell's sister Amy, the cigar-smoking imagist poet, wrote the literary critic Helen Bullis Kizera: "I remember my brother's telling me what an awful time he had to free himself from his inhibitions." Amy Lowell's remarks provide a brief but revealing glimpse of the psychic tensions that beset her brother. The struggle against inhibitions inculcated during his upbringing by the family and other authority figures in the world of Boston's Brahmins was an enduring one that reached particularly acute proportions when critical decisions regarding career and marriage had to be made.[1]

The inhibitions Lowell suffered from are not specified, but no doubt matters of personal style were especially important. The buttoned up, repressive atmosphere of Boston, which required respectability, deference to authority, and a guarded, understated presentation of the self, became burdensome for Lowell and his contemporaries. Consider, for example, Nathaniel Southgate Shaler's remarks about the geologist Charles T. Jackson: he "could not rest on his solid accomplishments . . . but must have a share of the approval of his fellows for his mental digestion, so to speak, with every meal. Boston, with its chariness of praise and its unreasoning shallow contempt for demonstrative persons, afforded him no outlet for his great talents, which closely verged on genius. Had he lived in Paris, where a personality is appreciated for what it is worth, and not in the atmosphere of decaying Puritanism . . . he would have won a great place for himself." The personal style that Jackson embodied was quite similar to Lowell's, and it seems clear that the two men suffered from the same sense of limitations. Both struggled in a changing environment to legitimate the more expressive lifestyle that seemed appropriate to their era.[2]

Lowell's challenge to the repressive features of Brahmin culture was supported by the example of Oscar Wilde, who lectured in Boston in 1882. The very next year Lowell contrasted Wilde's cult of personality with its absence among the Japanese people he had just visited. He was clearly drawn to Wilde's expressive presentation of self, his showmanship, and even his dandified appearance. Furthermore, Wilde's notion that wit was the best evidence of a man's individuality must have resonated strongly with Lowell. In identifying with Wilde, Lowell joined the rising generation of would-be expressive Bostonians, including the sixty Harvard undergraduates who sat in the front two rows of the Music Hall to applaud enthusiastically for their hero.[3]

Lowell's resistance to older Brahmin mores developed out of tensions with his father, whose values were those of upper-class Bostonians of the previous generation. As firstborn son, Lowell would have been expected to take up his father's role and responsibilities. Instead his younger brother Lawrence, the future president of Harvard and successor to his father as sole trustee of the Lowell Institute, stepped into that place. Lowell himself recognized this curious role reversal in 1883 when, in describing a family evening of gambling, he wrote ironically of how "even the eldest son, led astray by the prodigal, may fall." Lowell, in fact, was on the verge of leaving Boston for a long and relatively extravagant trip to East Asia which he took with great relish.[4]

The exceedingly close relationship between Lowell and his mother was largely responsible for turning Lowell into a rebellious spirit and his brother into a virtual firstborn. In his biography, Lawrence Lowell referred to "the daily stream of loving letters" that flowed back and forth between the two. By his sister Katharine's account, Lowell was "peculiarly fond of his mother" and "his tender solicitude for her was unfailing . . . we used to say that he could do anything with her and that she spoiled him." According to Katharine, their mother treasured her prayer book because her eldest son's baby teeth had bitten the tops off the pages. His sister also reported that Lowell took all of his mother's belongings when the estate was divided among the children.[5]

These secondhand reports are confirmed by some of the very few surviving letters to Lowell from his mother, written during a two-week period in August of 1894. Virtually all the letters are addressed to "My Darling

Percy." The first in the series begins as follows: "I watched you till you had quite gone last evening and my heart was in my throat. I love you so my darling that I could only commend you to his care who loves you more dearly . . . I leave you in perfect faith that he will do everything for you the person I love best on earth." Even allowing for the sentimental excesses of Victorian rhetoric, these letters articulate a powerful emotional bond between Lowell and his mother.[6]

It is not difficult to imagine the psychic dynamic that must certainly have been set in motion within the family by this very close relationship. In all likelihood the emotional intimacy between Lowell and his mother was for her something of a substitute for a similar relationship with her spouse. Under the circumstances, it would be quite normal for Mr. Lowell to have been resentful of the attention lavished by his wife on their eldest son, and for that resentment to spill over into his own relationship with their firstborn. This antagonism, in turn, could well have weakened Percival Lowell's identification with his father and caused him to reject the traditional role of the firstborn as responsible agent for the parents' interests. Instead, Lowell seems to have rebelled against this conventional role in favor of a more adventurous path through life that his mother was quite willing to indulge. As Lowell vacated the seat of the firstborn, Lawrence Lowell took his place and assumed the attendant opportunities and responsibilities.

Even though the evidence for this reversal of roles between the brothers is quite persuasive, it was no obstacle to Percival Lowell's acquiring some of his father's character traits, including the manly self-assertion highly prized by upper-class Victorian men. Despite Lowell's contrary career direction, he retained his father's State Street qualities, including an excellent business sense and a forceful, autocratic personality. Certainly the brothers and their father were accustomed to positions of command. In relations with subordinates, the Lowell men expected their views to prevail, and thus they had little sympathy for those, whether workers, immigrants, students, or observatory assistants, who might wish to share in making the decisions about matters that affected their lives. As Lowell's friend Barrett Wendell pointed out, the patriarch Augustus Lowell "was a man of such strong self-assertion that many people found him repellent. This very strength of personality endeared him to those with whom he was on affectionate terms." As will become clear, both Percival and Lawrence were heirs to the autocratic, commanding personality of their father, though they put this trait to quite different uses.[7]

For Percival Lowell, force of personality was both a central element of his identity and also an issue in his life. Ferris Greenslet, in his family portrait, likened an encounter with Lowell to being "suddenly deposited in a powerful magnetic field." Lowell's own writings, which might serve in this regard as his autobiography, were filled with discussions of identity, individuality, and personality. Lowell believed that the capacity of one human being to influence others was a hallmark of those advanced civilizations that encouraged the development of individualism and imagination. In this way, he had not the shadow of a doubt that individuality and a forceful personality were desirable traits.[8]

Nonetheless, Lowell diverged in significant ways from the lifestyle adopted by his father and brother. His exuberant personality, witty repartee, and charismatic posture were nicely expressed in the innovative, imaginative, and highly visible projects he undertook at far removes from Boston; his father and brother, by contrast, were very much the "proper Bostonians" of Cleveland Amory fame who believed in utter loyalty to family and ancestors and an understated presentation of self. Both father and second son were known for their work ethic, integrity, and balanced judgment. Both were properly married, in Brahmin terms, heavily involved in family affairs, and interested in a variety of philanthropic enterprises. Augustus Lowell had chosen his wife from one of the leading cotton families of Boston, while his son followed another Brahmin strategy by marrying a third cousin.[9]

Without question, personality differences led to tension between Lowell and his father. The evidence for this tension is subtle but unmistakable, though Lowell's previous biographers have ignored it. While Lowell's obituary for his father appears to be laudatory, a careful reading suggests a deep antipathy to certain of his father's traits. Lowell identifies his father as "impersonal"—a term he used to describe East Asians, whom he believed incapable of imaginative thought or creative action. In case his readers had missed the point, he added that the elder Lowell could give details on every plant in his garden but was innocent of any general framework that could explain the links between them. Lowell, the devout evolutionist, was thus at pains to clarify his father's rejection of the theory of evolution. Lowell then proceeded to denounce the "puritanism" that his father "inherited doubly distilled," and his consequent hostility to loafing and recreation. From this passage alone, it is clear that Lowell's own battle against puritanism had begun at home. To be sure, Lowell acknowledged that his father was a man of will who succeeded in directing business and cultural enterprises, and that the gruff exterior had hidden a tender heart. However, this last point was introduced only in response to Amy Lowell's prodding after she had read her brother's first draft.[10]

In addition to this memoir, there is evidence of conflict on essential matters between Lowell and his father over a period of many years. Most important, Augustus Lowell was staunchly opposed to his son's East Asian sojourn, no doubt because it necessitated his withdrawal from regular participation in family business affairs. As a result, he continued surveillance of his son's activities well past Percival's thirtieth birthday; the ensuing tension between the two men was evident in their mutual dependence on Lowell's sister Elizabeth and, later, her husband to serve as intermediaries between them. She, for example, reported their father's distress at not receiving business reports from Lowell, while Lowell asked Elizabeth's husband to explain to the "family" the significance of his East Asian venture.[11]

Interestingly, as Lawrence assumed Augustus Lowell's responsibilities following his father's death, the relationship between the two brothers also deteriorated. Even four years earlier, Lawrence Lowell had joked privately that the canals of Mars, seen through Percival's new telescope, were actually "the bars of a huge safe deposit vault" enclosing the planet and protecting Martian securities. During the next decade, Percival Lowell brought the conflict into the public realm. In the *Boston Evening Transcript* he expressed his unqualified opposition to the merger of MIT and Harvard, even though his brother had just been named to a committee to negotiate the terms of such a merger. This incident, however, was much less ticklish than the squabble between Lowell and the Harvard College Observatory over whether planetary news from Flagstaff would be circulated to other observatories as part of Harvard's communication service. Percival Lowell deliberately brought this issue to the attention of the *Boston Post* just two months after A. Lawrence Lowell was inaugurated as president of Harvard.[12]

These tensions highlight the two brothers' contrasting personal styles, which, in changed circumstances, divided them. Lowell himself was highly contentious, flamboyant, and eager to take advantage of the media, while his brother, more cautious, preferred behind-the-scenes negotiations. We may wonder as well whether the growing division between the brothers, rather than the press of business, explains Lowell's failure to attend his brother's inauguration as president of Harvard in 1909 and the absence of

Lawrence Lowell from Percival's funeral in Flagstaff in 1916. The evidence would thus suggest that a previously amicable relationship had cooled as Lowell came to see his brother in the father's role.[13]

Lowell's commitment to personal autonomy, which eventually drove a wedge between him and the male members of his family, was first recorded during his grand tour of Europe in 1876 and 1877, following his graduation from Harvard. He then wrote Barrett Wendell about the potential dangers of marriage. "One needs to take every precaution before he throws away his independence." Marriage might be a stepping stone to power or social standing, to which Lowell claimed he was indifferent, but it should not be an emotional trap. "Like but few, love none is a capital motto if one only has the courage to follow it out. Thus, will you most easily escape the passions which bind most men in their passage through life . . . If an unknown something is wanting, you had better retire within yourself and have as little to do with acting as possible." This formula suggests as an ideal the individual who shuns deep involvements with others. It sets forth the principle of self-interested behavior and implies a kind of aloofness from the community that ran counter to the traditional orientation of Brahmin leaders. An individual so committed to personal independence and relying so exclusively on his own resources would inevitably have difficulty taking into account the needs of community or family. At the same time, he would maximize his own freedom of action and potential for self-expression. While I would not go so far as Lowell's secretary, who claimed that "at heart he was a hermit," it is true that Lowell had a keen sensitivity to the protection of his own privacy against the invasive instincts of the small but prying Brahmin community.[14]

It turns out that his 1877 declaration of personal independence governed Percival Lowell's behavior and guided his decisions during many of the crucial periods of his life, and accordingly brought him into conflict with members of his family and social class. Clearly, marriage was only one of a number of events that could threaten his personal freedom. In 1891, for example, after Lowell's brother-in-law Alfred Roosevelt died suddenly in a train accident while Lowell was on one of his long sojourns in Japan, he was summoned by the family to return home as an act of solidarity in a time of mourning. To the request of William Lowell Putnam, his other brother-in-law, Lowell replied as follows: "I shall out of regard for the family's wishes in the matter sail on Oct 22—before that I am too busy . . . Every man must lead his own life. He will certainly never make a success of it if he allows others to guide him absolutely in the matter. We are not full of individuality to become automata and the very people, our own kind relatives, who most wish us to conform to their code, forget that far from promoting they did their best to thwart our career, when it succeeds." Interestingly, Lowell followed this assertion of independence with a request to Putnam to find a house for him in Boston. (He had been living with his family in the intervals between trips to Japan.) "I think such an arrangement will give the family more pleasure by my running in and seeing them and me more freedom than if I roomed at home."[15]

This acute sensitivity to the prospect of the family's infringing on his freedom was closely linked to feelings about the social atmosphere of Boston. In Lowell's view, the remnants of puritanism still tainted Brahmin culture. The city's atmosphere of repression, in this respect, was simply an extension of the family. Pleasure seeking of various kinds—sexual, aesthetic, or physical—was frowned upon. Lowell used the occasion of writing an obituary for Oliver Wendell Holmes, Sr., to denounce the excesses of Puritans past and present: "the most austere society the world has ever known, unless it be that of the same society's ancestors in which to be human was itself perilously near sin. So people wore masks shocked by the sin in sincerity."[16]

The repressive atmosphere that Lowell deplored was an affair of the entire Boston Brahmin class but was perfectly illustrated by the attitude of his parents. An atmosphere of repression prevailed at dinner parties and dances given by the Lowells, despite their two-year stint in Europe. As Robert Grant reported, guests were encouraged to converse, but not necessarily on any subject they might choose. On one occasion, when Lowell's father was away from the table, Grant had begun to recount a tale of a naked lady wheeled into a room in a bathtub. The return of Augustus Lowell silenced the narrator, who gathered from Mr. Lowell's stare that the topic was forbidden.[17]

The same sobriety and unexpressive personal style characterized those Bostonians whom Lowell encountered outside of the family. As a result, after living in Boston for five years following his graduation from Harvard, Lowell was frustrated and bored. He remarked to his sister that "here things are rather dull. Not that I quite go in for murder as a necessary stimulus, but people might learn to be lively and gay without being bad." The journey to East Asia and back so intensified his distaste for Boston social life that he gave public vent to his feelings in *Appalachia,* the journal of the

Appalachian Mountain Club. Commenting on a leper colony encountered at Kusatsu in the Japanese Alps, he avowed that the disease would be an ideal solution to the "bores—who ruin one's taste for humanity by wilfully cheapening the article! If they were only lepers, one could still meet them—in a delightfully distant way!"[18]

Unable to isolate Boston's bores in that fashion, Lowell proposed instead an antidote to local limitations through escape. As he saw it, travel broke the geographical and intellectual boundaries that constrained one at home. Fellow Bostonians' criticism of travelers could be understood, he said, as "the instinctive revenge of staid mediocrity upon discursive genius. What fails of comprehension is sure to visit rebuke. For to pass the bounds of the recognized is not unnaturally to transgress the limits of regard." Once out of Boston, the traveler would become a kind of reservoir, storing sustenance that would enable him to survive back in Boston's cultural desert. Accordingly, Lowell advised his sister Elizabeth to store up "for yourself riches above the reach of fickle fate . . . You are making, as it were, a friend of yourself. One to whom you can go when time or place shall send you from others . . . So school your mind that it may come to the rescue of your feelings." Clearly, he was basing this prescription on his own disillusionment with life in the Hub. And consistent with his insistence on personal freedom, he sought fulfillment not from friends and family, but from his own experience.[19]

Outside of Boston, he hoped to achieve the release from his inhibitions that was so difficult to accomplish in town. He explained to Elizabeth that he found Arab squalor picturesque and the "languor" of the natives quite appealing because he himself had none in his own makeup. He also found an antidote in his travels to the overwhelming sobriety of Boston life. He relished the "exuberant, thoughtless gaiety" of Kyoto during his brief visit there in 1883, and he later wrote Amy Lowell that he would "revel" in his large house in Tokyo; in his book *Chosön,* he described a joyous pilgrimage from Seoul to a country monastery. "Our advance was one continued frolic. The bracing air and the spirits of the travellers made decorum impossible. One by one we deserted our chairs and took to strolling. Then we exchanged conveyances all round. Then we played amateur jinrikshamen and dragged the girls in the kuruma [jinriksha] as fast as we could run, to their great delight and very near upsetting."[20]

In Boston, however, Lowell found very little sympathy for his struggle to overcome inhibitions and release animal spirits. Brahmin leaders, including his father, resisted Lowell's quest to integrate an expressive lifestyle with an acceptable career. Indeed, no less a figure than Harvard president Charles Eliot deplored the manner in which Lowell had violated the decorum proper to a Boston Brahmin. On the occasion of Lowell's refusal in 1894 to give adequate credit to the Pickering brothers for the establishment of the observatory (see Chapter 9), Eliot remarked in a letter to E. C. Pickering, the head of Harvard College Observatory: "Mr. Percival Lowell is undoubtedly an intensely egoistic and unreasonable person . . . Fortunately he is generally regarded in Boston among his contemporaries as a man without good judgment. So strong was this feeling a few years ago that it was really impossible for him to live in Boston with any comfort." Clearly, from Eliot's perspective, Lowell had violated the prescribed standard of behavior: good judgment and regard for the community. And while Eliot was vague about the nature of the conflict between Lowell and his contemporaries, the portrayal of Lowell as "egoistic" squares strikingly with Lowell's own commitment to individualism as expressed in his earlier letters to Wendell and Putnam and his identification with Oscar Wilde.[21]

Quite clearly, Lowell challenged the personal style that characterized upper-class men in nineteenth-century Boston. In so doing, he threatened to undermine the values and lifestyles that had prevailed over many years. Such a threat to the dominant culture could not go without response from the authorities, and the ensuing conflict further widened the cultural divide between generations of Brahmins. Although Lowell was not alone in undertaking to challenge current values, we shall see that the resulting conflict exacted a psychic toll manifested at various points in his life.

Lowell would suffer not only from the retribution of social authorities but also from the psychic effects of his own individualism. He acknowledged in *The Soul of the Far East* that self-consciousness, the source of the awareness of individuality, could become overexcited. Under such circumstances, the individual would experience "the haunting horror of his own identity." Consciousness would then become "an uninvited spectre at the feast." Lowell here called attention to the powerful impulse among individualists to turn away from others and engage in excessive introspection. Such a posture, in turn, brings about isolation from society and breeds a strong sense of loneliness. The vivid language that Lowell used to clarify the nature of self-consciousness suggests that he had experienced the loneliness and isolation he described.[22]

These factors, along with his high-strung nature, put him at great risk for nervous breakdowns. So did the loss of his mother, with whom he had been in intimate daily contact and who was his strong supporter in every one of his endeavors. This loss and the marriage of his sister Elizabeth had removed the two women who could mediate between Lowell and his father. No doubt, these changes had increased tensions between the two men and had helped to bring on Lowell's nervous breakdown in 1897.[23]

Like other men who suffered breakdowns, Lowell blamed the condition on a too-faithful adherence to the work ethic, which in turn flattered the victim. In Lowell's case, however, there must have been considerable truth to the explanation. Over a period of three years, he had been working at a frenetic pace to finish writing *Occult Japan* and launch the observatory. Meanwhile, he was experiencing not only his own doubts about observations he had made of Mars and Venus but also strong criticism from professional astronomers at rival observatories.[24]

The uncertain course of Lowell's disease, like other bouts of neurasthenia, seemed to call for different sorts of treatment at different times. During this and a subsequent nervous illness, Lowell lived briefly with members of his family and received advice about proper treatment. Lawrence Lowell, reflecting conventional wisdom, recommended that Lowell "take very little mental exertion" and "live in the open air and take a considerable amount of exercise." He even suggested facetiously that Lowell could approximate these conditions by becoming a mule on a cargo ship sailing from New Orleans to the Cape of Good Hope. This jocular advice was apparently not well received. Lawrence Lowell remarked that "giving such advice to such people I call casting pearls before swine." Nonetheless, Lowell did appreciate the information passed on to him from his brother about how W. R. Thayer, editor of the *Harvard Graduates Magazine,* had overcome his own bout with neurasthenia.[25]

After the onset of his illness, Lowell rested at his Beacon Hill home. From mid-September until mid-December, he lived in his father's house, apparently in a state of complete exhaustion, and under orders to have *"absolute rest,"* with no visitors and no work permitted. Lowell later claimed that his internment had further weakened him and thus blamed his father for delaying his recovery. During this time, Lowell was placed under a doctor's care and was told that he could not return to astronomy for more than a year.[26]

After some improvement, Lowell, in the company of Dr. Alfred D. Lindstrom, launched the next phase of his recovery by turning to travel, exercise, and sunshine. He tried Bermuda, the South, Maine, New York, England, and eventually joined Mabel and David Todd on their eclipse-viewing expedition to Algiers in 1900. Not until the spring of 1901—after nearly four and a half years away—did Lowell feel strong enough to return to Flagstaff.[27]

It is significant that Lowell first sought to cure his nervous breakdown at his own house as a way of protecting his independence from family intrusions. Only a complete breakdown brought him to rely on his father and on Putnam and his brother to provide advice and a place where he could convalesce. Given the tension between them, it seems likely that his father's death in 1900 contributed to Lowell's recovery. However, Lowell's partial collapse in 1913 suggests that his nervous instability was rooted in other factors as well. In any event, the personal crisis provided a kind of stage on which family dynamics were acted out.

The family relationships that turned Lowell into an ambivalent rebel are clear in retrospect. The intimacy between mother and son alienated Lowell from his father and turned him away from the responsibilities and opportunities that traditionally belong to the firstborn. In his place, Lawrence Lowell became the virtual firstborn, and Lowell effectively the prodigal son. Nonetheless, Lowell himself was heir to the authoritarian disposition so characteristic of male members of the family.

While parents and siblings provided a nurturing environment for Lowell on various occasions, they could not, for obvious reasons, fully support his efforts to challenge their own Brahmin culture. From the family point of view, his quest to develop a more expressive and individualistic style was more problematic than his career path. Accordingly, Lowell sought support for his diverse careers both within and outside of the family. By contrast, and with the exception of his sister Amy, Lowell found little support at home for his challenge to Brahmin values. He thus sought and found a sympathetic cohort to support this challenge among his peers, many of whom shared his antagonism toward the mores of the older generation.

2

◆ ◆ ◆

Lowell and His Peers

Lowell consorted with men and women of his class and generation who shared his ambivalence about old Brahmin values. Together they sought to enrich their personal lives by liberating themselves from their parents' confining culture in order to develop more interesting and demanding personal lives and careers. In this endeavor, as we shall see, Lowell served as a leader. Despite his staunch individualism, he joined contemporaries on several occasions to nurture new career options and opportunities to enjoy leisure time.

The new culture could be nourished only under the proper conditions. Lowell and his friends had to distance themselves from their parents to find a time and place to create the new culture. Various ways of achieving such distance were tried out, including unchaperoned travel and bachelor apartments. Through these strategies, the younger generation escaped the surveillance of their parents to consort together, experiment with new activities, and lay plans for more systematic and organized ways of setting their own course.

To be sure, men succeeded far more than women in winning their independence from the older generation. After all, men were less protected under the old regime than were women. And, of course, men could more readily establish a new culture than their female contemporaries by using the club system to launch new activities in a manner less threatening to the older generation. Nonetheless, Lowell's one-on-one encounters with women in Europe took place independently of the chaperons who, in earlier times, would have constrained these meetings. His reports suggest that his women friends were only too eager to collaborate in rejecting old social constraints.

Bohemian Bachelor

In establishing a community of male friends in various parts of the world, Lowell was following long-standing practice among Bostonians, not to mention their British counterparts. What was distinctive about Lowell's generation, however, was the widespread prevalence of an extended period of bachelorhood. In what Howard Chudacoff has called "the age of the bachelor," many men were choosing to delay marriage or to remain single all their lives. The population of bachelors, both in the upper class and more generally, thus grew rapidly after 1880. And whereas previously bachelors had been condemned for refusing to take on the responsibilities of family life, in the late nineteenth century they often established institutional affiliations and activities that came to constitute a subculture. Under these circumstances, the traditional bachelor values of individuality and autonomy were more widely accepted in the male population at large, especially among the younger generation. Thus, the tension between Lowell and his father, arising within a family context, was part of a more generalized conflict between the older generation's preference for family values and obligations as against the younger cohort's advocacy of individualism.[1]

The bachelor ideal of individuality, autonomy, and resistance to communal obligations and respectability overlapped substantially with the code of the Bohemian. Both groups rejected certain mainstream values, though not always the same ones. Many bachelors, for example, participated in traditional professions even when they did not marry. Bohemians, by contrast, favored the arts and rejected occupations that were tainted with monetary values. Practice, of course, did not always match principle. By the late nineteenth century, many so-called Bohemians were engaging in artistic activities on the side while supporting themselves through substantial investment portfolios. In such cases, the rebellion against mainstream values was at best a halfhearted one.

Most bachelors were no more systematic than Bohemians. Their enjoyment of single-sex activities did not obviate against mixed company on occasion. Indeed, for most men bachelorhood was a temporary state that ended eventually in marriage, though marital arrangements usually permitted the husband to continue to consort with bachelor friends at work and leisure. For the confirmed bachelors, ladies' nights at the men's clubs, dinner parties, and dances had their place in the otherwise all-male rounds of recreational and social activities. The true enemy of the bachelor—and of many married men—was the radical feminist, who demanded an end to male privilege and thus to separate male organizations.

Lowell selected as his friends Harvard contemporaries who sought to renounce their fathers' obsession with work and family, while embarking on richer and more challenging careers and leisure lives. To revitalize his generation, Lowell took the lead in establishing men's clubs to promote demanding physical, intellectual, and artistic activities. Club members found opportunities to test their interest in and commitment to the cultural and scientific careers that several of them launched. Lowell and his friends also threw parties that featured quantities of alcohol—and sometimes drugs—in an effort to turn Brahmin sobriety into gaiety. In the end, their artistic, scientific, and social endeavors challenged without displacing the smug and repressive values of State Street.

Lowell and his contemporaries were Bohemian in a rather limited sense. They played occasionally with the idea of the artist in a garret. Their lives, as Mark De Wolfe Howe commented, were "tempered with a mild flavor of Bohemia," by which Howe meant that the newly created literary and artistic clubs promoted an offbeat ambiance. Among other things, the clubs were hospitable to homosexual men at a time when no formal categories existed to separate men according to sexual preference. Indeed, among Lowell's best friends were William Sturgis Bigelow and Ralph Curtis, both of whom were probably gay. Lowell and company thus had a foot in both camps. They maintained their wealth and correct relations with mainstream Brahmins while developing an artistic subculture based on a circle of like-minded friends.[2]

The roots of this Bohemian bachelor culture must be situated in the Harvard College environment of Lowell's undergraduate years. In 1876 a group of artistically inclined students founded the Harvard *Lampoon*, the humor magazine that continues to play an important role on the campus in our own time. Among the founders were Lowell's three best friends, Barrett Wendell, Frederic Stimson, and Ralph Curtis, as well as Robert Grant. Though Lowell was not a member of the staff, he readily embraced their use of sophisticated humor to satirize Harvard life and establish an identity for themselves that would set them apart from mainstream undergraduates. For the rest of his life Lowell remained good friends with the *Lampoon* writers and sought to create a lifestyle that would reflect their mutual artistic interests.[3]

What is particularly striking about the lives of Lowell's three best friends is their commitment in postgraduate years to art, literature, and cosmopolitan values. Ralph Curtis returned after graduation from Harvard to live with his parents at the Palazzo Barbaro in Venice, where they entertained the likes of Henry James, John Singer Sargent, and Isabella Stewart Gardner. Curtis himself became a painter, befriended Sargent, and was the subject of several of Sargent's paintings and an Edith Wharton short story. Frederic Stimson, who had attended prep school in Switzerland, launched a career in the law while becoming a prolific writer of adventure stories that appeared regularly in *Scribner's Magazine.* Stimson later became America's first ambassador to Argentina. Barrett Wendell taught literature and composition at Harvard, published a pathbreaking history of American literature, and became the first Hyde Foundation professor of literature at the Sorbonne in 1904.[4]

Lowell and his friends created a kind of intellectual subculture that operated within Harvard and the Boston upper class, and also participated with their classmates in mainstream college activities. Following graduation, they pioneered in developing careers and lifestyles straddling the Bohemian world of art and the mainstream of business and finance. For several of Lowell's friends, the law became a stepping-stone to other careers or was combined with other activities. This casual approach to the profession, according to Stimson, would not have been tolerated in New York firms, where young lawyers were closely supervised by their senior partners. Robert Grant, confirming Stimson's claim, noted of his own law practice: "I was content to sit at my desk and let my partner make acquaintances outside, but to read law in cold blood when I had a story fresh from the mint of imagination in embryo seemed like wasting opportunity." Grant continued to write novels even after he was appointed probate judge of Suffolk County Court.[5]

The Bohemian credentials of Lowell and his friends, however, were established more clearly through their lifestyles than in their professional lives. To develop distinctive lifestyles, they sought to distance themselves from their parents by living away from home. Their new digs became the setting for serious dialogue about the value of what they were doing as well as their plans for the future. Such conversations revealed a sense of doubt bordering on hopelessness and sometimes despair about their current situation. Over time, however, Lowell and company developed a constructive approach to their dilemmas. Plans were made for engaging in new and more challenging activities to satisfy their adventurous impulses.[6]

Although Lowell himself continued to live at home until 1883, he took advantage of his friends' bachelor quarters to establish a certain degree of independence from his parents. Meanwhile, Stimson and Wendell were laying plans in the summer of 1879 for a "jolly comfortable" bachelor apartment. Stimson's ideas for decoration were inspired by the walls of Ralph Curtis's apartment in Paris, hung with nude paintings. According to Stimson, Curtis's apartment had an air of "charming artistic and intellectual idleness."[7]

Letters from the period testify to the nature of bachelor activities as well as the limited satisfaction they provided. The literary set placed a high value on serious and intimate conversations enhanced by the consumption of drugs and alcohol. When Lowell entertained his cousin H. C. Mercer during the summer of 1890, they listened to the music box, ate lychee nuts, and heard the mystic "tapping, tapping of the kakemonos," which produced dreamy visions of the "Orient." The exotic atmosphere Mercer's letter recreates suggests the possibility that the two men were smoking opium. Just four years later, in *Occult Japan,* Lowell remarked that "opium . . . is delectable, not more for the peculiar ideas it gives a man than for the poignancy of them."[8]

Conversations on "oriental" subjects were also fashionable during outings to William Sturgis Bigelow's cottage on Tuckernuck Island, south of Cape Cod. The invited guests, male only and including Percival Lowell, wore pajamas during the day while they swam, fished, and sailed. In the evening they dressed for dinner and at the table discussed such topics as Buddhism and East Asia.[9]

The Bohemian bachelor lifestyle initially intensified a sense of purposelessness and ennui that Lowell and his friends expressed in the decade after graduating from Harvard. During these years they were groping toward careers and marriages that were established only several years later. In the midst of this time of uncertainty, Barrett Wendell complained that his generation was suffering from overrefinement, which he called the "curse of the century." Clearly Wendell worried that he and his fellow Brahmins were not vigorous or hard-working enough to bear up against the competition from the more worldly industrialists of New York and Chicago.[10]

The anxiety about overrefinement no doubt contributed to the feelings of depression that Lowell and his friends experienced. A year after graduation, Lowell's "misanthropic" sentiments led him to shun all but his best friends and consider "migrating to another planet or ceasing to exist." The next year he sought companionship with Wendell, writing that "the blue can cheer the blue," and when Wendell's blues persisted, Lowell recommended "the life of an opium-eater" as a remedy. Meanwhile, Stimson and Wendell were also feeding on each other's gloominess. "I wish you and I were both dead," Wendell wrote. "I must find a grisette [French for whore] in a garret." Two months later Stimson confessed: "I too am rather blue." Of his law studies, he despaired: "Weary, weary winter, barren, barren term." Stimson considered various courses of action to alleviate his condition. He reported to Wendell that he and Lowell had just dined together and that Lowell had prescribed speculation on stocks as a way of bringing excitement to life. Stimson tried attending the opera every night of the week and drinking scotch; imagined a religious and transcendental flirtation with an Anglican woman as an antidote to boredom; proposed travel and study abroad; and, in 1880, still "without serious occupation," began experimenting with opium, but stopped when he became frightened. On two occasions Wendell spoke of suicide, but he rejected the thought after a cure in a Turkish bath, out of regard for friends, and not to appear weak.[11]

This gloominess was dissipated only as Lowell and his friends developed a greater sense of purpose. Marriage and careers frequently contributed to a sharp change of attitude. Stimson and Wendell were soon embarked on demanding careers in the law and literature, respectively, and were raising families. Lowell, by contrast, would continue his search for meaningful work into the 1890s and his bachelor status into the new century, while acknowledging that his "path in life" was "not the happiest." Meanwhile, these men and their friends abandoned their sometimes gloomy conversations in favor of various measures designed to address their problems.[12]

One of the most important of these steps was the revamping of Boston's club system in order to open up opportunities for a new, more adventurous lifestyle and the testing of new careers. Like his friends, Lowell frequented the Somerset for mainstream activities, including dinner and cards with men of the world, and then chose from among the Papyrus, Tavern, and St. Botolph clubs to enjoy the company of men of letters and the arts. Robert Grant explained that he "vibrated between the Somerset and Tavern clubs." Lowell himself preferred the St. Botolph where he socialized and gave a number of talks. The expressive ethos of the club was reflected not only in its choice of the patron saint of travelers for a namesake but also in Oscar Wilde's 1882 visit and its display of John Singer Sargent's famous portrait of Isabella Stewart Gardner in a low-cut dress.

Brahmin wits proposed that Sargent had "painted her all the way down to Crawford's Notch," referring to a famous landmark in the White Mountains of New Hampshire and also to Mrs. Gardner's close friend, the novelist Francis Marion Crawford.[13]

Lowell advanced his Bohemian bachelor lifestyle as well by joining or organizing other clubs to promote cultural, scientific, and sporting activities among like-minded friends. In 1881 he cofounded the Mathematical and Physical Club to bring together professional and amateur scientists in the Boston area. Of course, the members of the club were distinguished by their expertise in science, but we should remember that many Boston scientists, including those of the Peirce and Agassiz families, were themselves members of the social elite and could thus engage in "pleasant social intercourse" as well as intellectual interchange.[14]

In 1887 Lowell attempted to launch an Authors Club, which was to include fifty writers in the Boston area. The professional goal of this enterprise was to generate greater solidarity and prestige for Boston's authors (see Chapter 4). However, the Authors Club was also intended by its founders to extend and enhance the Bohemian bachelor world. This matter was clarified when the founders sent a questionnaire to prospective members asking whether women should be admitted to the new club. While no decision was ever reached on the issue, Lowell expressed his preference for a men's club that would admit women on designated "ladies' nights." "We will take our 'freedom of intercourse' and our charm of femininity consecutively—a compromise not compromising to either sex." Lowell's approach suggests that, though he hoped to establish an authors club that was more than a traditional men's social club, he still expected the new club to retain some of the latter's features—including risque references to women![15]

More bachelor than Bohemian was the Dedham Polo Club Lowell cofounded in 1886. He and his friends were among the upper-class Americans drawn to the sport, which had been imported from England ten years earlier. Harvard undergraduates had already formed a polo club in 1883. In 1888 Lowell emerged as captain of the Dedham team, which played the first game of polo in Massachusetts. In the next year, during a game with the Myopia Hunt Club, he had a memorable collision with George Von L. Meyer, a future secretary of the navy. According to several reports, both players had to be carried from the field.[16]

It seems clear that Lowell and his friends were drawn to polo because the healthful activity in the country helped restore the physical vigor of the players, who spent their weekdays in city offices. Also, by virtue of the skill and courage required, participants could gain a much-needed sense of adventure in an otherwise prosaic life. And the polo players must have found the sport to be a pleasant escape from the formalities of upper-class life in the city as dictated by Boston society matrons.

The eagerness to develop a vigorous lifestyle in the country was evident in remarks by Frederic Stimson, cofounder of the Dedham Polo Club. In the club diary, Stimson chastised his friends as "mere *city Clubmen* and *Summer sports*" (Stimson's emphasis). Alone at the Dedham Polo Club one day, he rode to the Country Club in Brookline, where "he found some *real men* who had ridden out or walked from town." These remarks suggest that Stimson and friends regarded vigorous exercise as one measure of manhood. Indeed, well before the creation of the Dedham Polo Club, Lowell and Stimson had explored the Boston suburbs and the Smoky Mountains on overnight horseback trips.[17]

By all accounts, the Dedham Polo Club was intended to be a country retreat for men. It was not simply that polo was a male sport; the social events surrounding the polo matches also were usually designed for men only. Club members, including Stimson, future Supreme Court justice Louis Brandeis, and Brandeis's law partner Samuel Warren rode their horses on long Sunday morning excursions through the Dedham area. In the winter, they used the clubhouse as a base for skating on nearby ponds and the Charles River. Club camaraderie and relaxation, as the club's diary, *The Vaccination Book* (from its origin as a medical record book), makes clear, were often enhanced by drinking. Stimson exclaimed with evident pride: "What an unconscionable quantity of sack to 1 and ½ d. worth of bread!" The ratio of money expended on food to the cost of drink for the average member was about one to two with Lowell weighing in just below the average. Following one of the club's drinking bouts, Lowell challenged Samuel Warren to see who could knock a polo ball down the streets of Dedham and back to the clubhouse first. In their fondness for drinking and uninhibited activity, the polo players resembled the "Bohemians" in their Boston clubs. Indeed, the polo players and Bohemians were often one and the same.[18]

In this male sport there was, however, an important place for women. Lowell, among others, promoted polo as appropriate entertainment for mixed audiences. Indeed, he wrote from Tokyo to urge his sister Amy to become "a matron of the sport" much in the manner of his friend Isabella Gardner, who was a frequent spectator. The polo club promoted parties for men and women, and on New Year's Eve of 1889 Lowell rented a train to bring his friends to Dedham for one such event, hosted by Mrs. Arthur T. Cabot. At midnight, according to the *Vaccination Book,* "ye neat-ankled damsels leap 12 candles." Others remembered the mock steeplechase in which guests jumped over the backs of chairs.[19]

After his return from Japan, however, Lowell lamented that the carefully drawn boundary between the sexes had been violated. As a result of the popularity of golf among men and women, the "sanctum sanctorum," the clubhouse, had been invaded by "femininity." Among the members at dinner there, Lowell discovered that he was "the only bachelor of life, as well as of arts, now surviving . . . Births of things are also milestones of life." Clearly the Dedham Polo Club and its clubhouse, which he had helped to establish as a male preserve in 1887, were no longer serving that purpose in the changed circumstances of 1894. Accordingly, Lowell prepared himself to abandon the club and enter a new phase of his life.[20]

It was easier to arrange a Bohemian bachelor lifestyle outside of Boston than within the city. After all, the bachelor life was intended to secure men's independence from a too regular involvement in social occasions for mixed company. On the Arizona frontier and across the Pacific, the population of educated Western men vastly outnumbered their female counterparts. Thus Percival Lowell lived largely in a world of single men in both places. His three closest male associates in Tokyo, the British writer Basil Hall Chamberlain, his American friend and cousin William Sturgis Bigelow, and his Japanese secretary, Miyaoka Tsunejiro, were all unattached and shared Lowell's preference for literary and artistic associations. Accordingly, despite Lowell's geographical separation from Boston, he was part of a social world that, in its broad outlines, quite resembled the one he had left behind. Moreover, as we shall see, Tokyo had social clubs, professional societies, and events around which men could develop a Western-style social life. And Lowell's observatory, situated on America's western frontier, was in its early years staffed almost exclusively by men. Not until after the turn of the century was Mars Hill equipped to accommodate women in a comfortable manner. Indeed one of Lowell's assistants, T. J. J. See, wrote to request a cook for "the bachelor observatory."[21]

Thus Lowell managed to replicate in Asia and the Arizona Territory the bachelor life that he and others had established in Boston. As at home, so in distant corners of the world men consorted together for business and pleasure, while spending relatively little time with wife and family. Mixed social occasions were less important and lengthy than time spent at clubs and in less formal settings with other men. But the associations of Lowell and his friends were distinctive all the same. They organized themselves to take up polo and mountain climbing to test skills and courage in close contests with competitors and nature. They also persisted in turning intellectual pursuits into social activities by helping to found the Mathematical and Physics Club, attempting to create an Authors Club, and joining the St. Botolph Club. Thus their club activity, in contrast to that of mainstream Brahmins, had an intellectual and Bohemian cast. It also featured the consumption of quantities of alcohol, which was intended to remove inhibitions and lead to a gayer and more emotionally expressive life. However, despite their rebellion against the routines and values of their parents and more conservative contemporaries, Lowell and his friends maintained their associations with and continued to depend on State Street financiers.[22]

Their club activities also provided a training ground for entering careers that were off the beaten path. While many of their peers took up the traditional occupations of investing and managing inherited wealth, Lowell and his friends used their club activities, both at Harvard and in Boston, to test their interest in science and culture. The clubs thus became launching pads for careers outside of the business world.

After 1894 Lowell's reliance on club life seems to have diminished. His disillusionment with the Dedham Polo Club, a major focus of time and energy in the previous decade, was only a minor reason for this change. Much more important, his passion for astronomy turned him into a belated convert to the work ethic. To be sure, he had compiled an enviable publication record on East Asia—four books and numerous articles and poems in ten years—but he seems to have attacked the subject in brief bouts of intense work followed by periods of devotion to polo and other leisure activities. By contrast, his passion for the canals of Mars was expressed in a frequently feverish work pace that left much less time for participation in club activities of any kind.

The death of his mother also drew Lowell away from club life. Deprived of his most intimate partner of the opposite sex, he seems to have experienced a strong need to establish closer relations with women of his own generation. Accordingly, as we shall see, he began to cultivate more intimate and long-lasting relationships with various women. This development, in turn, provided yet another opportunity to challenge the prevailing constraints of mainstream Boston, this time in the realm of sexual practices.

The Fair Sex

Lowell's relationships with women were driven by his desire both to satisfy deep emotional needs and to resist Brahmin efforts to impede his personal freedom. To achieve these sometimes conflicting aims, Lowell followed a complicated course. He abandoned an engagement at age twenty-seven to protect his independence, and he remained a bachelor until age fifty-three. In the intervening period, he found emotional fulfillment through his close relationship with his mother and then through a series of short-term relationships with upper-class women, conducted beyond the scrutiny of Brahmin society. So devoted was Lowell to his work that he found it impossible to establish an emotionally intimate and enduring relationship with those women who were his social and intellectual peers. Finally he married a woman beneath him in social rank and too old to bear children. Her adoration for and loyalty to Lowell bolstered his ego; at the same time, he incurred none of the family obligations that marriage to a Brahmin woman—and procreation—would have entailed.[23]

While Lowell found considerable emotional satisfaction in male friendships, he highly valued his intimate associations with women. Lowell viewed women as fundamentally more pure, tender, and charming than men, and therefore capable of providing greater emotional fulfillment. In the realm of art and literature, intellectual exchange with intelligent women could be especially gratifying. In such exchanges, Lowell was content to defer to the superior aesthetic sense of his partner. On occasions, he even came to acknowledge the pleasures of a softer, gentler society dominated by women.[24]

However, the prospect of sexual intimacy with women remained problematic. Lowell was caught up, along with other members of his generation, in the struggle to overthrow the sexual restraints imposed by his Puritan forebears. He gestured toward sexual liberation when, en route to viewing the solar eclipse in Algiers, he commented suggestively on postcards of nude women he sent to his partner David Todd. For a diary he kept in 1904, Lowell photographed nude statuary, and he also photographed in the nude a woman with whom he was conducting an affair. These gestures notwithstanding, Lowell avoided occasions for sexual intimacy and achieved little satisfaction from sexual intercourse. From his own accounts, he was timid, inexperienced, and quite reluctant to pursue long-term involvements. We are left with the impression that Lowell sought desperately and unsuccessfully to overcome the remnants of Puritan reticence in order to achieve an emotional intimacy that would mitigate his frequent bouts of loneliness.[25]

At age twenty-seven Lowell reached a defining moment in his life when he succumbed to a matrimonial scheme to match him with an eligible Brahmin woman. Given his wariness of marriage and his disapproval of Brahmin inbreeding, it is more surprising that Lowell entered the engagement than that he eventually withdrew from it. Cryptic comments in his published writings and letters from Lowell to his sister Elizabeth, who was then traveling in Europe with their parents, provide insight into this painful event. On July 19, 1882, he wrote, "I am so blue myself that I would I could speak with some of you." This letter was followed two weeks later by another in which he added, "It grieves me to have brought so much sadness to all of you. But I have had it too and still have for I never had the great happiness you all thought about on the other side. Everyone seems to have been happy while I alone was wretched." In this moment of crisis, the family returned to support him despite his protests against "sacrificing their plans for his welfare." The criticism that Lowell received in Boston society for breaking the engagement clearly reinforced his already strong resistance to long-term involvements with women of his own social class.[26]

Over the next two decades Lowell turned to a series of harmless flirtations, such as those he conducted in East Asia. He found in Japanese women a "naturalness" and delicacy, but passed up opportunities for intimacy because, he said, "I don't fancy them sufficiently to take to myself a fair Jap partner." Yet he apparently wished to suggest to friends that he was making sexual conquests at will. He remarked to Frederic Stimson that he intended to carry on with "Sweet Smile," a Japanese geisha, in a "distant, quiet, gentlemanly way. My well known antipathy to the sex fortunately renders this observation well nigh unnecessary." Lowell then noted suggestively that the evening with Sweet Smile would not close with a mere smile.[27]

In this way, Lowell must have established among Brahmin society a reputation as a womanizer. At any event, the future literary critic John Jay

Chapman was convinced of Lowell's prowess. Chapman was enamored of Minna Timmons and had been reading Dante with her. Timmons, along with Lowell, was a member of the Place aux Dames drama group. The stage was thus set for a drama of a different sort when all three parties were invited to a soiree at the Walter Cabot home in Brookline. There, emotionally overwrought, Chapman decided that Lowell had taken advantage of the opportunities offered by the Place aux Dames to seduce Timmons. To show that he did not blame *her,* Chapman beat Lowell with a cane. After Timmons had reprimanded him for his violent act, Chapman retired to his lodgings and deliberately burned his hand—so badly that it had to be amputated. While this incident developed from the fevered imagination of John Jay Chapman rather than from any evidence of Lowell's capacity for sexual conquests, it does suggest that his reputation as a lady's man was well established in Boston.[28]

Following his bout with neurasthenia in 1900, while on vacation from his astronomical work and away from Boston, Lowell launched at least three affairs, two of them serious relationships with upper-class American women from Philadelphia. In these affairs he would attempt both to overcome his Brahmin inhibitions and to find emotional intimacy. Until his marriage in 1908—and possibly beyond—Lowell took advantage of the increasingly liberated practices of certain of his more worldly upper-class contemporaries. In Europe he consorted with Ralph Curtis, who himself gravitated toward American expatriates, including Henry Adams and his companion Elizabeth Cameron, or "Scam," the estranged wife of Senator J. Donald Cameron of Pennsylvania, as well as the art connoisseur Bernard Berenson and his friend Carlo Placci, and Mrs. Cooper Hewitt, widow of Abram Hewitt, a steel manufacturer and former mayor of New York. This circle of intimates thrived on the sophisticated cultural life of Europe and tolerated extramarital sexual relationships. In that milieu Lowell could exploit his vacation time in Europe to play hard, just as he worked hard in America.

Adams and Berenson, like Lowell, were Boston cosmopolitans. They were discovering for Americans the brilliance of traditional art. Adams was just finishing work on his book *Mont Saint Michel and Chartres* (privately published in 1904), which celebrated the Romanesque and Gothic cathedrals of medieval Europe. Berenson, empowered by Isabella Stewart Gardner, was purchasing great works of Renaissance art to decorate her new mansion in Boston. The two men, like Lowell, were thus promoting an aesthetic response to the industrial revolution.

Their challenge to Brahmin social conventions was also appreciated by Lowell. Adams was deeply in love with Mrs. Cameron, though there is no evidence that either had yielded to the temptation to turn their love into a sexual affair. And Berenson had been unfaithful to his wife on several occasions. On both intellectual and social grounds, Lowell found this cosmopolitan company congenial, and vice versa.[29]

The worldly expatriate group of which Lowell was a part engaged in a seemingly endless round of dinners and teas at Europe's best restaurants, shopped in the most fashionable boutiques, attended plays and musical events, and pursued their affairs. Lowell himself seems to have been somewhat ambivalent about participating in this life. On the one hand, he was eager to challenge and overcome his inhibitions and to enjoy the good things of life, including sexual experience. On the other hand, his bouts of playing in Europe were a sideshow to the serious career that he would resume after his few weeks' vacation. Furthermore, the absence of physical or intellectual challenges made Lowell ill at ease in the expatriates' sybaritic environment. In this regard, Ralph Curtis was not too far off the mark in claiming that Lowell "revels in discomfort and cave-dweller's luxuries." Exception should be made, however, for the consumption of alcohol, caffeine, and tobacco. These three drugs, coming at the end of a meal, constituted, as Lowell said, "a well perioded conclusion." The cigar in particular was an indispensable item that provided "the only excuse for a dinner": "Everything . . . seems the better through the pleasing mist of the cigar."[30]

Along with smoking, Lowell had a consuming interest in fashion—both for himself and his female companions. Fashion, after all, was an ideal weapon to use against the Brahmin preference for self-effacement, restraint, and reticence, as Oscar Wilde had already demonstrated. On various occasions Lowell outfitted himself in a claret necktie and a "striped flannel suit which would have paled Solomon in all his glory." Curtis took pity on Lowell's tailor, who was wracked with insomnia under the pressure to build garments "on unheard of lines with 'personal ideas' as to how they fit." Lowell suffered as well, from "prolonged mental agony, in getting my new ones (clothes) to fit." However, the results were apparently worth the suffering: "I am now cladable in decent garments in the latest style. I am so proud of them I am afraid they will sink in mid-ocean."[31]

In shopping with women, Lowell seems to have enjoyed the fashions as well as the company. On "a great shopping day!" with a companion recorded as "N. N.," he noted that the search for frocks was succeeded by a chance spotting of "an orange sack in the window . . . We entered and it ended by our coming out with the sack and with the best hat I have yet seen, black with a crown of carmine." The day was completed with the purchase of "an excellent lace blouse in black to which we instantly made love . . . all three acquisitions became N. N. exceedingly well." He later noted that "orange is the *dernier cri* in Paris and one which I find penetrating . . . It is now what pink and purple roses were nine years ago."[32]

During his travels in Europe, Lowell was swept up in the worldly atmosphere of expatriate gatherings. In St. Moritz, then as now a fashionable gathering place, he was taken by Ralph Curtis "to call on Mrs. Cameron who was the same as always . . . In the evening I sampled the Hotel du Lac, St. Moritz Bad for dinner and found it German and with a certain Teutonic bequem [German for comfortable]; then later to the ball at the Palace where Miss Brown favored me with a smile and an outstretched hand as she was whisked off for a waltz." In Paris, still in the grip of Ralph Curtis, the pace picked up: "At Henri's lunching with Scam, Henry Adams, the C's [Curtis's] and I. Poor lunch, left to the discretion of the Waiter! . . . Ralph and I walked to the Palace Hotel for tea and to watch the hunting of the fair. Were at once hunted ourselves but not caught, the eye-shots not penetrating. Down past Colombin's which R informed me was excellent ground for such sport . . . To dine at Larue's on the rue Royale."[33]

In this worldly environment, Lowell moved beyond the casual flirtations that characterized his female relationships in Boston to pursue full-fledged affairs and to consider marriage once again. This flurry of romantic activity would suggest that in the wake of his mother's death and his bout with neurasthenia, he experienced a much stronger need for emotional intimacy than had been the case twenty years before, though he was still reluctant to tie himself down. And, as always, Lowell enjoyed the opportunity to flaunt his rebellion against the puritanism of Brahmin Boston.

Lowell met Miss Irva Struthers, a Philadelphian of social-register standing, in York, Maine, in the summer of 1903. In the course of the next five months he wrote some two dozen notes to her, made at least two trips to visit her in Philadelphia, and in the end proposed marriage. Struthers was a miniaturist who exhibited her work at the Pennsylvania Academy of Fine Arts on at least two occasions. Lowell played on this theme in the romance, suggesting that his knowledge of humanity (love making) and hers of art were areas of expertise that could be shared: "You should instruct me in the one and I should wake you to the other." She must have resisted this suggestion, because he later wrote, "Life is so short and you so rational." It is doubtful whether Lowell's marriage proposal, after such a short romance, could have been taken seriously. Miss Struthers, in any event, resisted Lowell's "impetuosity" and rejected his proposal on grounds that his nomadic lifestyle and secular inclinations threatened her commitment to family and faith.[34]

Just eight months later Lowell became involved with Miss Edith Pettit, also from an upper-class Philadelphia family, and Erna M. Stevenson, a thirty-year-old "spinster." Edith Pettit, like Miss Struthers, had a strong interest in art. Her portrait had recently been painted by Frederick MacMonnies, and she would later write a brief and compelling account of MacMonnies's transformation from sculptor to painter.[35]

Lowell's romance with Miss Pettit blossomed almost as soon as they met on board ship, on a trans-Atlantic voyage. She was a delightful companion to him, sitting at his table and venturing forth to private trysts by the ship's funnels. He was taken with her wit, charm, and beauty, as is evident from various journal entries. On the matter of wit, he exclaimed, "Would I could recall more for she speaks most cleverly to the point and she appreciates with a glance or a wrinkling of her nose what she does not answer." The playful character of the romance is evident in Lowell's entries about their promenades through the Swiss Alps. Seated on the grass, "she cut doubled blades of grass into little bits. It felt so pleasing to do it. And then she made a salad for a rabbit with knife. I had made a stone footstool for her foot." Pettit's outfits, and the figure they dressed, were also appealing to Lowell: "There is nothing lovelier than a lingerie blouse, nothing in which a woman is at once so dressed and so bewitchingly undressed . . . Hers was superb, the most exquisite of lace."[36]

Early on in the romance, however, it became clear that a permanent attachment was out of the question so far as Lowell was concerned. We gather from his diary that Miss Pettit had lost her virginity four years earlier and that her loss of innocence—both sexual and psychological—constituted an obstacle to a continuing involvement with her: she "would be a rare full flower but for her awakening." Later he remarked: "I miss Europa and if Europa were mentally miss I might fall." Despite his reluctance to become too intimate, the relationship was more than platonic. Following an injury to her ankle during a mountain walk, Lowell administered kisses liberally, along with much touching. He carried her over a mountain stream, an act that recalled for him Jupiter's assistance to Europa. As she leaned on him the next day, he acknowledged that the "touches ever so light intoxicate to forgetfulness." On August 22 he wrote that she was "sweeter than ever and her mouth divine." On that occasion and others Lowell insisted in his diary that the "philosopher" triumphed over the "man," indicating that his own inhibitions limited the degree of sexual intimacy. On at least one occasion, however, Lowell acknowledged that he was "not so philosophe." And on another he announced that he would put his "imaginings" into "reality," but he offered only "a half performance rather in earnest of what I would perform." Miss Pettit's ambivalence about becoming more intimate was evident to Lowell in her occasionally "provocative" behavior, which was designed to break down his resistance, and her praise for him at other times for being "true to [his] colors."[37]

It is clear from various entries in the diary that Lowell took great pleasure in "Europa's" wit and intelligence and deferred to her interests in literature and art. Both read *Stella,* one of Camille Flammarion's science fiction novels, and, at Miss Pettit's suggestion, *Man and Superman,* the avant-garde play by George Bernard Shaw, which Lowell found "decadent." The intellectual basis of their friendship was enhanced by visits to Giverny to join the two sculptors whom Miss Pettit had already befriended. Frederick MacMonnies had a year-round studio there, while Janet Scudder, who had studied with MacMonnies, was spending the summer there with a woman friend. Lowell and Pettit roamed the area on their own and joined the MacMonnies family on picnics. On several occasions the couple enjoyed the company of Scudder and her companion. As a result of this experience, Lowell commented in his diary that "women have a natural charm when they outnumber men." In short, his deep affection for Miss Pettit seems to have drawn him into activities initiated by her and to have demonstrated the pleasures of a feminine environment that he had not previously understood.[38]

Miss Erna Stevenson, according to Lowell, was a woman just over thirty years old who was craving an intimacy she had been willing to pass up earlier. In the Swiss Alps, Stevenson and Lowell interrupted a mountain walk to engage in sexual intercourse in a hut off the trail; according to his frank account of this occasion, he gave an "incomplete performance" despite her encouragements. Notwithstanding this unpromising beginning and Lowell's evident desire to avoid situations in which such opportunities might arise again, it appears that Stevenson was eager for further intimacy. In the park at St. Cloud, outside Paris, they photographed each other fully clothed, with Lowell mounted on a pedestal and Stevenson in a provocative pose. These playful moments, however, did not satisfy Stevenson's eagerness for an overnight stay; accordingly, at St. Germain, also near Paris, they shared a room and preserved the occasion in pictures of Stevenson, one taken of her naked, from the rear. The experience of sexual intercourse that followed the photo session was no more pleasurable to Lowell than the previous occasion in the Alps had been. This time, intercourse filled him with feelings of "regret" and "remorse" as well as "lowered vitality," which he apparently regarded as a normal aftermath. In place of sexual pleasure as evidence of his success in overcoming inhibitions, Lowell came away with a trophy in the form of photographs of his nude companion. However, this tangible evidence of his violation of Brahmin proprieties was a poor substitute for the sexual and emotional intimacy that eluded him.[39]

It seems clear that Lowell had derived a great deal of pleasure from his short and ardent courtships with Miss Struthers and Miss Pettit. He obviously valued their interest in literature and art as well as their witty and imaginative interchanges with him. He also found both women physically attractive, but was clearly uneasy about the prospect of sexual intimacy with either of them. In any event, both affairs provided occasions for Lowell to contrast the satisfactions he obtained from his love life and his work. In perhaps the most significant passage in his "Journal—1904," he announced, "I miss the only solace of life, work." And two days later, with a clear sense of relief, he exclaimed: "Work at last!" Similar reflections followed Miss Struthers's refusal of his marriage proposal; on that occasion, he wrote to lament his monklike existence as a scientist: "And so I go back to my only solace. Alone! Yes, always alone, though I had hoped for the opposite." He concluded: "Perhaps that [work] I may make a success; my own life is such a dismal failure." By this point, Lowell clearly must have forgotten his admonition in *The Soul of the Far East* that excessive self-consciousness brought on loneliness and isolation. At any rate, it seems clear that his obsession with independence rendered virtually impossible long-term intimacy with a woman and brought about the painful loneliness that he often experienced.[40]

These affairs serve as an interesting prelude to Lowell's decision to marry Constance Savage Keith four years later. In some complex way Lowell must have decided to abandon the idea of marrying a social and intellectual equal in favor of someone below him in social standing. At the time of his engagement, he had established a certain intimacy with Miss Keith and with Louise Leonard, both middle-class women who adored him. No doubt the unrestrained adulation they accorded him, as well as the promise of unstinting support for his astronomical work, were key elements in determining his choice. Precisely why he chose Constance Keith over Louise Leonard, however, is not clear.

In the year of his mother's death, Lowell had hired Wrexie Louise Leonard as his secretary, a position she was to hold until his own death twenty-one years later. In the next year, Lowell purchased his house at 11 West Cedar Street on Beacon Hill from an interior decorator and real estate entrepreneur, Constance Savage Keith, who was to become his wife twelve years later. Both of these women idolized Percival Lowell and devoted their lives to him. On his side, the commitment was less complete; he encouraged the affections of both women, which brought about a bitter rivalry between them, even while he was also involved with Struthers, Pettit, and Stevenson.[41]

Wrexie Louise Leonard grew up in a small Pennsylvania town, moved to Boston just before she hired on with Percival Lowell, and stayed there until well after his death. She handled much of Lowell's correspondence, including instructions to his assistants on astronomical and administrative matters at the observatory. She also typed and edited his manuscripts and arranged for exhibitions of observatory drawings, photographs, and spectrograms. All of these activities suggest a high degree of intelligence, as does the fact that Louise Leonard turned astronomer herself on occasion and contributed her observations to *Popular Astronomy*.[42]

In addition to their official relationship, there was an obvious intimacy between the secretary and the boss. In the early years, Lowell seems to have had an amorous interest in his secretary. From Algeria in 1896, Lowell wrote playfully to his cousin Eliot Clarke about Miss Leonard's eagerness to enter mourning in order to wear black, which particularly became her. Later letters from Lowell to Leonard comment on the garden, on mutual friends, and offer warm thanks for gifts exchanged; their tone suggests that he was fond of her, but more as a good friend or even a family member than as a lover. The one surviving letter from her to him reveals a strong emotional attachment on her part. She reported on the condition of a jack rabbit they had been observing together—"I wanted you to see him too!" and concluded that "the house seems very big and quiet without you." For a Christmas present that year, Leonard sent Lowell a sprig of pine with mistletoe attached because he "would like it better than all the 'greens' in New England."[43]

Meanwhile, Lowell was developing a close relationship with his future wife, who would offer, if anything, even more devoted attention to him than Miss Leonard. Ralph Curtis's remarks on Constance, following Lowell's death, provide the best evidence of her devotion: "What is his wife likely to do now? I find her profoundly boring. Her adulation of him was pathetically bourgeoise—like the respectful gratitude of a family retainer—a semi-freed slave, befo' de war!" Constance Lowell remained in mourning for thirty-eight years until her own death, and for the first seven years lavished attention on the marble mausoleum that was to house Lowell's remains.[44]

The attraction between Lowell and his wife was also physical. In 1904 Lowell noted that he had seen "the face of a young girl with the stary [*sic*] open eyes unpencilled of the eyebrows, the ugly feature of CSK, and yet I find myself in love with them for remembrance sake. I like to sit and gaze at them." At the same time, he admired—and hoped to exploit—Constance's business acumen. Indeed, less than a year after their marriage Lowell put his new wife in charge of the Boston office of the observatory. Strangely, Constance Lowell's well-established record as a litigant seems not to have deterred Lowell from naming her coexecutor of the estate. The fact that Lowell himself was simultaneously subject to two lawsuits suggests that he may not have regarded a proliferation of legal problems as anything out of the ordinary.[45]

Even after his marriage Lowell did nothing to end the rivalry for his affections between Louise Leonard and Constance Lowell. On several occasions he traveled with Leonard while his wife was elsewhere. Moreover, their struggle for possession of Lowell continued after his death. Within two weeks, Mrs. Lowell, assuming her role as executor for the estate, fired Leonard from the observatory staff. Five years later Leonard took revenge on her rival by publishing *Percival Lowell: An Afterglow,* a collection of Lowell's letters to her. Most of the letters had been written after the marriage, but none of them mentioned his wife.[46]

It would appear, then, that Lowell preferred the devoted attention of a woman of modest intellectual and social achievements to the challenge of a marriage of equals. No doubt Constance provided companionship without threatening his ability to give first priority to his work. Indeed, she was clearly willing to assist him in office chores. By contrast, a marriage with Struthers or Pettit would have enriched his leisure life but raised the prospect of diversion from work. Lowell evidently felt that in such a marriage he would relinquish control of his life—clearly an unacceptable prospect to him.

In his relationships with women Lowell was, thus, quite wary of compromising his independence. Until 1895 he found emotional satisfaction in his close relationship to his mother. Following her death, he engaged in a number of affairs that brought a certain emotional fulfillment and allowed him to challenge the sexual constraints of his parents' generation. Lowell's marriage to Constance Keith, in turn, affronted Brahmin notions of class, which transcended generational lines.

On a separate track, and much earlier in his life, Lowell also engaged in subverting Brahmin ways. By helping to establish new social clubs dedicated to a more challenging and individualistic way of life, he played a leadership role in the development of a new, more expressive Brahmin culture. Lowell and his contemporaries challenged themselves physically and intellectually, while also playing harder in an effort to have more fun. They hoped to liberate themselves from inhibition and at the same time prepare for nontraditional careers outside of the business world and away from Boston. This strategy would distance them geographically and culturally from the world of their parents.

3

Preparation of a Polymath

Lowell's quest for personal liberation was inextricably linked to his search for a satisfying career. By joining the family investment business, he had committed himself to living in Boston and interacting regularly with his family. Thus his decision to leave the investment business was the quickest and easiest way to break the hold of his family and expatriate himself from Boston.

Of course, many career options were available to Lowell. His choosing to launch a long, leisurely trip to East Asia clarified for all his alienation from various dimensions of traditional Brahmin life. In giving up investments, he not only eschewed the business world for cultural experiences but also left the family behind in favor of establishing a new community of friends in Tokyo; at the same time, he distanced himself from the familiar routines of Boston in order to explore an exotic culture.

And yet, in important ways Lowell would carry with him the world of Boston, business, and family that at first blush he appeared to have left behind. He would, in fact, depend heavily on his investment skills to maintain a portfolio to support his travels. Further, during his East Asian ventures, he would rely on family and friends to be an audience for the experiences he recounted to them in his engaging correspondence. Finally, after leaving Boston, he would join a small group of dedicated Bostonians who were amassing cultural artifacts in an effort to preserve traditional Japanese culture.

Lowell's decision to abandon one career in favor of another was made easier by the devotion to dilettantism among his Brahmin contemporaries. They were, after all, part of a transitional generation coming of age in the first stages of the industrial revolution, and as members of a small, privileged class of Bostonians, they could take advantage of their situation to pursue one or more fields of interest. However, money was neither necessary nor sufficient to explain their polymath tendencies. Many rich Brahmins preferred the investment business to developing other talents. Those who became polymaths were receptive to the message of the Harvard curriculum, which expressed the ideal of the gentleman scholar and prepared students for a variety of professions.[1]

Harvard

In June of 1889, Percival Lowell delivered the annual poem on behalf of the Phi Beta Kappa honorary society at the Harvard graduation ceremony. Fresh from a six-month stint in Japan and still five years away from the founding of his observatory, Lowell eulogized two Harvard professors, both graduates of the class of 1829. Benjamin Peirce, his former teacher, was a distinguished mathematician and astronomer who had also directed the U.S. Coastal Survey, and Oliver Wendell Holmes, during his tenure as Parkman Professor of Anatomy and Physiology at Harvard Medical School, had published a steady stream of poetry and novels. Lowell seized the occasion to remind his listeners of the achievements of two Harvard men remarkable for their breadth of learning. We may surmise that, as a polymath in the making, Lowell identified them as models for his own life.[2]

Indeed, he was already well advanced in that direction. Despite the rising tide of specialization, Lowell was creating for himself a blend of careers that perfectly reflected his varied interests. As a Harvard undergraduate, he had distinguished himself in both literary and scientific areas. Lowell won the Bowdoin Prize in his senior year for his essay, "The Rank of England as a European Power between the Death of Elizabeth and the Death of Anne." At the same time, he prepared his graduation disquisition on the nebular hypothesis. These two essays reflected the nearly perfect balance he had achieved between science and the humanities in his elective courses: eight in physics and math and six in classics and history. This undergraduate straddling of the two cultures clearly anticipated his postgraduate careers, first as interpreter of Japanese culture to the American public and later as advocate of the canals of Mars. In 1891 Lowell was elected to the American Academy of Arts and Sciences as a member of the section on literature and the arts. Eleven years later he was moved into the section on mathematics and astronomy. Later Lowell objected to this way of handling his shift of careers, maintaining that he should be allowed to keep his membership in both sections, as was the practice in the French Academy. Clearly he cherished his identity as a polymath.[3]

Lowell and his classmates were the last generation at Harvard to experience a curriculum that encouraged the development of competence in more than one field. To be sure, Lowell entered Harvard just as President Eliot was masterminding a campaign against prescribed courses in favor of electives. Nonetheless, the changes set in motion by Eliot took hold gradually, and despite the rise of electives, the prevailing values at Harvard remained those associated with the prescribed curriculum.[4]

The development of several talents was an appropriate response to the Harvard educational philosophy, which gave priority to mental discipline. This aptitude could be acquired only through immersion in different academic subjects. According to the reigning theory of faculty psychology, a good education ought to provide a harmonious and balanced development of the various faculties. This would prepare the individual not for any specific vocation but to become a well-rounded human being. As Francis Bowen, professor of philosopy at Harvard from 1835 to 1889 and one of the leading education theorists of the day put it, "The successful pursuit of a single art, or of the fraction of a single science, is but poor compensation for the loss of all versatility and alertness of mind. One may become a good accountant, an expert mathematician, and even a skillful lawyer, without being anything more than a fraction of a man."[5]

General education developed mental discipline that would, in turn, prepare students to pursue different fields with equal success. To illustrate this point, Joseph Henry, director of the Smithsonian Institution, urged the compatibility of careers in law and science. Speaking to one of his former students, he quoted Dr. Johnson's remark, "'He who can travel ten miles in one direction can travel the same distance in another,'" adding, "and this is particularly true when the roads in the two directions are of the same general character. The written laws of Nature and of Nations are both expressions of general principles, and he who has sufficient strength of mind to grasp the one, can also by proper application master the other."[6]

Among Lowell's predecessors at Harvard, Henry Adams was the perfect exemplar of the polymath. A novelist, journalist, historian, and art historian, Adams claimed that he learned nothing at Harvard. But his bitter reflections on Harvard's failure to ground him in mathematics and languages as preparation for the twentieth century should be regarded as a case of dyspepsia after the fact. The Harvard of the 1850s, as Bowen indicated, was training generalists, not specialists. In this regard, Adams showcased the strengths of a general education.[7]

We may properly regard Percival Lowell as the heir of Adams. Despite the fact that the prescribed curriculum was on its way out, Lowell and his classmates still spent half of their class time in prescribed courses. More important, Eliot supported electives, as others preferred prescribed courses, largely to promote mental discipline. No one as yet regarded specialization as a desirable goal in the curriculum; there were no majors, minors, or concentrations. Only honors students were held to taking a certain number of courses in a particular field.[8]

The prescribed curriculum introduced students to a coherent view of the world that was based on the concept of natural religion. This world, the product of a benign creator, operated according to scientific principles, and the pervasive lawlike behavior of the creation was accessible to intelligent men through the study of its various parts in particular academic fields. In this way, the academic enterprise retained a palpable coherence. Each field was essentially a branch of natural religion devoted to explaining some particular dimension of the creation. It followed that the boundaries between disciplines were weak and that crossing them was both easy and necessary if some understanding of the universe was to be achieved.

The cues students were taking from the classroom were reinforced by those they received in the Bohemian clubs of Boston. Lowell and his friends viewed amateurism, or dilettantism, in a generally favorable light. The terms usually carried a positive connotation of love for a particular activity and only occasionally referred to casual engagement with a field. Dilettantism appealed to young Brahmins because it promised variety. It justified taking up several fields at once or passing from one to another. The French novelist Paul Bourget, after visiting the Tavern Club, remarked on "the particularly healthy state of American 'dilettantism'—a term that carried no hint of derogation." Lowell's friends Frederic Stimson and Barrett Wendell praised a biography of John Hay, which "vindicated the 'dilettant' and unappreciated man of the world." Hay, as a consummate diplomat, novelist, and friend of Henry Adams, was another excellent example of the polymath.[9]

Given the educational and social climate at Harvard and in Boston, it is not surprising that serious students of Lowell's generation moved easily from one career to another. Teddy Roosevelt (Harvard '80), for example, considered a career in botany, studied law, herded cattle in Dakota, and wrote a four-volume history, *The Winning of the West* (1889–1896), before entering politics. Closer to home, A. Lawrence Lowell ('77), Robert Grant ('73), Frederic Jesup Stimson ('76), and Barrett Wendell ('77) attended law school and practiced law for varying periods of time before turning to other occupations. None of the five—nor Henry Adams, for that matter—quite matched Percival Lowell's bridging of the chasm between the humanities and science, but all exemplified in varying degrees the polymath phenomenon.[10]

Businessman

Percival Lowell demonstrated in business as in science a capacity for taking bold and sometimes risky initiatives. He was an astute businessman whose considerable zest for aggressive investments in both the scientific and financial worlds was matched by impressive managerial skills. Lowell's capacities were developed first in the business world, particularly between 1877 and 1883, and then tested as he assumed the directorship of his own observatory in 1894. In this role, he was responsible for directing a small staff, paying their salaries, and purchasing the increasingly expensive instruments required for astronomical work. Indeed, the existence of the observatory called forcibly to Lowell's attention the importance of creating and maintaining a substantial investment portfolio that would support its current and future needs. In the absence of children, the observatory would become Lowell's legacy.

As a businessman, however, Lowell clearly operated within a tightly knit kinship network that was committed to sustaining the economic well-being and cultural standing of the family and the Brahmin class. This required a division of labor between those concerned primarily with increasing financial assets and others devoted to cultural and scientific pursuits. Percival Lowell's brother-in-law William Lowell Putnam, who was a key player in the Boston financial community, devoted himself to the task of increasing the family patrimony, while Percival and Lawrence carried on their cultural and scientific pursuits. Putnam's marriage to Lowell's sister Elizabeth pooled the money and brains of two Brahmin families. Such pooled assets were one way in which the upper class addressed the relative decline of Boston capital in relation to the newly rich industrialists of New

York and Chicago. Putnam also encouraged the Lowells' involvement in an ambitious investment scheme, concocted by Boston and New York investors, that dramatically altered the financial orientation of the Lowells and other wealthy Bostonians. In this manner, Brahmins of Lowell's generation, despite a reputation for passivity in the financial arena, pursued an aggressive strategy to increase their assets and to encourage greater prominence for Bostonians in cultural and scientific affairs.[11]

In the fall of 1877 Lowell returned to Boston from his postgraduation grand tour of Europe with "no impulse toward a profession." Under the circumstances, he took the path of least resistance by following in the footsteps of his father and grandfather and, under their tutelage, learned to invest and manage the family assets. By virtue of this conventional choice, Lowell not only entered the investment business but also pursued an active social life with his Brahmin friends. Indeed, business, social life, and individual mood were seamlessly connected in the minds of Lowell's companions. As Frederic Stimson remarked, "The social market is dull, stock better. Harshaw @ 30 gives Percy $3,000 profit and me $600—Percy is boominger than ever." Thirty-five years later Wendell noted his friend's "heavenly" mood and "infer(red) him on the right side of the market."[12]

Lowell and his brother were each given $100,000 by their father, so in addition to having responsibility for family investments, Lowell also managed his own. According to his brother, Lowell was a "shrewd investor," which suggests that these original funds grew in value through his investing acumen. Indeed, at the time of his father's death in 1900, Lowell's assets exceeded $500,000—though it is not clear that all of the increase resulted from growth in his original investments.[13]

Lowell's decision to embark on lengthy trips to East Asia and Arizona after 1883 should not suggest that he had abandoned his work as a businessman. In fact, he recognized that his freedom to travel would be compromised by a decline in his investment income. Accordingly, he wrote Barrett Wendell during his first trip to Asia that "it's lucky my purse stretches across the Pacific or I should be bankrupt." Living off the income from his investments, made him acutely aware of the need to tend his portfolio during the periods when he was in Boston. To secure his finances, he continued to make regular visits to his Boston office; he also served as treasurer of the Lowell Bleachery in 1888 and as acting treasurer of Massachusetts Cotton Mills in 1894.[14]

Lowell was also aware of his own inferior financial situation in comparison with the newly rich industrialists. After visiting Teddy and Edith Wharton's Newport house in 1887, Lowell recounted that "I dined with seventy millions one evening, yet we were less in number than the Muses . . . I sat between Cornelius Vanderbilt and Mrs. Somebody Burden, somebody's burden indeed." It would appear that Lowell found compensation on this occasion for his financial inferiority by demonstrating his sense of cultural superiority. He remarked of Newport that "of all the inanities of which man is capable I believe that town can show the most utter [*sic*]. Plutus whom one might well mistake for Pluto reigns supreme." Vanderbilt and company's seventy millions would have dwarfed Lowell's own considerable holdings, just as he believed his charm and learning exceeded theirs.[15]

The largest share of responsibility for Lowell family finances fell on the shoulders of William Lowell Putnam, who was well respected by both his family (Lowell regarded him as "an excellent businessman") and such leading Boston investors as Henry Higginson and T. Jefferson Coolidge, with whom he served, respectively, as director of American Bell Telephone and A. T. and T. In addition, he was director of half a dozen other companies, including Waltham Watch Company and Tampa Electric, and served as an executive of such leading Boston financial institutions as the Massachusetts Hospital Life Insurance Company and the State Street Trust Company.[16]

Beyond their link to the Boston financial community, the Lowells also established a New York connection by virtue of the marriage of Lowell's sister Katharine to Alfred Roosevelt. In this way, Putnam and Lowell drew the Roosevelts into their financial dealings. They used Alfred's brother Kermit on occasion as a broker, and after Alfred died, they mentored their nephew James in the early phases of his career and discussed various investment prospects with him. Lowell, for example, advised his nephew against taking a position with the Third Avenue Traction Company because of its reputation for graft, and several years later subscribed to "the Bethlehem proposition . . . in order to help you on in your career." The nature of this "proposition," which Lowell later rejected, was never explained.[17]

While it is clear that Lowell accepted the overall strategy of Boston financiers and relied on his brother-in-law as a direct link to their operations, he was very much involved in decisions about the family's financial affairs. He was consulted about all major decisions, advised Putnam on investment opportunities, kept himself well informed about the financial situation of the country by reading various financial papers, and took an active role in the management of several companies. As a general rule, Lowell set guidelines for investment policy while Putnam took care of day-to-day matters. In return for Putnam's services, Lowell paid him a 2½ percent fee on the earnings from his investments.[18]

Lowell laid out his investment strategy in a series of letters to Putnam. Among his important principles was the value of concentrating decision making and assets. Lowell maintained that "money is made by concentrating not scattering one's interests" and that "a one-man power is best." He urged his brother-in-law to take the responsibility for buying and selling his assets, but this did not preclude a steady stream of his own advice about investment decisions. Letters to Putnam in 1903 listed promising stocks as well as poor investments, and set forth the reasons for Lowell's views. One of Lowell's important criteria for a good investment was that the company be run by a bold, imaginative leader. To derive maximum profits from an investment, Lowell urged on Putnam the underwriting of bond and stock issues. This strategy also promoted holding companies à la J. P. Morgan's, such as the one that Lowell, Putnam, and James Roosevelt considered forming in 1915. In addition, Lowell believed that investment decisions should take into account current economic and political conditions. In 1916 he evinced disappointment that "the war bonanza" had not yet boosted the value of his sisters' trusts.[19]

And of course Lowell was always eager to take advantage of Putnam's insider knowledge of good investments. It is not entirely coincidental that Lowell had invested almost half of his assets in A. T. and T. by 1903 or that he later invested in Tampa Electric, both companies for which Putnam served as director. And on one occasion Lowell urged his brother-in-law to "let me know all you know about the telephone situation. . . What you know undirectorily, but stockholderly so to say."[20]

Consistent with practices of other Brahmins, Lowell and family dumped virtually all of their holdings in New England cotton companies in favor of investments in other industries in other parts of the country. In the 1880s, leading Boston investors promoted the Calumet and Hecla Copper Mine of Michigan and such western railroads as the Atchison, Topeka, and Santa Fe. Lowell was interested in the mines but not the railroads. Even though the observatory was served by the Santa Fe, Lowell sold his stock in the railroad in 1897 and advised his sisters not to invest in it. By 1900, however,

Boston capital was moving to the South, where investors sought to merge existing southern railroads and power companies. For example, H. W. Poor and Company of Boston and New York underwrote a large bond issue for a Georgia railroad, much of which was purchased by Boston interests.[21]

The point men for these arrangements were H. M. Atkinson, a Bostonian headquartered in Atlanta, and P. S. Arkwright. Atkinson engineered investments in Georgia Railway and Electric, which proved so profitable that he organized two subsidiaries: the Atlanta, Birmingham, and Atlantic Railroad, of which he was president, and the Atlantic and Birmingham Construction Company. P. S. Arkwright served as president of Georgia Railway and vice president of the A. B. and A. Railroad.[22]

Lowell invested heavily in these three Atlanta-based companies. According to his will, fully 25 percent of his assets consisted of stocks or bonds of Georgia Railway and Electric. Lowell also served for several years as chair of the company's finance committee. As for Georgia Railway's subsidiaries, he wrote to Atkinson that "like Tarquin the Grand in the Sibylline Books—though for different reasons—I should now like to become possessed of 200,000 more of Atlanta, Birmingham and Atlantic Railway," and Lowell also invested heavily in and served as director of Atlantic and Birmingham Construction Company. In addition, as we have seen, Lowell bought shares in the Tampa Electric Company. While Georgia Railway and Tampa Electric were profitable enterprises, the two subsidiaries failed, and in the process Lowell suffered severe losses. Unfortunately no record exists as to the value of his stock holdings in the two companies, but his will indicates that bond losses amounted to slightly more than $200,000. Accordingly, there is no way to calculate how profitable the southern investment strategy actually was.[23]

The skills that Lowell brought to the investment and management of his property were extremely useful in establishing and maintaining the Lowell Observatory. Beyond generating an endowment to support the observatory, Lowell's business sense assured that the funds would provide the personnel, equipment, and setting for an effective operation.

The Lowell family ethos of concentrating money and power led directly to the governance structure of the observatory. Lowell adopted the sole-trustee model pioneered by his great uncle John in founding the Lowell Institute. According to this arrangement, the founder would designate a single male family member to govern the observatory following his death. Inasmuch as Lowell had no offspring, he would have to select a relative willing to devote time and energy to furthering the observatory's work. Even before his death, however, he had occasion to test the family's interest in the project. When he fell ill in 1897, Lowell turned to Putnam to manage the observatory.[24]

Of serious concern to Lowell was how the observatory would survive his death. To follow him, Lowell chose his cousin, the architect Guy Lowell, who served for ten years as sole trustee. Guy Lowell was in turn succeeded by Lowell's nephew Roger Lowell Putnam. The sense of family commitment to the project was evident even more clearly in the contributions of Lawrence Lowell. After the observatory endowment had been depleted through conflict over the provisions of the estate, Lawrence gave $10,500 to purchase the 13-inch telescope through which Pluto was eventually photographed, and then proceeded to write a well-received biography of his brother. Thus the Lowell family treated the observatory as a collective responsibility.[25]

In establishing the largest private observatory owned and supported by one individual in the United States, Lowell had taken on a heavy financial responsibility. To be sure, in 1894, when the observatory was founded, Lowell was a wealthy man. However, as we have seen, wealth is a relative concept, and in this case, it needs to be measured in relation to the cost of building and maintaining the observatory. (It is, however, meaningless to convert these expenses into modern equivalents, since the kinds and costs of instruments have changed so dramatically. And, unfortunately, the information at our disposal is rather sketchy.) It seems likely that Lowell had an annual income of about $25,000 in 1894, which, by the standards of the day, was a substantial sum. All the same, Lowell incurred considerable expenses. Start-up costs, not counting Lowell's own office and travel expenses, were $2,300 for the first four months. No figures are available from then until 1903. However, in 1896 Lowell purchased the 24-inch Clark refractor for $20,000 and funded what must have been an expensive expedition of seven individuals to Mexico City. During his illness, Lowell instructed his assistant Andrew Ellicott Douglass, through Putnam, that total annual expenses should not exceed $6,000 per year. Thus, even for a man with an income of $25,000, the creation and funding of an observatory was not an incidental matter.[26]

After his father's death, Lowell's annual income grew from $25,000 to slightly more than $100,000, which must have eased the strain of financing the observatory. He also began to build the value of his estate, which grew from $500,000 in August of 1900 to more than $2,000,000 at the time of his death. Financial records from the years 1903 to 1908 indicate that observatory expenses, including salaries, equipment, repairs, and new buildings, averaged a little more than $10,000 per year. This figure does not include Lowell's own travel and office expenses, nor the salaries of the secretaries and "computers" (mathematicians hired to calculate the orbit of Planet X) whom Lowell employed. And of course there were other significant costs, including $3,500 for his 1907 expedition to South America and $10,800 for a 40-inch reflector in 1909. However, even if we double the average annual expenses to $20,000, Lowell's income was more than adequate to support the observatory. Indeed, in 1906, Lowell seriously considered commissioning from George Ritchey an 84-inch reflector at a cost of $55,000.[27]

Lowell was faced with a financial problem of a different order following the introduction of the income tax in 1913. In negotiations with the Treasury Department, he insisted that a private institution like the observatory should be allowed to deduct expenses in calculating taxes. After an unfriendly ruling from the Treasury Department, Lowell pointed out to Secretary William Gibbs McAdoo that the decision favored public over private scientific institutions only because they were funded by many men rather than one. Meanwhile, Lowell mobilized the Brahmin network in his behalf, urging Senator Murray Crane of Massachusetts to seek the repeal or modification of the income tax, while his cousin William Sturgis Bigelow wrote to Henry Cabot Lodge to protest the discriminatory treatment of the observatory: "When an individual starts in by himself and does an absolutely first class thing off his own bat without getting under the skirts of the university for protection, this highly intelligent Democratic administration tries to squeeze him as if he were a 'Corporation.'" Lowell did not live to see the resolution of this issue.[28]

Granted the adequacy of his resources, launching the observatory from scratch, running it for twenty-two years, and putting it on a solid foundation for the future tested Lowell's skills in areas besides the financial. He had gained some experience in the management of cotton factories that helped to prepare him for directing the observatory, though he was never involved in the day-to-day running of the mills, as Barrett Wendell reported: "He did the honors of the [Massachusetts] mill with his grandest airs and I came back with a profound conviction that the inside ring of the cotton trade contains some soft and comfortable places." By contrast, A. G. Cumnock, the Scottish immigrant who served as manager of Boott Cotton Mills, wrote Lowell's father to report that "he [Lowell] has shown marked ability since you went away in all that he has touched."[29]

In the management of his own observatory, Lowell was particularly adept at selecting promising young astronomers whose work would be well regarded in the profession and who were willing to live in an isolated setting. It is significant that the only difficulties Lowell experienced with assistants occurred when he was incapacitated with neurasthenia. Of course, Lowell demanded the complete loyalty of his assistants, and he quickly fired A. E. Douglass after Douglass characterized Lowell's methods as unscientific. During the following sixteen years, however, Lowell hired and retained three asistants, Vesto Melvin Slipher, Earl C. Slipher, and Carl O. Lampland, who not only served under him but also stayed on for many years after.[30]

Keeping astronomers on the job required, in addition to salaries and housing, the purchase of increasingly expensive instruments. Lowell not only invested in two large telescopes but also supplied V. M. Slipher with a Brashear spectrograph that was the best that money could buy. He hired Stanley Sykes, an English-born machinist, to keep the equipment in good working order.

In dealing with his assistants, Lowell managed to strike a rough balance between insisting on their supplying data for his work on Mars and encouraging their own research projects. Observatory correspondence and Lowell's writings are full of praise for the work of his assistants. He generously granted leaves when they were ill and inquired after their health at frequent intervals. He wrote V. M. Slipher, for example, to ask if he had been able to procure a servant and if Lampland and maintenance man Harry Hussey and his wife were properly provided for. His genuine concern and charismatic personality generated real feelings of attachment among staff members at the observatory.[31]

Lowell not only directed his own observatory but also served as nonresident professor of astronomy at the Massachusetts Institute of Technology. Through this affiliation, he acquired instant academic credentials—at a nationally respected institution—that he hoped would counteract his reputation for unprofessional behavior. In his professorial capacity, Lowell delivered various lectures and occasionally exhibited his photographs of Mars.

While his professorship was an academic matter on one level, it should be regarded as a business proposition on another. Lowell's negotiations with MIT President Henry S. Pritchett in late 1902 occurred just a few months after all of Augustus Lowell's children had pledged funds for a new electrical engineering building in his honor. On June 30, 1902, Lowell sent a $10,000 contribution to Pritchett as his part of this gift. Immediately following the contribution, Lowell proposed an expedition, to be funded jointly by the Lowell Observatory and MIT, to determine the best astronomical sites in the world. Under the arrangement, Lowell would put up $8,000 while MIT would contribute $1,000 and supply the staff, which would be under Lowell's direction.[32]

The project was dropped after Pritchett's doubts about finding manpower and sufficient funds to implement it were reinforced by his astronomer colleague Simon Newcomb. Meanwhile, in October of 1902 Lowell sought a permanent affiliation with MIT, but bristled at the thought of accepting a lesser title than "Professor of Astronomy at the Institute, *honoris causa*." Pritchett, whose field was astronomic and geodetic science, was certainly aware of the controversy in the profession over Lowell's work; he accordingly rejected this proposal but compromised on the title "Non-Resident Professor of Astronomy." Exactly what role the Lowells' collective gift played in the appointment is uncertain, but it seems probable, given the timing, that President Pritchett wished to oblige a member of a family whose generosity might continue in the future. We should also note that William Lowell Putnam was then serving on the MIT corporation, the governing board of the institution.[33]

Lowell's shrewdness as an investor built a modest initial inheritance into a substantial portfolio, which could then be enhanced by the collective efforts of Boston colleagues. It is not entirely coincidental that their investment strategy should have targeted industries far from Boston, at a time when Lowell himself was using the income from his portfolio to build an observatory in the Arizona territory. It is also significant that Lowell's preoccupation with science was made possible by the newly developed strategy among Boston Brahmins designed precisely to free up men like him for cultural and scientific pursuits. The division of labor by which some tended financial matters so that others could engage in nonmaterial pursuits worked both to arrest the decline in assets among the group and to support new projects in other realms. This collective strategy also secured the future of the observatory by ensuring that family members would come to its support.

Traveler

In September of 1882, just eight months before his departure for East Asia, Lowell confessed to his sister Elizabeth, then traveling in Europe, that he "was born a traveler." Lowell's comment calls our attention forcefully to the transitional period he was about to enter and also to the central role traveling played in his life. Travel, as he put it, was a way of seeking the good rather than goods. It offered the prospect of more fully satisfying his zest for life and his yearning for adventure than the business career in which he felt trapped. It was both an escape from the limitations of the society he had been raised in and an opportunity to broaden his perspective by learning about other cultures and, through them, to understand his own way of life.[34]

Travel, however, could be broadening only if the traveler actually penetrated the material facade of a culture to engage its people, language, and customs. And a true cosmopolitan would attempt to discover geographically and culturally distant civilizations. Lowell was contemptuous of those who never ventured beyond Europe or who resisted the idea of transcending the Earth to discover and explore other cultures in the solar system. In this sense Lowell was the ultimate cosmopolitan, already preparing for the space age. However, in insisting on the value of intimacy between the traveler and the non-Western cultures he was seeking to understand, Lowell set a high standard that he did not always live up to.

In May of 1883 Percival Lowell abandoned forever his full-time involvement in the family investment business in favor of travel to Asia. That Lowell was mentally ripe for this transition is evident from comments he made to his sister Elizabeth about a recent visit to family-owned cotton mills. Lowell had treated the occasion as if it were a visit to a European cathedral. His spectatorial approach suggests that for Lowell the priority of managing cotton mills had been superseded by the traveler's quest to understand the meaning of cultural landmarks.[35]

It is hardly surprising that Lowell became an avid traveler, in view of the preparation he received as a child. From 1864 to 1866, during his formative years, the Lowell family established residency in Paris and sent their sons to a nearby French boarding school. At the end of this stint, Lowell had mastered the language and could communicate effortlessly in both written and spoken French. However, he was "so ill at ease in travel" that, while the rest of the family visited Italy, Lowell spent two months in a boarding school in Vevez, Switzerland.[36]

This case of nerves had clearly worn off by the end of his college days. Following graduation from Harvard, Lowell joined his cousin Harcourt Amory on a fifteen-month tour of Europe in 1876 and 1877. This trip was to set the stage for a lifetime of traveling for Lowell, including five trips to Asia in a ten-year period, ranging in length from three to ten months. Two of these trips included stopovers in Egypt, India, Burma, Indonesia, and China. Lowell was also one of the first Westerners to undertake more than a superficial visit to Korea, stopping there for three months in 1883 and 1884. After 1894 the periodic Asian trips were replaced by sojourns of two weeks to seven months in Flagstaff. The western desert jaunts were interspersed with semiannual European tours, as the work of the observatory permitted. It is no wonder Lowell sympathized with "the lust of vagabondage" that seized Harry Hussey, the observatory maintenance man. Lowell clearly had a strong case of it himself.[37]

The laundry list of countries he visited could be repeated for many wealthy Americans of Lowell's generation. In the wake of the Civil War, and with the simultaneous completion of the Suez Canal and the transcontinental railroad, as well as the invention of the packaged tour by Thomas Cook, globe-trotting became fashionable, and many wealthy Americans and Europeans traveled around the world. However, Lowell's traveling was of a different order in the sense that he had thought a great deal about the role of traveler and the larger meaning of this activity for his own life. He understood that not all bodily movements over the globe should necessarily qualify as travel. Indeed, Lowell's remark that true travelers journeyed for good and not for goods implied his own rejection of the tourist's superficial approach to foreign cultures. Accordingly, his books are interspersed with pejorative comments about "mere tourists" and "globe-trotters." The true traveler was something of an anthropologist, and would seek a deeper involvement in and understanding of the host culture.[38]

As early as his postgraduate grand tour, Lowell espoused this anthropological approach to travel. He wrote from Amsterdam to Wendell, for example, apropos of contacts he had made with Dutch people, that it was "pleasanter to see something of life and manners of a people than simply to pass through as the average traveler does seeing nothing but the outsides of things." Similarly, he remarked to his sister Elizabeth in 1886 how much better she would know the West for having stopped in Salt Lake City—as he had not—to visit the Mormons.[39]

Lowell connected travel with the opportunity to learn. Of course different types of travel were appropriate for different types of learning. The "age of bodily adventures" experienced by explorers of undiscovered lands, as Lowell remarked in *Chosön,* had already passed. In his era, travel was better suited to profound discoveries about cultures, based on intimate acquaintance. And discoveries of a different sort could be made in astronomy, which Lowell spoke of in both literal and metaphorical terms as a form of travel. Contemporary astronomers were obliged to travel to remote areas that had pure, steady air through which to view the heavens; their observatories were thus "monasteries in the wilds." "To sally forth into the untrod wilderness in the cold and dark of a winter's small hours of the morning . . . is almost to forget one's self a man for the solemn awe of one's surroundings. Fitting portal to communion with another world."[40]

The telescope, in turn, provided the means for the astronomer to continue his "journey" and to "visit" another world. "Since the seeing strange sights is the essence of all far wandering, it is as truly travel so the eye arrive as if the body kept it company." In effect, Lowell believed, telescopic exploration of Mars, like a journey by foot or train in Japan, revealed physical and cultural features of the landscape and its possible inhabitants. Lowell contrasted the relatively simple Martian polar expeditions conducted through the telescope with their frostbitten earthly counterparts. Extending the metaphor, he dedicated *Mars and Its Canals* to his hero Giovanni Schiaparelli, whom he considered the Columbus of astronomy for his discovery of the canals of Mars. "He made there voyage after voyage, much as Columbus did on Earth, with even less of recognition from home. However, the full import of his great discovery lay hid even to him and only by discoveries since is gradually resulting in recognition of another sentient world." In a similar vein, he considered his own words as "travelers" whose wisdom might reach the readers of his books through the printed page.[41]

This notion of the traveler as a knowledge seeker and knowledge giver was expressed most clearly in the search for a "cosmopolitan" perspective that Lowell had begun during his tour of Europe after graduation. At that time Lowell confessed that he had no great feeling for America. "As for what other people call patriotism, I have it not, except more or less towards England." This attachment to England, based on a fondness for old things, in no way interfered with Lowell's cosmopolitan agenda. He scorned the pallid contrasts between innocent Americans and jaded Europeans featured in the work of Henry James and Paul Bourget. "They think themselves cosmopolitan; in fact they are bornés [French for "narrow" or "limited"]." James and Bourget spent too much time exploring "social etiquette" and too little investigating the more profound dimensions of human existence. Appropriately, then, Lowell's books on Asia were devoted to achieving a better understanding of the essential elements of Western civilization by exploring the contrasts between West and East. Remarking on the backward way in which East Asians appeared to do things, Lowell noted that "it is because the far-East holds up the mirror to our own civilization . . . because by her very oddities, as they strike us at first, we learn truly to criticize, examine, and realize our own way of doing things, that she is so very interesting." She offers "vistas into the springs of the world's life, as a whole."[42]

In a similar fashion, Lowell urged on his readers the value and interest of planetary civilizations that were truly different from those on Earth. They would offer a greater "parallax" for "cosmic comprehension," or a "cosmoplanetary breadth of view." Reckoning with varieties of intelligent life would enable Earth-bound men to more fully understand themselves. Despite this theorizing, however, Lowell proceeded to depict the character of Martian civilization in a thoroughly mundane fashion. Mars, he supposed, must be run by a committee of technocrats resembling MIT professors, who presided over a more advanced version of the developing American technocracy. And he strongly opposed the new stellar research agendas of larger observatories, even though their efforts to gauge the boundaries of the universe could be regarded as more "cosmopolitan" than his own effort to investigate surface features on a nearby planet.[43]

Curiously enough, Lowell's cosmopolitanism was not inconsistent with the "broad-minded localism" he attributed to Oliver Wendell Holmes, Sr., in his 1895 obituary. Lowell admired Holmes's keen sense of place, which he demonstrated most clearly during a two-year stint as a medical student in France. Rather than adopt superficial French customs, Holmes remained true to his New England roots. According to Lowell, the universal appeal of Holmes's work was based precisely on this clear sense of time, place, and personal identity.[44]

In the course of his own comings and goings, Lowell gave serious consideration to the concept of travel as cultural engagement. He clearly admired those who could forgo material comforts to achieve greater intimacy with East Asian culture. In principle he advocated the Asian way of living simply and rising above bodily needs, in contrast to Westerners' dependence on modern conveniences. The former, he believed, achieved a kind of spiritual transcendence. And at least during his early weeks in Japan, Lowell aspired to measure up to this standard. He claimed in a letter to Elizabeth, "I am beginning to talk Japanese like a native (of America) and I take to ye manners and customs of ye country like a duck to water."[45]

However, despite his best intentions, there were limits to the engagement Lowell sought with East Asian culture. He never achieved either fluency or literacy in Japanese, nor did he adopt the ascetic lifestyle he so admired. In fact, Lowell tried to steer a middle course between globe-trotters, on the one hand, and on the other, foreigners like Lafcadio Hearn who "went native" by marrying Japanese women, converting to Buddhism, wearing Japanese clothes, and speaking the language. The lifestyle adopted by Lowell and other expatriates militated against such intimacy. They were professionals, most of whom taught, wrote, or practiced law, and they frequently saw other English-speaking expatriates. Nonetheless, they contrived by various ruses to live outside of the foreigners' compound in Tokyo, even when they were not employed by the Japanese government, and to interact on a regular basis with a special group of Japanese nationals.[46]

At the center of this lively circle were Basil Hall Chamberlain and William Benjamin Mason, both long-time residents of Tokyo, coauthors of a revised English-language travel guide to Japan, and neighbors of Lowell. Chamberlain had garnered a reputation for his knowledge of traditional Japanese literature and religion. Lowell was also closely involved with two Bostonians, Ernest Fenollosa and William Sturgis Bigelow, who were leading American authorities on East Asian art and disciples, like Lowell, of the Salem, Massachusetts, zoologist and expert on Japanese material culture, Edward Sylvester Morse. Indeed, Fenollosa eventually became curator of the Boston Museum of Fine Arts "Oriental" art collection, much of which he had himself assembled during a seven-year stay in Japan. Lowell's cousin Bigelow, a Boston doctor, became Fenollosa's sidekick and frequently joined him in the search for art objects. Lowell was also in contact with John Henry Wigmore and Gustavius Goward, who were, like Bigelow and Fenollosa, fellow Harvard graduates. Goward was secretary of the American legation in Tokyo, while Wigmore was an attorney and an expert in Japanese law. Lowell accompanied Henry Terry, professor of law at Tokyo University, on a visit to the interior of Japan in 1883, and he incorporated George Agassiz, Ralph Curtis, and John and Isabella Stewart Gardner into his circle during their brief visits to Japan.[47]

This Anglo-American group saw much of one another, but also mixed with Japanese nationals who spoke English, had traveled abroad, and either held or were training for positions in the educational or political establishment. Lowell met his Japanese friends through Chamberlain and Morse, among others. Morse, for example, introduced Lowell to Miyaoka Tsunejiro, who, as his secretary, accompanied Lowell to the United States with a Korean delegation and then traveled to Korea with him. Miyaoka later became a specialist in international law. Through Miyaoka, he met Takenaka Seikin, whose family Morse had visited in Kawagoe, and Takamine Hideo, director of the Tokyo Normal School. Morse spoke of a visit to the Takamine home where he enjoyed "a most delicious dinner in pure Japanese style," and he remarked that "Dr. Bigelow ate every course with genuine relish" while Mrs. Takamine played "Auld Lang Syne" and "Home Sweet Home" on the violin. Through Chamberlain, Lowell met Masujima Rokuichiro, the head of the English law school of Tokyo whose house Lowell rented in 1889. In addition, Lowell was introduced to Ariga Nagao, professor at the Normal School, statesman, diplomat, and historical scholar who was the translator of Fenollosa's lectures on Japanese art. It is clear that Lowell was drawn to his Japanese friends because they had adopted European ways. For example, he remarked that Miyaoka "is very clever and at least half a European. I should say about three quarters one."[48]

The lifestyle of the Anglo-Americans emerges quite clearly from the robust letters Lowell wrote to family and friends. They describe a mixture of serious scholarly activities, travel, and rounds of parties, teas, and art shows. Many in the expatriate community studied Japanese culture and presented their findings at the Asiatic Society of Japan, which had a scattering of Japanese members. Lowell and others attended monthly meetings of the organization, read papers, and published them annually in the society's journal, *Transactions of the Asiatic Society of Japan.* In addition, virtually all of Lowell's friends wrote well-regarded books and many also taught at institutions of higher education, especially Tokyo University. Other centers of cultural activity were the Yokohama Literary Society, to which Lowell lectured on several occasions, and the Tokyo English Club, a club for Japanese nationals who wished to learn English.[49]

On the social front, Lowell engaged in activities that brought him into relatively high-level contact with foreigners and Japanese, provided occasions for learning about Japanese culture, and created an outlet for his sociability. As Lowell explained, "I am making the acquaintance of the eligible or as Chamberlain might put it the fruitful of the foreign population." To manage these acquaintances, Lowell needed to be in the right place at the right time. In this regard, his charter membership in the Rokumeikan (literally, the Hall of Deer Cry) was crucial. The institution had been designed by Japanese officials to provide comfortable quarters where foreigners and Japanese could socialize on a level of equality—facilities that were generally not available in Japanese homes.[50]

Lowell's various homes, located outside of the foreigners' compound in Tokyo, were also conducive to advancing his social life, especially given their size and appointments. In 1893, for example, he rented a house with eighteen rooms. Nine years earlier, he reported that the grounds of his house contained "a fine tennis lawn, a famous rose garden not to mention a miniature Fuji in the same. My house personnel is luxurious—cook, boy . . . gate keeper, kuruma man." On the same garden theme, he wrote his brother-in-law that he was "now thinking *more romanico* of making a Fuji and a chain of hills along my avenue and a part of my garden. That is something like in the way of magnificence." The houses were apparently furnished in a Western style, since Lowell shopped for furniture at auctions in Yokohama.[51]

In this fashion, Lowell was prepared not only to enjoy the hospitality of others but also to offer some of his own. Indeed, he attended a steady round of parties with his small set of Japanese and Anglo-American friends. Cherry blossom season produced invitations even from the emperor and other dignitaries. Lowell noted a few impressions of his first visit with the royal family: "Very pretty little garden party. Most frightfully select . . . Grounds beautiful, Mikado and Empress not so beautiful though she is said to be on Mongolian standards." Seven years later, Lowell arrived at the emperor's gate for a cherry blossom party but was turned away for not wearing a frock coat. Just a few days earlier he had attended another party, this one given by a Russian diplomat named Sienkiewicz; he wrote Amy that "the flowers were all out in full force and so were the elite of Tokyo—old friends to be recognized, new faces to be introduced to. I enjoyed myself much . . . There was of course a band and dancing on the lawn. I introduced Ralph [Curtis] and he is now launched upon the giddy whirl." Lowell himself repaid the hospitality by offering his own cherry blossom party. However, the Japanese foreign ministry turned down his invitation en masse because, as Bigelow explained, their attendance would have implicated them in the illegality of Lowell's living arrangements outside of the foreign enclave.[52]

Theater, painting sessions, wrestling, and polo matches were also occasions for becoming acquainted with Japanese high culture, as well as for entertaining. Lowell invited Takamine, Ariga, and Takenaka along with Bigelow, for example, to a thirteen-hour theater party. He commented that "the acting is admirable and besides the usual interest is the opportunity to see customs and costumes which have passed, or are passing away never to return." As a polo player himself, Lowell eagerly invited his friends to a polo party so that he could take advantage of the opportunity to see how the Japanese played the game. They had the satisfaction of viewing "former daimyos to the number of the Muses playing like the Graces an old noble sport for one's special benefit." Lowell also reported on what he called the "Japanese Whistler's Afternoon Tea," at which artists gathered to paint designs on paper or fans free of charge. And after Bigelow introduced him to sumo wrestling, Lowell attended another match where he arranged to meet with the wrestlers beforehand in his box.[53]

Lowell created a life for himself in Tokyo that established a comfortable middle ground between the superficial activity of the tourist and unrestrained engagement with mainstream Japanese culture. By associating with knowledgeable Anglo-Americans, he was introduced to a select group of Japanese men familiar with Western ways. Together, they enjoyed the opportunity to show off or to become acquainted with certain elements of Japanese culture. Lowell, however, was not satisfied with the rather protected lifestyle of the expatriate colony. He apparently wished to experience the challenge of venturing into the interior of Honshu and Korea, to witness both the exotic lifestyles of those who had not been exposed to Western customs and the beauty of the landscape. In Japan, Lowell succeeded in gaining permission from the government to travel beyond treaty-port limits, but his efforts to journey outside of Seoul in Korea were thwarted by weather and the poor road system.

In departing from Tokyo, Lowell cut loose from his friends to confront Japanese and Koreans who spoke no English and were unfamiliar with Western ways. (On three of these trips, however, he did travel with an American friend.) Lowell also abandoned the comforts of his home, including a bedstead, in favor of local sleeping arrangements. Among other things, he would have had to accommodate unfamiliar transportation, food, and housing. In the process, of course, his model of the traveler who sought to experience a new culture on intimate terms was tested and found wanting. Lowell fell back on Western ways.

Lowell's greatest challenge, evident even in Tokyo, was adjusting to an

East Asian diet. From the outset he was incredulous that people could survive on rice and vegetables. As he explained in his book *Noto,* "We are not inwardly contrived to thrive solely on rice and pickles." To his brother he wrote: "Think of the means of subsistence in a land where there is no milk, no butter, no cheese, no bread, almost no meat and not over many eggs." Accordingly, before leaving Tokyo to journey inland, he bought beer and "a stock of canned provisions at a native shop in Tokio patronized by foreigners . . . The above-mentioned provisions are cooked by my boy who also looks after our comfort in other particulars." In 1889, when these items had become available in the interior, Lowell sent his "boy" to buy them while he was sitting "japonico" in an inn writing a letter. "No one," as he reported to Barrett Wendell, "lives wholly on Japanese food." However, when *Noto* appeared, one reader complained that the author should have eaten the native diet. Lowell's response suggested a certain sensitivity to the charge of avoiding Japanese food: "I might perhaps have said so without doing it but to do it, poetry or not poetry, never." Recognizing Lowell's resistance to a steady diet of East Asian food, the members of the Korean delegation Lowell had accompanied to the United States, who had, in turn, invited him to visit Korea, took with them a Japanese cook from Nagasaki familiar with European-style cuisine. Nonetheless, Lowell seems to have enjoyed eating Korean food as a guest at banquets. And so long as it was not part of a steady diet, he found a "Jap feast even to a European palate is delicious." Clearly, however, Lowell's dietary notions created a barrier between him and his hosts that was reinforced by the constant presence of his "boy." Moreover, in failing to overcome his aversion to Asian cooking Lowell essentially compromised his commitment to fully engage with the people and customs of the country.[54]

In two other respects, Lowell did somewhat better in accommodating East Asian ways. He always stayed at Japanese inns in the interior of Honshu, except when he visited the Western-style resort at Miyanoshita, near Mount Fuji, which he found more to his taste. In Korea he insisted on having his own house at the Foreign Ministry in order not to stay with the American minister Lucius H. Foote and thus to become identified with one or another Korean political faction. His accommodations, however, were always modified to suit his tastes. For example, when Lowell stayed in a Japanese inn, he padded his bedding by using two futons to sleep on and one to roll up as a pillow rather than the single futon allotted to most Japanese. (Apparently even that routine did not suffice, because five years later he noted that his boy piled up half a dozen futons to increase his comfort.) Though his house in Seoul was equipped with European chairs, he appears to have accepted a variant of native sleeping arrangements there, despite the fact that they were uncomfortable; on this "bedinstead," he complained, he got less sleep than he would have in a Western-style bed.[55]

By contrast, Lowell seems to have enjoyed trying out various modes of local transportation. He wrote his sister Elizabeth, "I have just had a most interesting journey and have tried every means of conveyance from the most aboriginal to nineteenth century kinds, having just got off the train. Truly a land of oddities where one sees a jumble of life." He found the Korean palanquin uncomfortable and was amused by the Japanese jinricksha, which he described to Amy as "a big baby carriage . . . drawn by a man." Because the jinricksha was uncomfortable for long trips, he approved of alternating stints in the jinricksha and the *basha,* or horse-drawn carriage. Traveling home on board ship, Lowell experienced a certain nostalgia for these modes of transportation while suffering the boredom of a long voyage across the Pacific.[56]

Lowell's response to the landscape was much less problematic than his reaction to East Asian people and their customs. He was, from his first ventures into the interior, impressed by the extraordinary natural beauty of the mountains and rivers. That beauty was enhanced by the sensitivity of the Japanese to preserving their forests and their landscape. He remarked that "the gods are wise enough to like trees, a religious belief I should like to see the effect of in America, where local fiends destroy instead of reverencing." Lowell was particularly taken with the contrast between the respect shown by Japanese farmers to existing contours of the land and the carelessness of Americans. The Japanese farmer, he said, conforms "to his surroundings and defiles not what he touches."[57]

But the landscape was appealing as well because it offered challenges to Lowell. He was eager to try his hand at mountain climbing and tackled some of the highest and most difficult peaks in Japan. His interest in volcanic mountains—he climbed Fuji, Ontake, and three active volcanoes: Shirane, Asama, and Bandai—became almost an obsession. These visits enabled Lowell to satisfy his scientific impulse to examine the scenes of current or past volcanic activity. In the process, he came into contact with an austere and exotic landscape, and as his description of the weird stones and the hot spring at Kusatsu makes evident, he found this landscape aesthetically pleasing. Moreover, the ascent of Ontake enabled him to witness trance-practicing sects, increased his "literary budget," and later on helped him to write *Occult Japan.* His venture up Shirane had become the subject of a talk to the Appalachian Mountain Club and was published in the club's journal. While Lowell also enjoyed his floats down the Tenriyugawa and Fujikawa rivers, the latter in the company of William Sturgis Bigelow, he did not register the same intense emotional response to rivers as he did to the experience of climbing and exploring volcanic mountains. In this regard, it is significant that Lowell was in charge of the mountain ascents, while on the rivers he was in the charge of Japanese guides.[58]

Lowell's adventures took him beyond the confines of the West to experience East Asian civilization and then to an encounter with Mars. These expeditions were meant to develop a more cosmopolitan spirit that would enable him and others to more clearly understand their own way of life. However, Lowell's capacity for experiencing different customs and routines was limited. He sought the company of the expatriate community in Tokyo, where he could mix with a few westernized Japanese, frequent Japanese spectacles, and enjoy the comforts of his own home and the companionship of Anglo-Americans. Nonetheless, Lowell's travels expanded his own horizons and, through his publications, those of other Americans. He presented new images of Korea and the interior of Japan, which had been *terra incognita* before his adventures. His travels also helped to liberate him from State Street routines and thus from ties to Boston, business, and his father. And in the process, Lowell tested his interest in the cultural and scientific pursuits he was about to take up.

4

◆ ◆ ◆

New Careers

Lowell's endeavors as businessman and traveler provided the essential underpinnings for his careers in communication and science. His investments gave him the financial wherewithal to sustain new undertakings. His journeys provided subjects about which he could communicate. But neither business nor travel was fully satisfying, because neither permitted Lowell to become the public figure he wished to be. As investor, he remained largely within the family circle; as traveler, he satisfied a personal need to expand his horizons. The mastery of communication skills and scientific methodology, however, enabled Lowell to break out of his isolation and satisfy a yearning for fame as well as fortune. In opting for a career as a communicator, Lowell put himself on stage and discovered the pleasures of interacting with a large and receptive audience.

As his interest in East Asia declined, Lowell found a subject of greater interest to both him and the public. Having already traveled among and written about a distant and exotic people, he committed himself to an even more sensational adventure, the exploration of life on Mars. His telescope would be his new means of "traveling," allowing him to look in on an extraterrestrial world. Lowell would again seek ethnographic evidence, but this time it would be gathered through the telescope rather than the naked eye. Despite these continuities, though, the exploration of Mars required the construction of an observatory and the mastery of a new discipline. Already committed to science, Lowell would now become a true believer in the scientific method.

Writer, Speaker, Photographer

Lowell possessed many of the attributes of an effective communicator. He was gifted with a vivid imagination and great clarity of mind, and he mobilized these aptitudes effectively in shaping a unique career as a writer, speaker, and photographer; in this endeavor, he found solid support from his family, friends, and the network of Brahmin cultural institutions devoted to the arts. As Lowell recognized, however, the cultural world into which he had been born was in decline. Increasingly, New York was becoming the center of art, publishing, and other cultural activities. In response, Lowell proposed in 1887 to revive Boston's primacy in letters by founding the Authors Club, which would give more recognition to Boston writers of his generation. Despite the club's failure to take hold, Lowell's effort underscored his conception of writing as a public activity that required communal support.

In his biography of Percival, Lowell's brother Lawrence reported that "literature in the broad sense" was the "field he proposed to enter." It was, in effect, an excellent field for Lowell, because by adopting it he was able to demonstrate his extraordinary versatility. He wrote with equal ease travel books, works of science for the general public, specialized scientific papers, poems, lectures, and narratives. He illustrated several of these written works with excellent photographs that he himself had taken, and he frequently delivered captivating lectures on Japan or astronomy, as the occasion demanded. In this way, Lowell had a foot in both cultures at a time when the divide between them was growing; indeed, he thought of his scientific projects in important respects as essentially "literary" in character. And he might have chosen literature as his principal career had he not been discouraged by some of his friends, who were offended by precious tendencies in his style.[1]

Granting Lowell's astonishing versatility as a writer, it is nonetheless clear that he was most comfortable in casting his work—prose and poetry—into the form of romance aimed at a male audience. Working within this genre, Lowell often featured the heroic adventures of men who pitted themselves against obstacles, human and natural, and confronted extraordinary or mysterious events. While in his travel volumes and some of his scientific works Lowell was the hero of the plot, others played this role in certain of his narratives. However, in adopting the somewhat antiquated form of the romance, he deliberately rejected the currently fashionable literary realism, with its attention to surface detail. For his purposes, the romance was superior to realism because it strongly emphasized the inner qualities of individual characters. No doubt Lowell's emotionally charged prose style, in which the "I" was often at center stage, put him at odds with professional scientists who sought to report their findings in spare, impersonal, and unemotional prose.[2]

In fact, Lowell's emergence as a writer was anything but a lonely and heroic struggle against obstacles of any sort. Along the way, he found support from family, friends, and Boston's cultural institutions. Lowell came from a literate family that placed a high value on learning. His communication skills had also been sharpened by a classical education at the Noble School and Harvard. Lowell's love of language and precocity as a writer were demonstrated at age twelve, when he retired to his room at the family home in Brookline to commemorate the shipwreck of a paper boat in 100 lines of Latin hexameter. By the end of his college days he had become self-conscious about writing, as evidenced in his praise of Barrett Wendell's letters as "philosophical" for interpreting the events they recounted.[3]

It is no coincidence that Lowell's career as a writer was launched as he began his travels in Asia. His experiences there provided excellent subjects first for letters to his family and friends and then for his lectures, writings, and photographs. Before the end of his first year in East Asia, Lowell announced to his sister Elizabeth that he was not only "expected" but "urged" by his Korean hosts to write a book about his experiences. And despite Lowell's protest that he had a "wholesome dread of publication," his correspondence with family and friends prepared him for this undertaking. He wrote in letter books, so that he could keep copies of the letters he sent home. This practice enabled Lowell to tell the complete story of his travels by writing parts of it to different members of the family, who could then share the letters. At the same time, he was creating a record that could be used in writing his books. Indeed, many passages in the letter books later found their way into published works.[4]

Lowell's correspondence with Stimson and Wendell became an important occasion for practicing and reflecting on writing styles. For example, Lowell wrote Stimson that he had reread his note "as a good preparation . . . for my style." And, later: "I admire, my dear boy, your style and your ideas very much . . . and respect highly your literary criticism, but what do you mean by so decidedly preferring mine of the 29th Dec to that of the 12th Jan." As for Barrett Wendell, Lowell noted his "description of ye wedding of Georgeosis Peabody Gardner Bachelor of AB (of the arts of enticing ye lovely maiden)," which he found "delicious." To Wendell Lowell enclosed his essay on Korean hats, which, as he remarked, took Carlyle's essay on clothes as its model. A revised version of the letter on hats became the text for Lowell's chapters on this subject in *Chosön*.[5]

Once back in Boston, Lowell took advantage of family and friends to improve his writing. Indeed, after deciding to write a book on Korea, he had threatened his mother with "the exquisite ennui of reading it before it goes to print and then you know we can have corrections and improvements by the family." Sometimes, in fact, Lowell came to family occasions armed with a manuscript from which he proceeded to read aloud. Even more important for Lowell's career was the regular interchange with William Lowell Putnam, in conversations and correspondence, about any and all aspects of his work on Japan and Mars.[6]

A number of institutions also provided valuable opportunities for Lowell to test his work on an audience. As we have seen, he tried out his ideas on the membership of the Yokohama Literary Society and the Asiatic Society of Japan before committing himself to print. He also benefited from his family's control of the Lowell Institute to deliver lectures once his projects were nearing publishable form. In addition, Lowell had ready access to the *Atlantic Monthly*, whose founder was cousin James Russell Lowell. And, once a book manuscript was complete, he successfully approached Boston's leading publishers, Ticknor and Houghton Mifflin.

This web of associations was augmented by Lowell's membership in the St. Botolph Club, which provided not only an excellent setting for Lowell's talks on Japan and Mars but also potential recruits for the Authors Club Lowell hoped to form. The idea for creating such an organization emerged from conversations with Stimson, who had recently been invited to join the New York Authors Club and offered to propose Lowell as well. Instead, Lowell and other members of the St. Botolph decided to create a Boston counterpart to the New York organization as a way of enhancing the status of Boston writers. In Lowell's view, the city's existing literary groups were not effectively serving this function: "Papyrus being also journalistic—the tavern generally Bohemian and the St. Botolph everything but the commonplace." In effect, the new organization would be less distracted by social concerns and insignificant literary activities.[7]

Lowell, as chairman, played a leading role in attempting to launch the club, while a committee of five, led by T. W. Higginson, a frequent contributor to *Atlantic Monthly,* was charged with considering its feasibility. The committee invited the establishment of Boston letters, James Russell Lowell, Oliver Wendell Holmes, Sr., William Dean Howells, and Thomas Bailey Aldrich, to endorse the project and thus to give it visibility. Then a circular was to be sent to approximately fifty Boston authors, including Lowell, Wendell, and Stimson, who had each published at least one book and been nominated by an appropriate party. (Similar requirements governed the selection of members for the New York club.)[8]

A circular and questionnaire designed to discern interest among prospective members were sent to a list of sixty-six names, while Lowell located a room on Park Street to serve as a meeting place for the club. The project, however, failed to excite the interest of its prospective members. Only twenty-four of the nominees expressed an interest in joining the Authors Club, while twenty-eight never bothered to answer the circular. How much the response was affected by reports in the press of false claims by the club's founders is uncertain. At any event, the *Boston Advertizer* alleged that the four distinguished authors had not been contacted before their names appeared on the circular. In support of this claim, Holmes, interviewed by the *Advertizer,* confessed that although he was eager to support young writers, he had known nothing of the club. Other established writers soon let it be known that they were not interested in the project. Furthermore, according to the paper, the club had no criteria for choosing the fifty authors who were to be invited to join.[9]

While the charges of the *Advertizer* were vehemently denied by the rival *Boston Evening Transcript,* the press controversy put a damper on plans for the club. In the last analysis, however, lack of interest among the authors themselves was responsible for killing it. The committee of five announced that a meeting to deal with organizational matters would be put off until the fall; no record exists of any effort to revive the club before 1899, when several women writers took up the task.[10]

The demise of the Authors Club effectively ended Lowell's organizational approach to enhancing the prestige of Boston men of letters, but his failure in no way deterred him from establishing his own career as a writer. In his work, Lowell straddled the growing divide between science and the arts by adopting a philosophy of writing that emphasized the originality of the writer in his role as thinker. He argued that "the fact gives us but a flat image. It is our reflections upon it that make it a solid truth. For every truth is many-sided . . . thus true seeing is done with the mind from the comparatively meagre material supplied by the eye." A creative author could write about any subject whatsoever; moreover, by applying different treatments to the same subject, he could address his thoughts to different audiences. "Only the arrangement [of facts] differs with the character of the book. You string them into a necklace for the world at large. You pigeon-hole them into drawers for the scientist." For Lowell, then, a work could be addressed either to the educated public or to professional scientists, depending on the treatment of the material.[11]

Accordingly, Lowell conceived of his popular scientific works as essentially literary. He reported to Putnam, for example, that he "tried to make it [*The Solar System*] both scientific and literary." What this meant for both Lowell and his critics was that such a book would present scientific truths in a nontechnical fashion. Indeed, Lowell believed that popularization of science was a greater challenge than writing for a professional audience. It could be accomplished only by he "who is master of his subject and is both the best test of, and the best training to, that post."[12]

Lowell's success in crossing disciplinary boundaries and making scientific subjects available to the general public may be judged by the response of his contemporaries. No less an authority than Lafcadio Hearn remarked: "It is strange that Lowell should have written the very best book in the English language on the very old Japanese life and character, and the most startling *astronomical* book of the period,—'Mars'—more interesting than any romance." And in fact many of the critics who read Lowell's books insisted on their contribution to both science and literature. The *Atlantic* reviewer of *Chosön*, for example, remarked that Lowell possessed "a rare combination of qualities, a threefold strain, poetic, scientific and human running at the will of an exceptionally active intelligence."[13]

We may well ask how Lowell succeeded as well as he did in keeping a foot in both camps of the two cultures. An important element of his appeal to readers was the dramatic and personal way in which he presented his work on Mars and East Asia. No doubt, his enthusiasm for the literary romances of Robert Louis Stevenson, Edward Bulwer-Lytton, and Frederic Stimson played an important role in Lowell's conceiving of science as a story that could, as Hearn pointed out, rival the adventures recounted in novels. Lowell's books thus appealed to an audience of men and boys who sought in literature, as in life, male heroes with whom they could identify. The romance provided the perfect vehicle to deliver these stories to the public in an attractive form.[14]

The romance worked for Lowell's readers because they experienced a vicarious sense of participation through identification with the narrator or hero. Lowell insisted, for example, on the importance of the scientist writing about his own discoveries, to provide the reader with "an aroma of actuality" that would in turn make him a "codiscoverer." He also appealed to his readers by using homely illustrations to illuminate larger issues. For example, Lowell described a meteorite as a "messenger" that could unravel the whole story of the development of the solar system from an original nebula. Its message was effectively presented in the form of "celestial hieroglyphics."[15]

Lowell's own adventures as scientist and explorer were the true subjects of his books, to which factual information was subordinated. He wrote in 1909 that "to be a pioneer is to stand alone with nature . . . Yet think what it is to pass one's day in a thought-world where the thinker roams alone; to grapple with problems the very terms of which are beyond ordinary comprehension and the solution appreciated only in years to come; to contemplate in lonely ecstasy, after still lonelier despair, the revelation come with months and more of pondering."[16]

This theme of the lonely pioneer pitted against the elements and against his fellow men was to dominate Lowell's most important work addressed to the public. The great moments of scientific discovery were frequently recounted in this vein. One excellent example is his oft-quoted passage on flashing lights from Mars that he concluded were "the glint of ice-slopes" rather than signal lights. These lights, "after their travel of one hundred millions of miles, found to note them but one watcher, alone on a hilltop with the dawn." In a similar fashion, Lowell spoke of what he called the "romance" of seeing double canals by which he meant their strangeness: "Nothing astronomical I have ever seen has been so startlingly impressive as my first view of a double canal." The personal drama was often enhanced by Lowell's reminders that he was waging a lonely battle against prevailing scientific opinion to establish his own views. While this emphasis on the personal adventure in doing science was well received by the educated public, Lowell's astronomy colleagues were clearly disturbed by his emotionally charged rhetoric and egocentric focus.[17]

Of course, travel accounts lend themselves to the romance form even more readily than popular science. Indeed, in two of his East Asian books, Lowell takes center stage as the subject of a series of unusual adventures he and the reader experience together. In *Noto* he recounts his own hair-raising attempt to cross Harinoki Toge, a high pass in the Japanese Alps, while it was still snow covered. In *Chosön,* he presents his first glimpse of Seoul as one which "completely realized the fancies of my boyhood dreams." For the rest of his visit Lowell conveyed this same sense of surprise at the unexpected and exotic ways of the Korean people.[18]

It was not necessary, however, for Lowell to assume center stage for the romance to succeed. He was also capable of writing superb narratives in which other heroes played the central role. Two *Atlantic Monthly* essays, for example, related, respectively, the assassination of Japanese minister of education Mori Arinori and the failed Korean coup d'état of 1884. To be sure, the drama of both narratives was enhanced by Lowell's emotional involvement with the two heroes, who were both Westernizing reformers. Furthermore, he had personally accompanied Hong Yong Sik, the first Korean postmaster general, to the United States as secretary to the 1883 Korean mission and thus could appreciate his willingness to sacrifice himself in order to allow fellow revolutionaries to escape. "Thus died a brave and loyal soul true with his life to the principles he had publicly professed, and which he deemed it cowardly and wicked to abandon." Lowell thus succeeded as a narrator by dramatizing the dangers to which the heroes were exposed and explaining the moral stake in the outcome of each drama.[19]

In addition to blurring the boundaries between science and literature, Lowell wrote prose and poetry almost interchangeably. In this respect, he was following in the footsteps of his New England predecessors Holmes and James Russell Lowell, who used both forms, and whom he was fond of quoting to Basil Hall Chamberlain. Moreover, astronomers of his generation, including Giovanni Schiaparelli, routinely wrote poetry. Indeed, for Lowell, writing verse was the logical extension of a prose style that featured alliteration, punning, metaphors, and witticisms.[20]

Lowell's propensity for verbal fireworks is nicely illustrated by a letter he wrote to Frederic Stimson in 1884, when both men were preparing for careers as writers. Lowell's subject was the photographing of a beautiful Japanese dancer. "Flirtation's photography or photographic flirtations: alliteration's artful aid . . . I discovered subsequently that she was a virgin—to photography." Double entendre and alliteration were then supplemented by classical allusions as Lowell described the process of developing his picture of the dancer. "Thenceforth Pygmalion was but a shadow to me in intensity . . . the laboratory dark as Erebus, black as the crime perpetrated therein . . . the lovely image growing into life. In her bath too—of pyrogallic acid. Our friend Pyg. worked in the garish day. Such indecent exposure would have killed my little girl . . . Would it be indecent to add that I left her washing in her tub. Scientifically I had accomplished that most difficult of results, the establishment of a negative." This self-conscious verbal gamesmanship was as likely to offend, however, as to impress his readers, particularly when it was sustained through a book-length work.[21]

It was but a small step from writing of this kind to composing verse. This was especially the case because most of Lowell's verse was written in iambic pentameter, the standard form for narrative or epic poetry. And, appropriately enough, virtually all of Lowell's poetry was narrative in character and often recounted episodes or feelings previously presented in his prose work. His Phi Beta Kappa poem, "Sakura no Saka" ("Cherry Tree Blossom"), develops a central theme of his *Soul of the Far East,* while "Ontake," published by the Authors Club, essentially compresses the story of *Occult Japan* into a ten-page narrative poem. Similarly, "Mars," composed in 1894 but never published, recounts the discovery of signs of intelligent life, as does the book by the same title, which appeared a year later. However, as William Sheehan has suggested, the poem permitted Lowell to exercise his imagination in ways that would have been unthinkable in his prose.

> Or better still in some strange built town
> To mingle with the Martial men
> Enlarging my Earth restricted ken,
> Amid their pageant seeming show. . .
> Learning a little of what they know
> Grown by my travel in some wise then
> Myself a cosmic denizen.[22]

In similar fashion, Lowell's Phi Beta Kappa poem personalized his experience of viewing cherry blossoms in Japan.

> Was it that light which through my senses thrilled
> Was it that cloud-form's swelling curves that filled?
> My soul's vague void? Who guess the riddle can—
> Why beauty should so penetrate a man!

It is instructive to contrast Lowell's florid style, highly emotional language, and syntactically complicated sentences with the more compact and cooler imagist poems of Amy Lowell, whose interest in Japan was sparked by her brother's travels.

The one hundred and sixty streets in the Sanno quarter
Are honey-gold
Honey-gold from the gold-foil screens in the houses,
Honey-gold from the fresh yellow mats.[23]

We may suppose, then, that Percival Lowell found poetic expression both rhetorically challenging and effective for conveying a personal reaction to events he could otherwise describe in scientific works.

Lowell's self-consciousness about his own and others' writing styles eventually brought him into conflict with friends and associates in Japan. In 1893, Chamberlain wrote Lafcadio Hearn about Lowell's "quite too dreadful poem (so-called) on Azuma-yama—a mountain which sates its fury

'On some Japanese students sent
From Tokio to investigate
That mountain's interesting state'!!!

etc. etc. etc." This poem, initially published and then attacked in the *Gazette,* an expatriate newspaper, caused Lowell's friend W. B. Mason to remark that he "should not have thought it possible to commit literary suicide with such thoroughness and dispatch." After much soul searching Chamberlain advised Lowell to devote his time to writing prose rather than poetry.[24]

The precious quality of some of Lowell's writing and his tendency to labor over it excessively were nowhere more evident than in his public campaign in the *Japan Weekly Mail* to correct an anonymous letter from an Australian woman who had wished to inform readers of opportunities for sugar planting in Queensland. Offended by her prose, Lowell published a seven-page letter of corrections to the original forty-eight-line letter. His objections ranged from grammatical to semantical points; in response to the author's word choice, for example, in the sentence "Single specimen trees have been known to yield 20,000 superficial feet," Lowell wrote: "Why superficial? 'Square' feet is an even more technical phrase and a terser one. Besides the choice of a word for its sound is one of the well appreciated details of the literary art. Sounds stick." In answering Lowell's charges, the editor of the *Japan Weekly Mail* wondered whether Lowell's "three columns of criticism spun out of a pin's point of solid matter" had not violated a sense of artistic proportion. "We cannot all go to Corinth, any more than we can all conjure with phrases after the fashion of MR. PERCIVAL LOWELL . . . It only remains for the less gifted to feel thankful and bewildered." Clearly, the editor's irritation with Lowell's priggish response to a harmless advertisement was intensified by resentment of his wealth and leisured life.[25]

This incident, along with various critical responses to his poetry, no doubt contributed to Lowell's decision to abandon the literary career he had been seeking to establish. His decision was also likely hastened by Lafcadio Hearn's replacement of Lowell as the preferred interpreter of Japanese culture to American audiences.

Lowell's prowess as a prose writer, if not a poet, was equaled or exceeded by his reputation as a public speaker. Indeed, like many nineteenth-century authors, he made little distinction between writing and speaking; his lectures were effectively dress rehearsals for articles that would then be gathered between the covers of a book. By all available testimony, Lowell was a dynamic public speaker. In addition to a way with words, he had a commanding presence at the rostrum, developed in part from participation in the Place aux Dames theatrical productions under his mother's direction.

Lowell's reputation as a public speaker was first established on the basis of lectures delivered at the Lowell Institute in 1894 and 1895. His presentations on Mars caught the attention of Edward Everett Hale, who applauded "the charm of Mr. Lowell's lectures," and acknowledged "his humor, his ready wit, his complete knowledge of the subject with which he deals." These qualities, as well as his ease in public speaking, made Lowell's appearances notable. Eleven years later, his audience at the institute was so large and enthusiastic that the lecture series had to be repeated. One spectator wrote to express her regret that the course had ended, while the *Boston Globe* praised his "graphic and vivid style," his "quiet and delicate humor," and noted as well the "enthusiastic crowds" that thronged the lecture hall.[26]

Word of Lowell's prowess as a public speaker soon reached the territory of Arizona, where he was in frequent demand on such occasions as Memorial Day, the Fourth of July, and graduation exercises at the Normal School in Flagstaff. Accordingly, in 1903 he wrote his brother-in-law: "Can you send me by return mail any commemoration addresses (good ones) you can lay your hands on: also any literature bearing on the subject? There is a prevailing impression locally that nature constructed me for addresses. I wish they could also construct the addresses."[27]

Not inappropriately, the man who conjured up images through words also had a great facility for picture taking. Indeed, photographic images played a central role in presenting Lowell's view of East Asia to the American public. At a time when few Americans had visited Korea or the interior of Japan and no previous photographic images of Korea had been published, *Chosön* offered a rich visual experience for its readers in the form of the twenty-five reproductions of photographs taken by Lowell himself. Similarly, *The Soul of the Far East* and *Occult Japan* put a photographic face on the interior of Japan, which was far less known to the public than Tokyo and Kyoto. In this fashion, Lowell's images of East Asia, as those of William Henry Jackson of the American West, introduced Americans to a previously unseen part of the world.[28]

We may surmise that Lowell took up photography at this time, as part of what he called "a very catching epidemic," referring, no doubt, to the popularization of photography following the introduction of the hand-held camera and the dry plate. The hand-held camera replaced the cumbersome old boxes on tripods, while the dry plate enabled photographers to develop pictures at their leisure rather than having to bring along a darkroom. Lowell himself continued to use a tripod, but he enjoyed the convenience of the dry plate and accordingly fitted up a darkroom in the Tokyo house he was renting just in time to develop his Korean pictures. Though he found developing pictures a laborious process, he nonetheless completed an album of fifty-three photographs of Korea for the king of that country, to whom he had been presented during his stay in Korea. He also enclosed and annotated a number of the images in his letters to family and friends.[29]

The artistic quality of Lowell's photographs is evident on first inspection. The images are sharp, an effect that is enhanced by the contrast between light and dark areas. Lowell was a master at framing his pictures so that the focal images were nicely framed by interesting backdrops of mountains or city walls. He strove to capture a broad range of subjects, including royalty and common folk, landscapes and cities, interiors of homes and courtyards, street scenes and shops. Even though most of his subjects had never before seen a camera, they appear remarkably unposed in the photographs. Lowell was equally successful in capturing the dignity of Korean officials and the activities of common people in the streets. As a result, he managed to preserve and disseminate images of clothing, buildings, and practices in both Seoul and the Japanese interior that had never before been seen by Western viewers. Lowell also had a sense of how the camera might help to penetrate the secrets of Japanese culture. For example, he sent his sister Amy "some brocade pictures for your august eye." One of the photographs, as he explained, portrayed a woman who had been possessed by a spirit: "See how happily pictorially the two personalities are suggested by the sliding screen."[30]

In shooting these images, Lowell apparently never hesitated to risk creating a disturbance if he considered the subject worthwhile. His quest to photograph a beautiful geisha in Tokyo, which required prior permission from her father, became a story in itself. Even greater difficulties attended the photographing of a Korean hat shop: "Horse and man gathered at great expense by the photographer. Crowd on the sides gathered at no expense by themselves. They typify one of the ills that photographers are heir to—the uninvited sitter." Lowell violated custom by obtaining images of "virtuous Corean Washerwomen." He broke the law, designed to protect the privacy of Koreans, to shoot the interior of a courtyard in Seoul. Under the circumstances it seems remarkable that such an intrusive operation should have succeeded so well in conveying the vitality and dignity of life in Korea and Japan.[31]

"In the Valley of the Clothes," north of Seoul, 1884 (courtesy Museum of Fine Arts, Boston)

Lowell achieved considerable recognition for the quality of his photographic work on Korea. In reviewing *Chosön,* the *Nation* judged the photographs as "no mean addition to our stock of knowledge" and concluded that they were "the most valuable portion" of the book. Following his return to Boston in 1884, the Boston Society of Amateur Photographers awarded the photographs its prize for "highest excellence."[32]

The single most striking feature of Lowell's work as writer, photographer, and public speaker was his versatility. Few contemporaries managed to skillfully exploit so many different media to bring their ideas to the public. Without question Lowell possessed verbal skills of a high order and a capacity to write polished prose with considerable ease—a talent that sometimes got him into difficulty. His work was particularly notable for addressing both professionals and amateurs and in this way joining two cultures. No doubt Lowell's success in reaching a broad audience was related to his rejection of the literary realism of Howells and James in favor of the romance. By presenting his work as a personal journey, Lowell managed to encourage his readers (largely male) to participate vicariously in his various projects. Nonetheless, it is striking that a scientist like Lowell should reject realism, with its emphasis on the exact description of material reality. Lowell, of course, argued that he sought larger truths, but we can well wonder whether he did not sacrifice accuracy in detail in the process.

Finally, Lowell's work was sustained by a network of individuals and institutions that included his family and friends, the *Atlantic Monthly,* and the Lowell Institute. He came of age, in a literary sense, as Boston was losing its primacy to New York. He fought unsuccessfully to prevent the migration of cultural hegemony, and, having failed, he joined the winners. In 1905, he and his brother left the Boston publishing company Houghton Mifflin, which, like the New England cotton mills, he had come to regard as "old fogy," to publish with "broad-minded" Macmillan in New York. In this way he joined William Dean Howells and others in acknowledging Manhattan's primacy as a cultural center.[33]

Scientist

Percival Lowell fashioned a unique scientific career that nicely reflected his versatility and his capacity for risk taking. After abandoning a psychical research project in which he investigated a form of subliminal consciousness,

Lowell founded and managed his own observatory and thus controlled its research agenda and financial affairs. In the process, Lowell became the first astronomer to locate an observatory in the high desert country of the American southwest and to exploit the good seeing available in that region. Following its establishment, he took advantage of the location to investigate systematically the components of good seeing, including atmosphere, instruments, and the visual acuity of the observer. Lowell also became the first astronomer to approach the question of extraterrestrial life in a scientific manner and in this way to span the two cultures of humanities and science by using the telescope to gather evidence of both the natural and the cultural features of Mars. These features, he thought, would enable astronomers to gain insight into an extraterrestrial culture, in much the same way that he had used artifacts to understand and convey East Asian life. Meanwhile, he also collected previously unrecorded botanical specimens from the canyons of northern Arizona and shipped them to the Arnold Arboretum in Boston.

The highly ambitious observational programs that Lowell carried out in Japan and Arizona were based on his theories intended to illuminate the central facets of East Asian and Martian culture. However, working at the limits of seeing both optically and culturally, Lowell could never achieve definitive proof of his Martian canal theory, any more than he could prove the impersonality of the Japanese people. Accordingly, he adjusted his definition of "proof" to accommodate the difficulties he faced. His later search for a trans-Neptunian planet required only conventional evidence for proof; unfortunately, however, he failed in that endeavor.

Lowell's interest in science began as a boyhood flirtation with astronomy, which led to his acquiring a 2½-inch telescope and a number of popular astronomy texts. That interest continued during his undergraduate years at Harvard and he wrote his senior essay on the nebular hypothesis. Even after he left college and took up the investment business, Lowell found time to pursue his interest in science by joining the Mathematical and Physical Club. He also remained in contact with his college mathematics professor, Benjamin Peirce, until Peirce's death in 1880. And as evidence of the high regard for his scientific knowledge, he was appointed by the Harvard Board of Overseers, with the consent of the appropriate departments, to the committee on Mathematics, Physics, and Chemistry in 1882, 1883, and 1889, and as a visitor to the Harvard observatory in 1893, 1894, and 1895. Thus, even as a Boston-bound investor and during his travels in

Japan, Lowell exhibited an ongoing interest in and knowledge of the natural sciences. Not until 1891, however, did he opt for science as a career. Then, significantly, he announced that "science is the thing to investigate and write about." Thus, a scientific career appealed to Lowell because of the opportunity it offered to display his versatility.[34]

Lowell was drawn to the field of astronomy because he could use his entrepreneurial skills to create and direct his own institution and thus leave a lasting mark on the field. Lowell's autocratic style, inherited from his father, found a perfect outlet in the management of his observatory. He resembled, as John Lankford has suggested, American captains of industry of the late nineteenth century who were far more comfortable giving than taking orders. Lowell's combative instincts served him well in the struggle to assure the survival of his own small observatory in the highly competitive world of American astronomy. Challenged by competitors in large-scale observatories, Lowell drew on his skills as a communicator to defend his methods and his findings.[35]

At the same time, Lowell approached astronomy in the manner of European aristocrats who made careers in science, usually as amateurs. Like them, Lowell assumed a hierarchical division between the generalist, engaged with theory, and the specialist, who gathered evidence in the field or lab. Consistent with aristocratic values, Lowell regarded the creation of theory as superior to data gathering. Accordingly, he assigned a good part of the observational work necessary to support his claims to specialized paid assistants; once the evidence had been compiled, however, it was he who would categorize and interpret it. Of course in one important respect Lowell deviated from this model. He put in countless hours at the telescope, even though he was quite contemptuous of mere data gathering. Lowell's attitude was a perfect reflection of mid-nineteenth-century practices, when upper-class amateurs frequently directed the work of professional assistants.[36]

Lowell brought to science the adventurous approach that had characterized his other careers. He persisted in denouncing conservatives who followed old routines and failed to exploit the full potential of the field. Accordingly, at a time when highly specialized professionals were taking over astronomy and engaging in massive data-gathering programs, Lowell promoted science as at once a romance, a cultural activity, and a poetic enterprise that required a strong imagination. He congratulated A. E. Douglass on his discovery of surface features on Jupiter's satellite Ganymede but regretted that an imaginative naming scheme for these features would be wasted, "so unpoetic is the average astronomer." In heralding the Lowell expedition to the Andes under David Todd's direction, Lowell proclaimed Todd "not only an astronomer" but a man who could "elevate science to culture." According to this distinction, the astronomer was a technically competent specialist who relentlessly gathered data, while the man of culture used data to verify hypotheses about the planets and extraterrestrial life. Mere astronomers, Lowell concluded, were not capable of this higher level of thinking. They showed, as he explained to Douglass, "a lack of jungle penetration off the beaten paths."[37]

Given the intellectual failings of mainstream astronomers, Lowell would seek help from mavericks and specialists in other fields. Accordingly, he enlisted the sociologist Lester Frank Ward to shed light on the social patterns of Martian inhabitants. In this way, Lowell hoped to redefine astronomy so as to include allied disciplines. And he must have agreed with T. J. J. See, his former assistant, who lamented in a letter to Lowell that there were so few "philosophers" among astronomers.[38]

Lowell sought to explain this deficiency he and See had identified in most astronomers. He noted on one occasion that "scientists all over the world" are "of a second class in society" because "all the intellect go[es] into science leaving nothing for common sense." No doubt he was here rephrasing the distinction between the specialist and the generalist, but this time calling attention to its basis in terms of social class. Generalists, many of whom, like Lowell, had been educated at elite institutions, were better prepared to engage the truly important questions of science, while specialists, more likely drawn from the middle class, were absorbed in lesser tasks.[39]

In addition to importing a romantic view into a field governed by cautious approaches, Lowell took a number of other risks. He entered astronomy rather late in life, at a time when small observatories, like the one he created, were about to be displaced as important centers of astronomy by large-scale institutions often associated with big universities. With limited resources, Lowell had to compete with larger, richer, and more powerful rivals. In this respect, ironically, Lowell was mounting an old-fashioned operation. His research program was based on traditional concepts like the nebular hypothesis and the existence of extraterrestrial life, and he would explore these concepts, at least initially, with only the most conventional instruments. Until after the turn of the century, his observatory had no reflecting telescope or spectrograph and rarely used the camera.

Lowell was also taking up an issue that was troubling in various ways.

Linguistic ambiguity at the outset surrounded the Martian debate. The Italian astronomer Giovanni Schiaparelli had first identified the lines he had seen on Mars as *canali,* without specifying whether he meant canals or channels, both legitimate meanings of the Italian word. Beyond linguistics, an investigation of the existence of life on other planets would necessarily test the limits of seeing, since the surface features of Mars were visible only during near approaches of Mars to the Earth and in the best of atmospheric conditions. Furthermore, Lowell turned to the nebular hypothesis to support his claims of life on Mars at the very moment when it was being strongly challenged by various authorities. According to the hypothesis, planets evolved over time to develop a life-supporting environment (see Chapter 8). Meanwhile, he complicated matters by conducting scientific research two thousand miles from his Boston home—although compared with the scientific laboratory in his Tokyo living room, the Flagstaff observatory was close.[40]

While Lowell had little respect for practicing scientists, he had a high regard for science itself as the most potent weapon for revealing the ultimate truth about the universe. He viewed science as an instrument for penetrating the superficial features of any subject in order to identify its fundamental elements. The scientific process, Lowell believed, had an especially important role to play in cases where the true causes of things were obscure or were obscured by men for their own purposes. In this regard, he was particularly suspicious of theologians who offered erroneous explanations of natural phenomena. In other cases, such as the likely evolution of the Earth, over time, to its own death, he felt that the truth was too painful for most individuals to swallow. Thus, given the sentimental disposition of the public and the built-in resistance of religious groups to the unfettered operation of the scientific approach, the scientist operated in a hostile environment.[41]

To prevail in the face of resistance, as Lowell saw it, the scientist had to become a kind of muckraker; instead of political graft, his inquiries would reveal the sometimes obscure dynamics of the universe. Accordingly, Lowell spoke of science on different occasions in the following terms: "the coldly critical eye of science," "strict science," "prying scientific curiosity," "ruthless science." His language suggested in a Baconian fashion that the universe yielded its secrets slowly and unwillingly, and that only by virtue of the power of science to penetrate the surface features of the subject—and the resistance of ordinary men—would truth be manifest.[42]

In venturing forth to Japan, Lowell would use science to understand the social dynamics of a people he believed too underdeveloped to think scientifically. In this venture, he would clarify the nature of science as the "innate desire to learn from the cosmos." Less developed people, like the East Asians, expressed this desire in the form of crude myths and artistic renderings of the natural world. Lowell noted with a certain condescension, for example, East Asians' indifference to learning the "reason of things." Scientists, however, could penetrate the storyline of myths so as to discern elements of truth in them. In this way Lowell recognized a "thread of truth running through" the Asian version of Korean racial origins, a story that was in other respects "a web of fiction." Moreover, even though he supposed Asians to have no understanding of their actions, they had mastered techniques similar to those used by Western scientists. In this regard, Lowell was particularly impressed with Shinto practitioners who readily put each other into a trance. "As science the commentators' elucidations are chiefly comic, but their practical outcome was immense."[43]

Only in truly advanced civilizations and in recent years had men learned to think scientifically. Now that a few societies had entered the scientific age, their prospects were grand in Lowell's view. Science—and especially astronomy—would do nothing less than liberate mankind from the slavery of ignorance; the telescope in particular had the power to literally and figuratively enlarge the viewers' perspective on their place in the universe. "We stand upon the Earth, but in the sky, a vital portion not only of our globe but of all of which it too forms a part. To feel it is to enter upon another life." Thus, astronomy could directly contribute to Lowell's cosmopolitan agenda by clarifying the relation between Earth dwellers and the universe.[44]

Lowell intended to use astronomy in much the same way that he had exploited his travels in Japan to further this cosmopolitan agenda. Astronomy, from this perspective, would provide a pioneering investigative tool to engage geographically and culturally distant instances of intelligent life, the first and nearest of which would be that on Mars. Until Lowell's time, astronomers had participated in endless speculation about the possible existence of a plurality of worlds, without ever anchoring their speculations in hard evidence. Lowell's observations of Mars, by contrast, were accompanied by publications in which he set forth a scientific approach to investigating the possible existence of intelligent life on other planets. To carry conviction, those seeking evidence of extraterrestrial life would have to demonstrate, first, that a particular heavenly body had an atmosphere (water, oxygen, and so on) that could support life and, second, that the evidence of such life, in the form of artificial markings on the planet's surface, could be identified.[45]

Lowell's interest in optimizing "seeing" no doubt arose in equal measure from the challenge of searching for evidence of life on Mars at the limits of seeing and his love of travel. Happily for him, Lowell entered the profession at a time when astronomers were seeking superior observing sites, especially in high desert locations. This quest coincided with the availability of resources to fund these observatories and more efficient modes of travel to transport people and equipment to the sites. Lowell, after testing the atmosphere in Japan, joined with W. H. Pickering to found the Flagstaff observatory in Arizona, and when conditions there proved less than perfect, he tested sites in Latin America and Africa. Even after settling definitively in Flagstaff in 1901, Lowell continued to advocate the placement of observatories in similar locations elsewhere in the world.

The debate surrounding the visibility of the canals on Mars raised questions about the selection of appropriate instruments and the possible distortions of telescopic images by the eye and the brain, as well as the identification of good observing sites. Accordingly, Lowell investigated systematically the combined effects on good seeing of instruments, the human eye, and atmospheric conditions. The relative merits of large versus small telescopes and the results of tests to measure brain and eye distortion of images became once again controversial subjects in the astronomical community as a result of his Martian project.[46]

Despite the popular association of Lowell with the canals of Mars, he first defined the program of the observatory as the investigation of the solar system in general and Mars in particular. Over time, practice made good on this promise. Following intensive research on Mars in 1894 and 1896, Lowell and his assistants turned increasingly to the generation of data about other planets. In 1896 Lowell had already determined the rotation periods of Mercury and Venus and was reporting spoke-like lines on the surface of Venus. Lowell and his assistants later explored the nature and durability of Saturn's rings, the composition of Jupiter's atmosphere, the rotation of Uranus and the satellites of Saturn, Mimas and Enceladus, and the oblateness of Uranus, among other solar system phenomena. In this fashion, the Lowell Observatory's program served to redress the balance between planetary and stellar astronomy that had been upset by the substantial shift of establishment observatories toward stellar research.[47]

Lowell's decision to embark on the search for Planet X further demonstrated his versatility within the field of astronomy. His research on Mars, Venus, Jupiter, and Saturn identified surface features of planets and used surface features to make determinations of rotational periods. In contrast, the search for Planet X was based theoretically on an understanding of the gravitational forces that operate within the solar system. In undertaking this project, Lowell drew on his knowledge of celestial mechanics, which, in turn, required complex calculations of the displacement of Uranus's orbit.[48]

The grand prospects of science could be realized, in Lowell's view, only if scientists married data gathering to scientific theory. He insisted that astronomical discovery occurred on two levels: those of particular objects and grand truths. Of course both were essential to progress, but Lowell gave particular emphasis to scientific theory. As he explained, "Coordination is the end of science, the aim of all attempt at learning what the universe is." The gathering of many facts would be useless unless they could be explained in relation to some larger framework or theory. Only then would they have any real significance. The usefulness of a theory, he said elsewhere, is that "it will not only fit all the facts, but at times go out of its way to answer questions." Based on theory, it would be possible to develop a set of a priori principles that would describe the behavior of phenomena. Thus, general principles had to be given a very important place in science.[49]

In Lowell's scheme of things, the elaboration of a grand theory would be followed by identification of the principles or laws according to which the system might operate. From those principles it would then be possible to deduce the characteristics of the system and the configuration of its particular elements, to conform to the laws of behavior. His emphasis on deduction was equally true of Lowell's approach to East Asia and the solar system. In the former case, for example, Lowell took as his starting point what he saw as the impersonality of the Japanese people, manifestations of which he tried to identify in every dimension of Japanese life. In a similar fashion, he assumed the evolutionary character of the gaseous mass from which the solar system had originally evolved and then, on the basis of natural laws, tried to demonstrate the timing and character of changes in this mass during the course of its evolution.

Lowell's colleagues in Japan and Flagstaff found his deductive approach "unscientific," for similar reasons. Chamberlain protested to Lafcadio Hearn that Lowell's plan was to "argue down deductively from some general notion, such as the supreme virtue of 'modernity,' the 'impersonality of Orientals,' etc.—he has only 3 or 4—and then bend the facts to suit the preconceived idea, seasoning the whole with verbal fireworks." Eight years later, A. E. Douglass made precisely the same charge about Lowell's astronomical work to William Lowell Putnam, then acting director of the observatory, saying Lowell's "work is not credited among astronomers because he devotes his energy to hunting up a few facts in support of some speculation instead of perseveringly hunting innumerable facts and then limiting himself to publishing the unavoidable conclusions, as all scientists of good standing do." Chamberlain's criticism had anticipated Douglass's complaints, but while Chamberlain's views were known only to Hearn, word of Douglass's criticism got back to Lowell, who promptly fired his assistant.[50]

Good science requires the development of hypotheses that can be tested in the "real world." Lowell, however, frequently dealt with such concepts as "impersonality" and "life on Mars," for which empirical evidence was difficult or impossible to obtain. Under these circumstances, the normal process of testing the theory could not be implemented in a way acceptable to ethnologists or astronomers. In the absence of reliable data, Lowell was inclined to rely heavily on interpretation of the data available. Thus he insisted on the centrality of "imagination" to the generation of good scientific results. Of course, when he spoke of imagination, Lowell had in mind something quite different from the creative impulses of a novelist or a painter. As he explained, imagination was the faculty that enabled a scientist to generate a hypothesis or theory and then to establish relationships between it and the observations he had made. Imagination was absolutely essential to the scientific process, since uninterpreted facts would not contribute to the scientist's goal of understanding the nature of the universe. However, Lowell acknowledged that imagination had to be kept in check; otherwise there was a risk of generating theories that were not supported by facts or did not help to explain them. "Without the discernment to perceive relations no ascent to first principles is possible and without the support of quantitative criterion, one is in danger of becoming giddy from one's own imagination."[51]

Imagination was useful not only in interpreting data but also in reasoning about conditions and phenomena in the solar system that were beyond the limits of seeing. A good theory built up from reliable data could be used, Lowell believed, to suggest the probable appearance of certain surface features on a celestial body or atmospheric conditions invisible through the telescope. Accordingly, Lowell frequently resorted to using what he called "the mind's eye" or "the eye of analysis" or even "the finger of the sign board of phenomena" to ascertain the existence of undetected bodies or conditions where no observational data were available.[52]

Lowell's conception of "proof" was also affected by the difficulty of obtaining reliable data to confirm his hypotheses. With regard to Mars and Japan, "proof" in the form of hard data would be impossible to come by so long as Martians were invisible through the telescope and the grammatical structure of the Japanese language did not readily yield information about the "impersonality" of the Japanese people. Accordingly, Lowell set the standards for proof so as to assure himself and others of the validity of his propositions in the absence of an overwhelming mass of data. He claimed, for instance, that "proof is nothing but preponderance of probability. All deduction rests ultimately upon the data derived from experience. This is the tortoise that supports our conception of the cosmos. For us, therefore, the point at issue in any theory is not whether there be a possibility of its being false, but whether there be a probability of its being true . . . The test of a theory is, first, that it shall not be directly contradicted by any facts and secondly, that the probabilities in its favor shall be sufficiently great."[53]

Among various difficulties with this conception of truth were Lowell's failure to acknowledge here an idea central to his theory of writing: namely, that "facts" were always subject to interpretation; in the process, they might serve to support not only the theory at hand but also some other one. For example, the lines on the surface of Mars would either support or not support the theory of extraterrestrial life, depending on whether they were considered to be canals or cracks. Lowell also failed to mention what he would do with evidence that neither contradicted nor supported a particular proposition. Furthermore, Lowell's notion that probabilities had to be "sufficiently great" to declare a theory true left unanswered the question of what constituted "sufficiency" in a particular case. This vagueness was compounded by other statements, such as: "We need not to add example to example or proof to proof" in cases where there were no contradictory phenomena. Again, it was unclear how many facts would be necessary to convince an intelligent scientist or lay person of the truth of a proposition. Of course, such problems were magnified when Lowell was dealing with extraterrestrial life and tended to diminish when he turned to projects, such as the search for Planet X, in which more conventional evidence would suffice to settle the issue.[54]

Despite his willingness to accommodate theories for which evidence was weak, Lowell insisted on the importance of scrupulous data gathering in his work on Mars. He had been rather careless in his East Asian projects, where he pretended to speak authoritatively about language and religion without having acquired a profound knowledge of either. In his Martian research, however, Lowell sought to eliminate the "personal factor" from observations and thus make them more credible. In his extensive data-gathering process, Lowell advocated what he called "mental effacement," just as he insisted on "mental assertion" in the creation of theories and the interpretation of data. Accordingly, he argued that one person using one instrument should make a series of drawings on Mars's surface rather than have several people alternate or change instruments. "As the same personality enters all of them, it stands, as between them eliminated from all, to increase certainty of deduction." However, as between observers, Lowell believed that personal factors could not be entirely eliminated. Indeed, it was to be expected that two astronomers looking at the same body—even from the same location and with the same instrument—would not see that body in exactly the same way. Thus Lowell regarded the differences between Schiaparelli's findings and his own as evidence that the observations were honest. Had they duplicated each other, the second observer might rightly be suspected of "conscious psychic reproduction." (Curiously, he did not apply this logic later to Louise Leonard's drawings of Venus which duplicated his own.)[55]

The great danger in observations like the ones he was making, Lowell admitted, was that "we easily see what we expect to see, but with great difficulty what we do not." He admitted that observers were affected by "individual idiosyncrasy" or "a prevailing idea of the time." He added that "fashion is as potent here as elsewhere." Here was a partial admission of what critics were saying about his own Martian observations. His defense tread a delicate line between arguing that the canals were real because he and Schiaparelli saw them and that they were not real because they saw them differently. Lowell tried to split the difference.[56]

Lowell launched his scientific career by traveling in and writing about East Asia. He remained a scientist for the rest of his life, though in 1894 he put aside the study of Japan in favor of Mars. Even his astronomical work, however, had a cultural dimension inasmuch as the surface features of Mars were interpreted as evidence for the existence of a highly developed civilization on Mars. In this way, as he wrote Todd, they would "elevate science to culture."

The definitive refutation in our time of Lowell's image of the Martian surface tempts us to ignore his important contribution to thinking scientifically about the existence of extraterrestrial life. To be sure, he misinterpreted the evidence he found and adjusted his notion of proof so that propositions for which evidence was scanty or uncertain could be considered "true." All the same, Lowell must be credited for his pioneering venture in locating an observatory on a high desert plateau to maximize good seeing; equally bold and fruitful was his attempt to become the first scientist to verify claims for extraterrestrial life with telescopic evidence. Lowell established a useful framework of investigation even if his results were unpersuasive.

While Lowell never gave up his work on Mars, after 1907 he shifted his attention increasingly to the search for Planet X. In so doing, he sought to clear his reputation for sensationalism so that he could be remembered as a sound scientist. The search for a new planet was also consistent with Lowell's original plan to establish an observatory committed to solar system research. His eagerness to dedicate his observatory to planetary astronomy at a time when other institutions were moving relentlessly toward stellar research was a characteristically bold stroke on Lowell's part. With the advantage of hindsight, we can say that the fruits of the decision justified him in taking this path. Unfortunately, Lowell's focus on his Mars project—and the interest of current historians in that work—has distorted the more complicated agenda of his observatory in those early years.

At first blush, Lowell's career search seems not much more than the curious saga of a talented gentleman amateur with independent resources, who tried out seriatim a number of fields in which he was interested. Certainly there is considerable truth to this version of Lowell's odyssey, but there is more to be said about it. Of particular interest is the fact that Lowell never abandoned skills and activities in which he was proficient but continued to use them as he moved on to other fields. In this way, it would appear, he was unconsciously seeking to exhibit his extraordinary versatility. And no doubt astronomy, at least as he practiced it, was precisely the right career for him.

As observatory director, Lowell had charge of the business and administrative sides of the operation, and as the sole source of funding, his expertise in financial matters would be essential. By locating his observatory in the distant territory of Arizona, he absented himself from Boston on a regular basis and traveled for reasons of good rather than goods. Once in Arizona, Lowell discovered that there were cultural and natural attractions beyond the good seeing that had drawn him there in the first place. Lowell also used his skills as a communicator to persuade the larger public, as well as scientists, of the reality of the canals. Even though he ceased to write poetry, he had the satisfaction of hearing from critics that his scientific work was poetic. For all these reasons, he enjoyed his work at the observatory much more than his research on East Asians. Also, from Flagstaff Lowell believed he would be in touch with a form of intelligent life that was vastly superior to the most advanced version on Earth—and therefore light years beyond East Asian intelligence.

This account of Percival Lowell's private life reveals a complex individual who, from his early years, had determined to set an independent course, which would inevitably bring him into conflict with the communitarian ethic of Boston. As a charismatic personality, he offended the sensibilities of upper-class Bostonians of his father's generation who preferred a more cautious, understated posture. By age twenty-eight he had already alienated Brahmin society and fled to Japan as a result. Lowell defied his father's efforts to saddle him with a business career, preferring instead the lifestyle of a leisurely man of letters. Finally, he tested the sensibilities of the community by conducting a series of premarital romances outside of Boston, marrying beneath him, and failing to leave heirs.

And yet Lowell remained a Brahmin even in the old-fashioned sense. He was a staunch partisan of club life, a loyal family man and friend to his Harvard chums, a skilled investor, and a permanent resident of Beacon Hill. He belatedly embraced the work ethic and found himself, despite persistent efforts, unable to enjoy his leisure life. In all these ways, he remained recognizably Brahmin.

His unhappiness stemmed from both personal and social factors. No doubt Lowell's generation was particularly hard hit by the confusions that resulted from rapid social and cultural change. The old Brahmin values were still strongly inculcated in Lowell and his friends, yet cultural signals of a newer sort suggested the desirability of enjoyment as well as professional success. Lowell attempted to have both, but we judge from his comments and those of his contemporaries that his private life was stressful. Ralph Curtis remarked to Mrs. Isabella Stewart Gardner that Lowell "got neither the fun nor the glory out of life, which he deserved." He was too much the Brahmin at heart to fling aside his inhibitions in favor of pleasure seeking. He was morbidly introspective, convinced that he didn't get the recognition he deserved, and subject, as we have seen, to bouts of severe and extended nervous instability. In fits and starts he put himself in the hands of Ralph Curtis and sought to enjoy the worldly life of the Continent. He made valiant efforts to achieve emotional and sexual intimacy with upper-class women, but quickly returned each time to his work. His struggle for independence succeeded too well, and he was unable to find emotional satisfaction of an enduring character. In the end, he made a marriage of convenience with a woman who was socially beneath him and unable to appreciate his wit and intelligence. In this way, the old values of work and industry triumphed over his efforts to achieve a pleasurable life.[57]

The tensions in Lowell's personal life had important implications for his efforts to find a suitable career. His rejection of the role of firstborn, of his father, and of Boston authorities suggests a psychological disposition toward distancing himself both literally and figuratively from Brahmin conventions in all matters, including career. And in fact, after a few years, Lowell did dispense with that most conventional of Brahmin careers—investments—and at the same time embrace the prospect of traveling to distant lands.

Lowell's strong inclination to maximize his personal independence also played an important role in his decision to become first a man of letters and then a man of science. To achieve personal independence, Lowell gravitated toward professions that would permit him to work at his own pace, under his own direction, and with limited responsibility for others. His selection of male and female friends, many of whom were themselves intellectuals, both reflected and encouraged his interest in culture and science. Furthermore, the fact that a number of Lowell's male friends pursued two careers at once, or moved from one career to another, provided excellent models of versatility that served to promote his own inclinations in this direction. Finally, Lowell's membership in Boston clubs enabled him to keep in touch with developments in both science and letters and thus opened the possibility of careers that straddled the divide between the cultures.

Lowell's career path, in turn, contributed to the emerging cosmopolitanism and individualism of his generation in Boston. The new ethos of expressive individualism justified Lowell and others in breaking with their forebears to invest in distant companies and travel more frequently to distant lands. In this way, they satisfied a yearning for greater independence from the demands of the community while, at the same time, contributing to its welfare. Lowell, his cousin A. Lawrence Rotch, founder of the Blue Hill Observatory, and Isabella Stewart Gardner, the wife of another cousin, all expressed their individualism by creating cultural and scientific institutions they could shape as they saw fit and left an important legacy to future generations.

Despite his persistent quest for independence, Lowell's career search required substantial support from his family and associates in Boston. Only his financial collaboration with Putnam enabled Lowell to put aside his business affairs while in Japan or Arizona. The willingness of Lowell's brother-in-law Putnam, his brother Lawrence, and his father to manage the observatory or support it financially during his illness and after his death played a crucial role in the survival of the project. Lowell's literary career, as well, was shaped by the critical audience of his family and the opportunities to lecture at the Lowell Institute. These institutions, moreover, enabled future Bostonians—and other Americans—to profit from the legacy of their founders. Consistent with these values, Lowell maintained a staunch loyalty to both his family and the Boston community even as he expatriated himself to Arizona. He worked in Boston to establish the Authors Club, which would have enhanced the city's cultural prestige, and in Arizona to bring Brahmin-style enlightenment to the wilderness and Arizona specimens to the Arnold Arboretum.

In these ways, Lowell satisfied some of his own needs and interests while at the same time finding a solution to the dilemmas of a new generation of Brahmins. His efforts helped to give greater visibility and clout to a Boston upper class increasingly challenged by the wealth and power of the new industrialists. Old money alone would not rescue the Brahmins from quiet oblivion; they needed to turn their intelligence and creativity to influencing the artistic and scientific future of the republic. This was essentially Lowell's project, and—as we shall see—his efforts to bring it to fruition were partially successful.

II

◆ ◆ ◆

Lowell as Spencerian

Percival Lowell's work was shaped by his lifelong commitment to the realization of Herbert Spencer's cosmic evolutionary project. Lowell embraced Spencer's concept of a cosmos governed by the law of evolution to explain the development of both the natural and the social world. He applied Spencer's evolutionary scheme during his travels in East Asia and again as he launched his search for extraterrestrial life. In this manner, Lowell built his evolutionary system in a backward fashion. By examining the relative position of Westerners and East Asians in the racial hierarchy of world civilizations, he told the end of the story first. Then he retraced his steps—and the world's—to consider the origin and development of the solar system from a nebulous cloud. As a result, Lowell obscured his intentions and confused readers interested in locating the source of his ideas.

Lowell's work was not only derivative but also theory driven, and thus it reflected in important ways deep-seated intellectual preferences on his part. To be sure, Lowell undertook painstaking empirical investigations of the canals of Mars over twenty-two years and studied trance-practicing Shinto sects in Japan for more than a year. And he was certainly following accepted scientific practice in using his data to confirm a theory. What distinguished Lowell from some of his counterparts, however, was his plan to use the data to support Spencer's all-embracing evolutionary scheme. In attempting to link data and theory, Lowell frequently engaged in thinking of an imaginative sort. Thus he confirmed Spencer's theory to his own satisfaction, but he acquired a reputation among his peers for selectivity in collecting and interpreting data.

Lowell's enthusiasm for Spencer was motivated by aesthetic as well as practical factors. Not only did Spencer's theory offer support for the existence of extraterrestrial life, but it also demonstrated the comprehensive, law-abiding, and purposeful character of the evolutionary process. In addition, Lowell was clearly entranced with the idea that a single process governed the development of inorganic, organic, mental, and social phenomena. And he reveled in the logic of a system in which all phenomena could be expected to develop along certain preordained paths. Such a system provided assurances of a predictable and purposeful future, even though Spencer had dispensed with the idea of God. After all, the system was governed by the law of development, which had already produced advanced industrial societies from the chaos of a primitive nebula.

Spencerian theory also appealed to Lowell because it gave him license and authority to investigate a variety of phenomena. Happily for him, Spencer regarded general astronomy—or cosmology—as the master discipline that, by virtue of its breadth, encompassed all other disciplines. According to this logic, complete knowledge of specialized subjects required prior understanding of more general areas. The generalist grasped the underlying forces and features that shaped the universe and therefore had insight into the character of more specific and derivative phenomena. The generalist was, thus, more capable than the specialist of crossing disciplinary boundaries and grasping at least the basic principles of phenomena in subsidiary fields. This process could not be reversed, however, since more specialized or local knowledge was frequently obtained without attention to grand theory. In this way, Spencer's theory was a great boon to Lowell's adventurous pursuit of various careers. So long as the human and natural worlds were governed by the same laws, he could bridge the gap between science and culture. Evolution would thus reinforce the effects of Lowell's Harvard education, and his invention of new careers incorporating older ones, to facilitate the development of new interests and the crossing of disciplinary lines.

These intellectual advantages of Spencerian theory were just as important to Lowell as its usefulness in providing justification for the existence of extraterrestrial life in general and life on Mars in particular. Indeed, Lowell was a Spencerian long before he accepted the existence of life on Mars. Once he had accepted that idea, Spencer's version of the nebular hypothesis, clarifying as it did the common origin of all planets, their similar composition, and the identical processes through which they were developing, supported Lowell's pluralism. It followed logically from those premises that evolution would produce life-supporting environments on all planets at some point in their development.

In a similar fashion, Spencerian theory could be useful to Lowell in his understanding of East Asia. Lowell's travels were characterized by a desire to grapple with the essential elements of the civilizations he encountered and to see them in relation to some larger evolutionary picture. He eschewed the gathering of endless details of everyday life and concentrated instead on identifying the level of development East Asians had attained. Given this approach, Spencer's theory could serve to clarify the way in which the West was superior, and would likely remain superior, to East Asia. At the same time, Lowell, through Spencer, could call attention to appealing attributes that characterized less developed societies.

In many respects, Lowell's Spencerian approach also enhanced his ability to contribute to the debates on two distinct but important issues in the late nineteenth century: the impact of industrialism and the origins of the solar system. Lowell's visits to East Asia coincided with the recognition of serious defects in the industrial system, including environmental degradation and labor-management strife. But by juxtaposing modern American society and traditional East Asian society, Lowell, following Spencer, highlighted the superiority of "progress" over "tradition." With regard to the solar system, Lowell had been committed to Spencer's version of the nebular hypothesis since his Harvard days, a theory that supported his claim for the probability of extraterrestrial life. Accordingly, Lowell kept abreast of cosmogonic arguments and developed a revised nebular hypothesis as a way of heading off critics of that theory and securing the theoretical basis for extraterrestrial life. In the process, however, he damaged his credibility with academic specialists by exercising a Spencerian prerogative to speak authoritatively in fields other than astronomy.

Lowell's untimely crossing of disciplinary boundaries suggests that the Spencerian approach had outlived its usefulness. After all, Lowell first took up Spencer's ideas in the 1870s, when both he and they were just coming of age. At that moment, Spencer was completing his *Synthetic Philosophy* and Edward Livingston Youmans was publishing it in *Popular Science Monthly.* By the time Lowell got around to applying Spencer's ideas in the late 1880s, serious questions were being raised about important assumptions and approaches in his system. Indeed, the systematic, predictable quality of Spencer's approach came to seem increasingly confining to many intellectuals. They yearned for a freedom of choice that Spencer seemed to deny. Furthermore, with the rise of specialization, academics attacked the sweeping character of the system and the privileges it accorded to astronomers. These academics continued to promote interdisciplinary projects but insisted on the right of specialists to speak authoritatively within their discipline. In this environment, Lowell found it difficult to sustain his cosmic evolutionary project. Moreover, the objections of scientists to his theoretical claims anticipated those that would be directed at his observational work.[1]

5

Cosmic Philosopher

Percival Lowell's books on both East Asia and the solar system strongly reflected Herbert Spencer's cosmic evolutionary perspective. Like Spencer, Lowell assumed the unity of the cosmos and this, in turn, required the invention of cosmology as a metadiscipline of scientific inquiry. He believed that the laws of development governed the evolution of phenomena in the natural and social worlds and would thus account for the entire history of the universe, from the origin of the solar system to the rise of human history. By embracing this ambitious scheme, Lowell became one of the last and most audacious exemplars of a characteristically nineteenth-century mode of inquiry. Among his predecessors were Alexander von Humboldt, whose five-volume *Cosmos* sought to explain the operation of the universe. But it was Herbert Spencer's *Synthetic Philosophy* that became the model for Percival Lowell's lifework. To understand his project, then, we must first recover the lost intellectual world of Herbert Spencer. Only in this manner can we grasp the framework that undergirded Lowell's work, enabled him to treat subjects as diverse as East Asia and the solar system, and supported his belief in the existence of extraterrestrial life.[1]

Cosmic Philosophy

Cosmic philosophers were a special breed of evolutionists whose earliest work preceded the publication of Darwin's *On the Origin of Species* (1859) by some fifteen years. What most distinguished them from Darwin was their application of evolution in a far more sweeping and ambitious fashion than he would ever have found acceptable. These evolutionists believed that nature operated in a lawlike fashion to bring complexity out of an original uniform matter in the inorganic as well as the organic and human phases of the history of the universe. By recounting the operation of the process in these various realms, evolutionists provided a comprehensive explanation of natural and human history. They treated the evolution of worlds from their formation as planetary bodies to their deaths. Such projects obviously required their authors to cross established disciplinary lines and to become involved in such important scientific controversies as the debates about the age of the Earth, the origins of life and of intelligence, and the existence of extraterrestrial life.[2]

In addition to its sweeping character, cosmic philosophy was distinguished from Darwinian theory by the fact that its practitioners frequently ignored or downplayed natural selection as an agent of change, and proclaimed the inevitability of progress in a more unambiguous fashion than Darwin. They were convinced that nature had programmed evolution in such a way that increasingly advanced forms of life would emerge to meet the challenge of changing environments. Strangely, they also agreed that "progress" would inevitably lead to death as the evolutionary process ran its course in particular organisms or planetary bodies.[3]

Four men who profoundly influenced Percival Lowell were among the most important exemplars of this mode of inquiry. As we have already indicated, Lowell drew heavily from the work of Herbert Spencer, but he also read Spencer's friend and associate, T. H. Huxley. Moreover, Lowell used the ideas of Ernst Haeckel, Huxley's German friend and fellow zoologist, many of whose popular works dealt with evolutionary issues. Finally, John Fiske, Spencer's American disciple and author of the four-volume *Outlines of Cosmic Philosophy,* helped to introduce Lowell to Spencer's ideas.[4]

Lowell was drawn to the work of these four men because he shared the ideology of scientific naturalism they promoted. Like them, he wished to be part of the crusade for universal knowledge based on the application of science. What this meant, in effect, was that their work would privilege the role of matter and energy, as opposed to spiritual, aesthetic, and ethical forces, in shaping human history. In the spirit of science, they sought to understand the development of life and intelligence in terms of matter and energy rather than by invoking God. In addition, all four men addressed the educated public as well as more specialized audiences. Indeed Spencer and Fiske, who lacked formal academic and disciplinary credentials, were almost obliged to do so, but even Huxley and Haeckel, well trained in zoology, courted the general public. In effect, all were possessed of a certain missionary zeal to spread their particular brand of scientific naturalism.[5]

To understand the nature of cosmic philosophy and its impact on Lowell, we need to clarify the underlying assumptions and methodology in the work of Spencer, Fiske, Huxley, and Haeckel. For all of them, evolution was a triumphant process by which intelligent life and human history emerged over a period of millions of years from the primitive gaseous mass of a developing solar system. The assumptions behind this edifying story of progress were succinctly stated by John Fiske in the introduction to his *Cosmic Philosophy:* "The term 'Cosmos' *connotes* the orderly succession of phenomena quite as forcibly as it *denotes* the totality of phenomena." The project was to be both all-embracing and uniformitarian in character. That is to say, Spencer and Fiske assumed that all developments occurred in a sequential and gradual fashion rather than a catastrophic one, leading from, as Spencer argued, "indeterminate uniformity to determinate multiformity." Huxley, who battled with Spencer over details, nonetheless enthusiastically endorsed cosmic evolution in his 1869 presidential address to the Geological Society, because it "embraces, in one stupendous analogy, the growth of a solar system from molecular chaos, the shaping of the earth from the nebulous cubhood of its youth . . . to its present form; and the development of a living being from the shapeless mass of protoplasm we term a germ." The published version of this address was carefully marked by Lowell.[6]

In the minds of cosmic philosophers, evolution was a story of progress, which worked itself out in identifiable stages. To be sure, changes in the cosmos were incremental in character, but their accumulation over time resulted in the qualitative transformation of bodies and organisms subject to these changes. This tale of progress through identifiable stages was clearly articulated in the organization of Spencer's *Synthetic Philosophy.* He left aside the inorganic phases in order to attend in great detail to the development of life in simple and then more complex forms. The organic phase culminated in the emergence of intelligence, which led, in turn, to the rise of simple and more advanced societies. This orderly procession, the opposite of entropy, fulfilled Spencer's a priori claim that increasingly complex and differentiated forms would follow the initial formlessness of the universe.[7]

The progressive direction of universal history was not without its reversals, however. Given its materialistic basis, death was always closely intertwined with life. As Huxley explained, protoplasm, the basis of life, "is always dying, and, strange as the paradox may sound, could not live unless it died." In this way, he said, "life and death are irrevocably united." In complementary fashion, John Fiske noted the devastating effects of gravitational pull on all planetary bodies, leading to the destruction of life-supporting conditions over time.[8]

Philosophers of the cosmos encountered a methodological problem resulting from the rapid expansion of scientific knowledge and the increasing specialization of scientists. Under these conditions, it would be difficult for any one person to master the wealth of new information produced by specialists in various fields. At the same time, the rapid growth of knowledge seemed to demand a synthetic approach to organize and make sense of the new findings. Herbert Spencer sought to meet this need, while at the same time formulating a hierarchical conception of disciplines that would justify his comprehensive approach.

In his *Principles of Psychology,* Spencer viewed fields as distinct though interrelated methodologies for understanding the universe. Consistent with synthetic principles, he considered all disciplines to be branches of cosmology, each occupying a place in a hierarchical scheme according to the degree of specialization. For example, "The astronomy of the solar system is a specialized part of that general astronomy which includes our whole sidereal system; and becomes specialized by taking into account the revolutions and rotations of planets and satellites." Geology, in turn, should be understood as "a specialized part of this special astronomy." Continuing in this vein, Spencer regarded biology as subfield of geology, followed in order by psychology, sociology, and ethics. From Spencer's point of view, the physical capacity of body and brain necessarily shaped the thinking process, which in turn set limits on the nature of social interaction.[9]

In sum, cosmic philosophers were exceedingly ambitious in their efforts to create a unified science capable of explaining the operation of the universe. Such a bold enterprise required a new conception of scientific activity that would be embodied in the discipline of cosmology. As the parts of the cosmos were closely related to each other by virtue of common origin, so cosmology as a single master discipline would seek to comprehend the laws of the cosmos. Other disciplines, devoted to exploring particular aspects of the cosmos, would subserve cosmology in explaining the operation of the system. In that he accepted these assumptions and this methodology as the basis for his own work, Lowell could be considered a cosmic philosopher.

Origins of Lowell's Cosmic Philosophy

In 1891, three years before Percival Lowell's decision to build an observatory to search for life on Mars, he wrote William Lowell Putnam from Tokyo, "Your cherished scheme for my life labors is in a fair way to a blossom." The reason for Lowell's optimism was that, while reading James Croll's *The Philosophical Basis of Evolution,* he had "thought of some things to be said about the philosophy of the cosmos and propose to enclose them in a cover." The nature of the project was then clarified in the balance of the letter and in marginal notations in Croll's book.[10]

Lowell's philosophy of the cosmos matured gradually from his Harvard undergraduate days to the publication of his evolutionary works in 1908 and 1909. During his senior year at Harvard, Lowell was introduced to the evolutionary perspective through reading Herbert Spencer's essay on the nebular hypothesis. On that occasion he likely read other essays included in the same volume that promoted an evolutionary approach, such as "Progress: Its Law and Cause" and "The Development Hypothesis." From these essays, Lowell developed an enthusiasm for the explanatory powers of the doctrine in various realms of knowledge, as was revealed in his books on East Asia and his response to James Croll's work. And, a few years later, his grasp of Spencer's theoretical approach was reinforced by reading Fiske, Huxley, and Haeckel.[11]

Lowell committed himself to writing a philosophy of the cosmos only after he had reflected on the philosophical underpinnings of evolution. From his reactions to Croll, it is clear that Lowell shared Spencer's commitment to replacing religious dogma with scientific knowledge wherever possible. Accordingly, as a basic premise, Lowell insisted that "I do not intend to touch on religion or ultimates, it being more scientific to write something of what we may know than nothing of what we may not." Like Spencer, he would put aside questions about the origin of matter and life that traditionally opened the door to theological explanations. In place of an "uncreated creator," Lowell accounted for the origins of the universe through "an initial act call it the evolving spirit. All else is then determined. What this is we do not know." Rather than exploring the nature of these beginnings, Lowell, following Spencer, would amass reliable knowledge from which he could induce the laws governing the operation of the natural and social worlds. In this way, his philosophy would be "scientific."[12]

The scientific ethos embraced by Lowell and Spencer was premised on the idea of an objective external reality that human beings had the capacity to comprehend. This premise ran counter to Immanuel Kant's claim that ideas are innate and thus the source of all knowledge about the external world. Lowell, following Spencer, justified his own partial repudiation of Kant in paradoxical terms: "The only ideas which are born with a man are just those he was not born with at birth." He resolved the paradox by pointing out that "the ideas called innate are merely the transmitted results of ancestral experience. They originated each at a definite time in the evolutionary past and have been handed down by heredity since like legs and noses." Here, Lowell was restating Spencer's Lamarckian notion of the heritability of habits in the form of instincts that would bring about gradual progress in the human race. Thus, what legitimated Spencer's concept of "innate" ideas, as opposed to Kant's, was that Spencer's were based on experience.[13]

Both Lowell and Spencer were more specifically concerned with Kant's insistence on the innate character of the ideas of space and time: a claim that would jeopardize their evolutionary project. To make evolution a workable philosophy, these categories would need to be conceived in a radically different fashion. Accordingly, Lowell argued in Spencerian fashion that "space and time are not forms of the intellect but essential attributes of the cosmos." He demonstrated the reality of these concepts—their objective existence outside of the mind—by imagining two separate solar systems that could be made to look like one if distances and sizes of the components were properly adjusted. In effect, the actual universe consisted of objects and distances that were distinct but might appear otherwise.[14]

Time, even more than space, is an indispensable element in the worldview of an evolutionist. Accordingly, Lowell attempted to clarify the nature of this concept as an objective reality. He argued, "Time is cosmic. The cosmos is a product of time by virtue of its essence. Everything is produced by virtue of a long ancestry first and variation second. The second, exists only on the shoulders of the first." Thus, change of any kind was conceivable only in relation to a condition that had existed at some prior moment. The process by which the cosmos evolved from simplicity to complexity required the passage of time.[15]

If innate ideas were invalid, equally so were a priori ideas. Both Lowell and Spencer, intent on defending their scientific ethos, insisted on a reciprocal relationship between ideas and experience. Accordingly, as Lowell remarked, "Every principle must be derived from experience. To take a priori ideas is tacitly to deny evolution." Thus, if ideas were a product of experience, they could come into existence only as human beings experienced their world, which, in turn, required the passage of time.[16]

Having established the philosophical premises for evolution, Lowell next described in general terms the operation of the evolutionary process. His account indicates that he was borrowing heavily from orthogeneticists like Spencer and Haeckel, who viewed internal, structural features rather than external environmental factors as the determining forces in organic development. As Lowell explained, "We may liken evolution to the breaking out of a path. Bit by bit the path is made by each newcomer each progressing rapidly and inevitably to the point last reached and then breaking out a little more for himself." In this sense, the direction of evolution would be determined in large part by the potential of the original organism to change over time. "Only in a potential sense *of the second order* [Lowell's emphasis] can the world be said to have been created at once. Each germ had the power to go a very little way of its own initiative, time and many consequent germs being necessary to eventual development."[17]

Thus Lowell sketched the philosophical foundation for the evolutionary approach to the cosmos he had used in his East Asian books. In the process, he weighed deeper philosophical issues with some care, though his thoughts were highly derivative. The more than ten-year delay in returning to this project might suggest that it was not a pressing issue. However, we should remember that, although not always explicit, the evolutionary assumptions of cosmic philosophy continued to undergird Lowell's search for life on Mars. In 1907, after presenting in exhaustive fashion his empirical findings in *Mars and Its Canals,* Lowell took up cosmic philosophy in his usual energetic fashion.

Lowell's Philosophy of the Cosmos

Percival Lowell adopted an essentially Spencerian approach to understanding the universe in his evolutionary books and articles. The basic principles that he articulated were borrowed from Spencer but shaped to suit his own purposes. Consistent with Spencer, he regarded evolution as an all-encompassing approach to understanding the manner in which the inorganic, organic, and social worlds in their present state had come into existence; at the same time he emphasized the theoretical support that evolution could provide for the existence of extraterrestrial life and incorporated into Spencer's model new, more plausible scientific hypotheses, such as the origin of the solar system through an encounter of celestial bodies.

Lowell's approach to evolution can be encapsulated in five basic and interrelated points, all consistent with—and often duplicates of—Spencer's assumptions and methodology. First, Lowell understood evolution to be a comprehensive, law-abiding, and progressive process that accounted for change in the inorganic, organic, and social worlds—from planets and plants to planetarians. Second, planetary evolution occurred in well-defined stages that were governed by the gradual cooling of the planet. This dissipation of heat brought about first progress and then decline. Third, life was widespread because organisms were made of the most common elements and were adaptable to widely varying environments. Furthermore, the development of life led necessarily to intelligent life and the creation of societies. Fourth, since evolution was progressive, purposeful, and inevitable, it would over time bring about more highly developed worlds. However, the operation of evolutionary law would lead inexorably from birth to death. Fifth, the writing of a philosophy of the cosmos must be based on a hierarchy of disciplines, with astronomy ranked at the top.

That Lowell fully participated in the ambitious schemes of the cosmic philosophers is most evident in an *Atlantic Monthly* article he wrote entitled "The Revelation of Evolution," which was ostensibly published to commemorate the centennial of Darwin's birth in 1809. According to Lowell, Darwin's focus on the development of the species was only "part of cosmogony's master-key . . . For in truth the rise of the organic is but the latest chapter in a serial of the sky." In that sense, he said, Darwin had not utilized the full potential of evolution, which was "the mainspring of the universe. Grand in its simplicity, it is the one fundamental fact on which all we know depends. From its influence nothing can escape; for it has fashioned everything, from nebula to man." In Lowell's recounting of the history of evolution as an intellectual construct, the real hero was Spencer, who had realized far more than Darwin the great possibilities inherent in the concept.[18]

Long before the centennial of Darwin's birth, Lowell had commited himself to Spencer's bolder version of evolution. Dramatically, he concluded *Occult Japan* with the assertion that "the spirit of the Cosmos is itself one great *idée fixe* working itself out." In effect, the current condition of the universe could be attributed to a programmed evolution from an original formless state. In *The Soul of the Far East,* he made the same point in a somewhat different way: "All life, whether organic or inorganic, consists, as we know, in a change from a state of simple homogeneity to one of complex heterogeneity. The process is apparently the same in a nebula or brachiopod, although much more intricate in the latter." And, of course, as we shall see, individuals and races were subject to the same law of development.[19]

An important consequence of the common origins of all matter in the solar system and its lawlike behavior was the ability to extrapolate from known conditions on Earth to conditions in other parts of the solar system. For example, since progress in a planet's development was governed exclusively by size, a large planet like Jupiter would logically be in a much earlier phase of the evolutionary process than the Earth. Accordingly, Lowell remarked that Jupiter was "on the road from Sun to world." The same kind of supposed lawlike behavior enabled him to infer from his close studies of Mars "corroboration of the whole evolutionary process from beyond the confines of our native earth. That the inorganic should develop into the organic on a single planet might perhaps be accidental, but not on two." In this way, the evolutionary approach encouraged its practitioners to enlarge their perspective. They could envision the prospect of complex developments on other planets at different points in time that would inevitably include the development of intelligent life.[20]

Like other evolutionists, Lowell found it convenient to conceive of continuous and gradual change as leading from one stage to another, with each succeeding stage representing progress toward fuller development. Lowell first employed this approach in his account of the development of societies from savage to civilized. Then he used precisely the same kind of analysis to describe the development of planets. Every planet of sufficient size would move through six stages, as he put it, "from sun to cinder." In the three initial stages, the planet would evolve from its original condition as a gaseous body to one with a solid surface contoured into continents and ocean basins by the uneven collapse of the crust. Once the surface temperature of the planet fell below the boiling point, the formation of planetwide oceans, from condensation of the steamy atmosphere, and the combination of chemicals into new organic and inorganic matter would begin. In these damp surroundings, organisms of the simplest sort would find an environment conducive to their existence. The gradual receding of the oceans would bring on the terrestrial stage, in which the planet would depend on heat from the sun. Seasons would develop in most areas of the globe, along with clear differentiations in the climate at different latitudes. At this point, life of a different sort would begin to develop, as water creatures adapted to take advantage of opportunities offered only on land.[21]

Following in the footsteps of his mentors, Lowell strongly emphasized the important role of heat and its dissipation in the process of planetary evolution, which, in turn, raised the matter of planetary size. Lowell reasoned that the accretion of relatively large quantities of matter would be necessary to generate enough heat over time to sustain life. Only bodies of a certain minimum size possessed the quantity of heat, which Lowell termed "evolutionary capital," required to set in motion the evolutionary process. The "spending" of that "capital" through planetary cooling Lowell termed the deus ex machina; it changed the character of the planetary body and thus the planet's environment. Bodies that dissipated their heat too quickly would not be able to sustain evolution.[22]

Lowell, along with Huxley, disputed Baron William Thomson Kelvin's claims that the Earth had cooled too quickly to accommodate the pace of geological and biological evolution proposed by Sir Charles Lyell and Charles Darwin. While Huxley argued that more time was available than Kelvin's calculations had indicated, Lowell took the opposite tack. He maintained that "the relative pittance astronomy has been willing to dole them [the geologists] out . . .—a paltry 50 or 100 millions of years" was perfectly adequate if only geologists had a proper understanding of the early history of the Earth. Lowell's maddeningly vague explanation assumed a more rapid development of the Earth's surface, thus leaving more time for the organic phase of history. He could more easily have refuted Kelvin by using Baron Ernest Rutherford's recent discovery of radioactivity as evidence of the ample supply of energy to support the evolutionary process. Apparently, however, Lowell had not seen Rutherford's study.[23]

Following Huxley and Haeckel, Lowell was convinced of the inevitable emergence of life on sufficiently large planets as a result of the presence of essential elements and the great capacity of living things to adapt to new conditions. Lowell cited various recent reports showing that life existed in the depths of the oceans under great pressure and at the poles despite extreme cold. From that perspective, cold climate and thin atmosphere were no barriers to the existence of life on Mars.[24]

The history of the Earth thus modeled the stages of development theory and provided evidence that life arose easily under the proper conditions. Evidence from the telescope and the spectroscope, moreover, suggested that Mars had preceded the Earth in its own development and must be home to life of an advanced kind. That it was intelligent life could be ascertained both by telescopic investigation and by the logic of planetary evolution. For intelligent life to arise, however, organisms would have had to develop minds, and this, in turn, required a proper environment. Following Huxley and Haeckel, Lowell argued that under certain conditions, minds would arise from matter, of which they were a "manifestation." Since the matter out of which life developed was common, it followed that minds would also be an abundant commodity. They would come into existence as the changing terrestrial environment created new challenges, which in turn stimulated the development of higher capacities to meet those challenges. Lowell thus argued that brain tissue developed through molecular action to assist organisms in responding to a more complicated environment.[25]

Minds, which were originally the product of matter, would be able to control and exploit matter in the age of intelligent life. Intelligence would first become a tool for survival; as Lowell put it, when "the organism develops brain it is able to circumvent the adversities of condition." Intelligence was also an instrument to "subjugate" the environment through the exploitation of resources. Evidence indicated, however, that intelligent life on Earth was not the measure of capacities elsewhere. Lowell believed that Martians, driven by the necessity for water, had developed new capacities, particularly the ability to cooperate. Tragically, however, the adverse tide of planetary evolution, evident in environmental changes that outpaced advances in intelligence, would lead inevitably to their demise.[26]

According to Lowell, the development of mind was the fundamental goal of the evolutionary process from the beginning. Such a conviction on his part indicates that the process was endowed with a purpose that could be achieved through progressive changes in organisms and their environment. He viewed the evolutionary process as "a great cosmic drama" that involved "the building up of the infinitely little into something far finer than the infinitely great." Here, Lowell referred in almost mystical language to the molecular action and chemical combinations that brought about changes in organisms and enabled them to take on more complicated tasks. These changes, in turn, were apparently the product of a purposeful agent. As Lowell explained, "Modifications of it [life] follow any and every change of environment, but nature strives to the last gasp to bring forth this [mind], her highest product."[27]

While Lowell never suggested divine presence in the system, he clearly endowed "nature" with the powers of divinity. Viewed from his perspective, which echoed Haeckel's, the evolutionary project had a clear direction from inorganic to organic life and from lower to higher forms of organic life, culminating in the development of mind and the rise of advanced societies. The key moment in this story of progress was the movement from terraqueous to terrestrial life. "An upward step in evolution occurred when life stepped out upon the land. While less directly favorable to life, the land was endowed with more possibilities for organisms capable of turning them to account. Brain was needed, and brain evolved." Further challenges would likely lead to further advances, as the case of Mars indicated. "Man is merely this earth's highest production up to date. That he in any sense gauges the possibilities of the universe is humorous." Thus, the outcome of evolutionary struggle was essentially preordained by the work of a purposeful and progressive "nature" clearly programmed to achieve the triumph of mind.[28]

The lawlike behavior of matter in its evolutionary course, however, brought about an inevitable decay and death as the end product. The solar system and all its members, as Lowell explained, are governed by "inexorable permutations" that "lead from one state to the next, till the last of all be reached." Thus, evolutionary advances would all be for naught; no matter how perfect intelligent life, it would be inadequate to the task of meeting the challenges of the future. In the end, nature, which had striven so hard and purposefully to bring intelligent beings to their current lofty state, would not be able to rescue the product of her efforts. However, while intelligent life disappeared on one planet, it would be emerging on another in some other part of the solar system or universe.[29]

Lowell's ambitious project, like Spencer's, assumed a hierarchy of disciplines. However, nowhere in Lowell's work do we find a full-blown articulation of the hierarchy privileging a metadiscipline like cosmology. Rather, Lowell, like Huxley, offered a partial account of disciplines according to which certain of them, by virtue of the breadth of their approach, had greater authority than others. Lowell made this point in telling fashion in

The Evolution of Worlds, where he maintained that geology was a subfield of astronomy. This hierarchy followed naturally from the fact that the Earth's history was the product of "cosmic causes that have been at work upon it," which in turn required "cosmic laws to explain them." In this sense, geology was a "local" science that could be authoritative only in describing events and conditions arising after the Earth had already been created. In Lowell's view, misinterpretations were possible if events on Earth of "cosmic occasioning," and over which "astronomic cause presides," were seen as "local." On this matter, Lowell's claims will inevitably seem to us to be self-serving, since they justified him in posing as an authority on issues outside of his field. In fact, however, they should be taken as evidence that he had thoroughly absorbed the cosmic evolutionary perspective from Spencer and Haeckel and, with it, the hierarchical notion of disciplines that privileged astronomy over geology.[30]

Percival Lowell belatedly brought to completion the "philosophy of the cosmos" that he identified as his life's work. In writing his evolutionary books and articles in 1908 and 1909 he continued his bold assault on the most important secrets of the universe, from the origin of the solar system to the rise of intelligent life and human societies, originally launched with the publication of *The Soul of the Far East* in 1887. In accomplishing this project, Lowell was heavily influenced by T. H. Huxley, Ernst Haeckel, and especially Herbert Spencer. Like his mentors, Lowell accepted the concept of the unity of the cosmos, which could be investigated only through the creation of a meta-discipline like cosmology. He believed that evolution held the key to revealing the laws of development in the natural and social worlds and was convinced that the new cosmic perspective could provide support for his own projects, including the identification of a racial hierarchy among civilizations and proof of the existence of extraterrestrial life.

6

Image-Maker

> Gooley!—I have found a marvellous book—a book of books!—a colossal, splendid, godlike book. You must read every line of it. For heaven's sake don't skip a word of it. The book is called "The Soul of the Far East," but its title is smaller than its imprint.
>
> LAFCADIO HEARN TO GEORGE GOULD, 1889

Lowell's writings on East Asia did indeed have a powerful impact on the educated reading public, as Hearn's letter suggests. Viewing East Asian culture in an evolutionary perspective, Lowell addressed the most important issue of his era: namely, the future of the industrial system. His books were occasions for exploring the relative merits of traditional and modernizing societies and for affirming at one and the same time the value of East Asian artistic achievements and the superiority of Western industrialism. This ambivalent response to industrialism expressed Lowell's distaste for the excesses of the system rather than a desire to restore a traditional way of life. The popularity of his view, however, was fleeting. In the wake of the depression of the 1890s, Americans turned to his disciple, Lafcadio Hearn, who better reflected their growing distaste for an industrial system in disarray.[1]

A serious consideration of Lowell's work on East Asia requires a proper understanding of the context in which his books, poems, and articles were published. We know, for example, that interactions between individuals from different societies are subtly conditioned by their own conceptual frameworks, from which they view the encounters. Thus, travelers' comparisons of other countries with their own are frequently based on ideas that were internalized well before the journeys began. In Lowell's case, as we have seen, Herbert Spencer's sociological views and, in particular, his ranking system, provided Lowell with a conceptual framework that focused attention on certain dimensions of East Asian and Western cultures and clarified their significance. In this way, Spencer's scheme enabled writers like Lowell not only to assert the superiority of Western ways but also to develop a systematic view of the diverse races around the globe. Accordingly, we shall chart the influence of this evolutionary perspective in shaping Lowell's approach to East Asia.

Another context for Lowell's book was the stream of travel volumes being written by other Americans in the ten years before and after Lowell wrote. Among dozens of travelers who wrote anecdotally about their experiences in Japan, only a few tried systematically to understand the history and culture of the country. Lowell and Hearn, along with William Elliot Griffis, whose work first appeared in 1876, stand out in this way. The American public had been wedded to Griffis's *The Mikado's Empire,* and Lowell's books competed for attention with it. In emerging as the leading interpreter of Japanese life for the intelligent reading public in the United States and England, Lowell partially displaced Griffis, as Hearn would later displace him. Lowell's work is thus one of the keys to understanding the evolution of the American image of Japan from 1885 to 1895, as well as to understanding the popularity of Lafcadio Hearn in the years that followed.[2]

Travel volumes, however, are the product of a conversation not simply between travel writers but between civilizations. Travelers frame their judgments of foreign countries in terms of the values and concerns of their own society as they emerge in the encounter with a foreign culture. Even without Spencer, Americans would have approached Japan with concerns emanating from their own industrial revolution. The encounter with Japan, which had just emerged from its feudal period, offered an excellent occasion not only for discussing the prospects and problems of postfeudal Japan, but also for achieving a sharper perspective on changing values and practices in industrializing America. In this fashion, Lowell's travel books provided a vehicle for comparing and contrasting two societies in the throes of rapid change.

Spencer's Ranking System

Spencerian evolutionary concepts provided criteria for ranking civilizations that enabled Lowell to systematize his comparisons of East and West. Using Spencer's system, he could compare societies by placing them on a continuum from so-called "savage" to developed, or mature. Development was defined as the achievement of complex social organization involving a high degree of economic specialization, which, in turn, could be equated with the capacity of a society to dominate its environment. In line with Spencer's theory, a society's location on the continuum was not permanent, since all societies, like inorganic and organic matter, evolved through the same stages, though at different rates.[3]

Embedded in this apparently benign scheme of evolution, however, was a not-so-visible racial component that derived from Spencer's disciplinary approach to understanding societies. According to this scheme, racial maturity was gauged, first, by examining the physical development of the race (biology), which laid the foundation for mental development (psychology) and, later, for the creation of social structures (sociology). Thus, by the logic of the system, racial advancement was correlated most directly with the capacity of the brain to respond effectively to stimuli from the environment. In the beginning, Spencer maintained, all men had been endowed with a brain of the same size. However, in Lamarckian fashion, those exposed to more stimuli developed increased brain capacity, which was inherited by their offspring. In principle, all races were capable of growth; in practice, however, races in less complex environments, such as Africa, parts of Asia, and the Americas, had not had the opportunity to develop the large brain necessary to negotiate complex environments.[4]

According to Spencer, the stimulating environment produced by competition between societies for economic, military, and political dominion would bring about the mental development of their members. At least in the initial phases of this competition, a society would gain an advantage over its rivals through internal cooperation. Such cooperation, however, would become stultifying over time if, as was usually the case, it led to patriarchy and authoritarian rule. Members would become accustomed to obeying orders rather than thinking for themselves and would thus learn to be good imitators rather than initiators. The only remedy for the resulting arrested development, which plagued all homogeneous societies, was to break "the cake of custom" by encouraging individual autonomy. The impact of individualism would be felt in all realms, but especially in the economy, where it would lead to specialization and economic growth through industrialization. In this way, specialization, industrialization, and individualism were closely correlated.[5]

Finally, such advanced societies would be more imaginative, and therefore innovative, but not in the artistic or poetic sense. Spencer argued that imagination was the ability to see beyond experience to new conceptions based on an original combination of aggregates that might explain the experience. In defining imagination in this fashion, Spencer clearly made it an attribute of civilized people who, unlike savages, had varied experiences to reflect on, as well as abstract concepts with which to organize thoughts about that experience. In this sense, true imagination had to be distinguished from "reminiscent imagination," defined as the ability to recount past events.[6]

While Spencer purported to offer a value-neutral analysis of progressive versus stagnant societies, he had, in fact, loaded the dice. He assumed that the norm for progress was Anglo-Saxon society in the second half of the nineteenth century. Accordingly, other societies would be ranked high or low depending on whether their political, economic, and social development approached or deviated from the Anglo-Saxon standard. For ranking purposes Spencer distinguished three stages of development, ranging from savage to mature societies. In the middle were the "half-civilized" societies, including China and India, with their well-developed artistic and literary traditions but stagnant and authoritarian social systems. Lowell launched his study of Japan with this Spencerian scheme as his guide.[7]

Griffis's Image of Japan

Lowell's work must be understood as part of a continuing effort to assess the American way of life by contrasting it with various foreign cultures. In the antebellum period, the nation was already debating the question of industrialization and considering an appropriate stance toward nonwhite cultures—issues that were to concern Lowell and his contemporaries. American leaders hoped to shed light on the effects of industrialism by contrasting a preindustrial America with an industrializing Europe. The contrast would help to determine whether industrialism would be likely to advance or undermine republican institutions.[8]

America's westward expansion provided an opportunity of a different sort to define the nation's way of life, this time in contrast to the cultures of "barbaric" races encountered on the frontier. Americans debated the extent to which the inclusion of Mexicans and Indians within the republic might endanger its future. The discussion of the character of Indians led occasionally to a consideration of "Mongolians." The Indians' ancestors in Asia were regarded as half civilized and "imitative" in character. In this sense, they were ranked above blacks but below whites, very much as they would be by Spencer and company after 1865. Thus, the method of contrasting, in these cases and others, served several functions: it helped to clarify the superiority of American ways, to explore issues of importance to the American public, and to provide an opportunity for self-definition through studying the problems of excessive civilization or barbarism.[9]

In attempting to identify themselves in relation to other nations and races, antebellum Americans ignored East Asia except as they linked Indians to their continent of origin. Not until the immediate post–Civil War period did cultural relations with East Asia emerge for Americans as a significant issue. The intelligent reading public in the United States developed an image of Japan that was largely the product of Griffis's *The Mikado's Empire.* Every authority on the period acknowledged the impact of the work, from 1876 until well into the twentieth century, and before Lowell came along, it had no serious challengers. The first of Griffis's volumes provided a detailed history of Japan, from its origins to the Meiji Restoration, while the second presented a lively account of the author's one-year experience as a teacher of English at a boys' school in Fukui, along with commentaries on aspects of Japanese culture.

In important ways, Griffis's work was shaped by the American Civil War and the Meiji Restoration. It reflected the predominantly progressive political sentiments of middle-class Americans in the North. The defeat of feudal society was a common theme in Japan and America, and Griffis gave a rousing account of this development in Japan with which his readers could easily identify. He was also writing just as the American industrial revolution was getting under way and appeared to be rather unproblematic. The American public greeted enthusiastically the prospect of increased production that the revolution promised. Accordingly, Griffis lamented the poverty of Japanese towns: "All the houses of wood, the people poor, the street muddy, few signs of wealth, no splendid shops." Asians, like Westerners, would benefit from economic development. Finally, Griffis and his American audience embraced a pre-Darwinian, liberal Protestant ethos that precluded a more secular, scientific account of Japanese civilization.[10]

Griffis's work reflected the confidence of Americans of this era in the progressive character of their way of life. Griffis himself was a devout Christian who was eager to see the Japanese as good material for conversion. But he was, of course, much more than that. He must have surprised himself in discovering a sympathy for many aspects of Japanese culture, ranging from religious practices to housing and education. Too often, however, he deserted description for prophecy, urging Japan on in what he regarded as her quest for her rightful place among the modern nations of the world. "We hail the brightness of the rising of this first among Asiatic nations to enter modern life, to win and hold a place among the foremost peoples of the earth." His heroes in Japan were the modernizers, including Fukuzawa Yukichi, the leading advocate of Western ideas, and Nakamura Masanawo, who translated John Stuart Mill. He praised the visit of the Iwakura mission to the Congress of the United States on March 6, 1872, because it "marked the formal entrance of Japan upon the theatre of universal history." Even his own involvement in Japan could be portrayed as a contribution to the modernization of the country. "The heart kindles to think of the young life of this Asiatic empire being fed at the streams of the science and languages of Christian nations." Japanese students, the samurai from Choshu and Satsuma, the mikado, and the missionaries were united in promoting this noble cause.[11]

Against them were arrayed various forces seen as tending to perpetuate poverty, ignorance, superstition, and debauchery. Some of Griffis's most critical remarks were directed against Westerners whose debauchery and exploitation of the Japanese in the treaty ports gave Western ways a bad name. Even more dangerous, however, was the Buddhist hierarchy in the country. Griffis could, on occasion, recognize the attractions of Asian religions. In his Fukui home, Griffis admired an amulet placed at the entrance by the previous occupants; enroute to Fukui he observed his traveling companions praying in a Shinto shrine and commented, "The act was as touching as it was simple."[12]

Nonetheless, he could never quite shake the conviction that Buddhism was superstition and, as such, was responsible for many of Japan's problems. "It seems curious, even in Japan, to see men dressed in foreign clothes, praying before the gilded and hideous idols, bowing down to foxes and demons, and going through all the forms of paganism. Clothes do not make a Christian," he acknowledged, though he instinctively connected Western dress and Christian practices. He explained away the Buddhist belief in the annihilation of the soul as a function of the terrible poverty in Asia that rendered life unbearably painful. He blamed the Buddhists for the mistreatment of women in Japan, saying that according to their "dogma, ecclesiastical law and monkish asceticism, woman is but a temptation, a snare, an unclean thing, a scape-goat, an obstacle to peace and holiness." His conclusion about the future of Japan was implicit in these remarks: "Nothing can ever renovate the individual heart, nothing purify society and give pure blood-growth to the body politic of Japan but the religion of Jesus Christ." In this respect, the Japanese were no different than Americans. Notwithstanding their distinctive past, the Japanese were destined to achieve the same progressive Christian spirituality as Americans.[13]

Progressive spirituality went hand in hand with the establishment of a modern political regime. Griffis applauded Japan's victory over feudalism through the restoration of the mikado. By contrast, except for chapters on games, household life, and religious practices, he had little to say about traditional Japanese culture. Among the missing topics were art, theater, literature, gardens, and the language. No doubt Griffis recognized that too much attention to traditional culture would undermine his assertion of the progressive direction of Japanese life. He also ignored the Spencerian ranking system, which questioned the capacity of semi-civilized people to evolve into modern societies. In effect, Spencer's system subverted Griffis's claims and refocused attention on traditional culture.[14]

A Changed Environment

Griffis's book brought to a close one era in Japanese-American cultural relations even as the Philadelphia World's Fair of 1876 was opening a new one. The Japanese exhibit on that occasion would reveal what Griffis's book, published in that very year, had ignored: that Japan was a country capable of producing beautiful art. Newspaper reports praised the artistic brilliance of the Japanese exhibit, especially the lacquerware and pottery on display there. The fair thus brought home to middle-class Americans the beauty of handcrafted items prepared by Japanese artisans, and in so doing, it presaged a major shift in the preoccupations of the American reading public. While Griffis delighted in Japan's revolt against feudalism, the proponents of Japanese art focused on the cultural traditions that made possible the creation of handcrafted items—traditions that were closely linked to the feudal regime. Recognizing this, proponents of the arts were inclined to advocate preservation of tradition rather than modernization.[15]

This inclination was given a powerful impetus by strong economic trends that developed in the five years after the publication of Griffis's book. During the industrial revolution, educated Americans had been increasingly disturbed by clashes between labor and capital, by the ugliness of factory-made goods, and by the fouling of the environment. Such concerns recalled Jeffersonian anxieties about modernization. Now, however, the triumph of industrialism offered more tangible evidence of its social costs. The ensuing backlash was particularly strong among members of the elite, who resented their displacement by the new rich. This rising sensitivity to the harmful aspects of industrialism created a receptive audience for the different approach modeled in Japanese life.

The earlier enthusiasm of the American elite for modernization in Japan was tempered by another consideration as well. It was clear by 1880 that industrial and technological advances provided resources for military establishments and the acquisition of overseas territories. Slowly, the full implication of promoting these trends in Japan was dawning on the American elite: Japan, a weak but friendly nation, could become a power capable of rivaling Western countries. Such a possibility gave pause to the advocates of modernization.[16]

These important developments deeply influenced Griffis's successors, Percival Lowell and friends. They shared strong reservations about both the industrial order and organized Christianity. These doubts caused them to reject the idea of turning Japan into a modern Christian nation. As Boston cosmopolitans who had much in common with British men of letters, they were eager to experience another culture, and perhaps to like it better than their own. They cherished the differences between Japan and the West that kept their host country from becoming a duplicate of their homeland. Among the attractive differences was the Japanese talent for artistic endeavors. These men had a stake, then, in Japan's continued adherence to traditional ways. They genuinely admired the Japanese artistic tradition, and they worried about the subversive effects of modernization. Unlike Griffis, they were thus inclined to view and to portray Japanese and Western civilizations as polar opposites.[17]

Not surprisingly, Boston served as a center for both the criticism of industrial culture in America and the articulation of skepticism in the religious domain. Members of the new generation of Brahmins had been nourished on Herbert Spencer's work and, as a consequence, had little regard for the formal religion of their fathers. Lowell, Edward Morse, and William Sturgis Bigelow were either indifferent or hostile to the Christian faith in which they had been raised. Their mutual friend B. H. Chamberlain spoke of a "violent hostility to Christianity" as one of the issues on which they agreed. On rare occasions Lowell's own writings confirmed

Chamberlain's observations. Reporting on the response of the Asiatic Society to his talk on Shinto, he remarked: "The fair missionarinesses think Mr. Lowell so nice! which makes me feel like a villain but if they can't see satire I stand absolved." Lowell thus rejected out of hand Christian uplift of the Griffis variety and regarded its exponents as sentimental and effeminate.[18]

Boston also served as a center of anti-industrial thought, which was frequently articulated through the promotion of things Japanese. Although New England had pioneered in the development of industry, the enthusiasm for it diminished as the early industrialists passed from the scene and their children and grandchildren recognized its limitations.

Boston possessed a well-developed network of cultural and scientific organizations through which news of Japanese traditions could be disseminated. The Lowell Institute, for one, exploited public interest in Japan by devoting a lecture series to the topic in 1882. The trustee of the institute, Lowell's father, invited Edward Sylvester Morse, one of America's leading zoologists, to deliver the lectures. Morse had gone to Japan in 1877 in search of brachiopods but had become interested in material culture while he was teaching zoology at Tokyo Imperial University. He had delivered lectures on this topic in various parts of the country before he addressed an overflowing crowd at the Lowell Institute that included Percival Lowell and William Sturgis Bigelow. Lowell and Bigelow were so taken with Morse's campaign to preserve traditional culture that they soon followed him to Japan.[19]

Lowell's approach to Japan was dictated in part by the preoccupations he shared with this cosmopolitan circle, and in part by a desire to move beyond the work the others had already accomplished. He was interested in the history and traditions of a country only recently emerged from the Middle Ages, though his scholarly inclinations were less developed than those of Chamberlain. He admired Japanese art, but he did not possess the passion for collecting of Edward Morse or Ernest Fenollosa. He had a virtual obsession with exploring remote areas of East Asia that was frustrated by the fact that much of Japan and Asia had already been explored. Coming a generation too late, he envied Chamberlain and company, who had had the opportunity to map Japan for Western audiences. His frustration spurred him to an obsessive search for novelties to report to his audience and to periodic expressions of disappointment at the rarity with which he encountered them. He lamented the absence of old-fashioned adventures in Korea and the fact that other Europeans had preceded him to the Japan Alps and the Noto Peninsula, and he noted with satisfaction the citation in

Chamberlain's guidebook of his discovery of a remote valley in the Japan Alps.[20]

While Lowell undervalued his role in presenting unknown landscapes and peoples to Western audiences, he clearly found his vocation in Japan in a different sort of discovery. He came to the realization that Westerners had not yet grasped the uniqueness of Japanese culture in contrast to the West, and he took as his mission the creation of a coherent portrait of Japanese culture that would, in turn, clarify the distinctiveness of Western civilization. In this way, he would be able to offer a "scientific" account of world civilizations, taking as its point of departure Spencer's system of ranking cultures.[21]

Art versus Science

Lowell took up the central questions raised in American minds by the Japanese experience: Could one appreciate Japanese art without conceding superiority to the civilization from which it emerged? Would a modernizing Japan challenge Western hegemony? Lowell answered these questions in a manner satisfactory to admirers of both Japanese art and Western science. In the process, he found a way to present Japanese civilization as a coherent enterprise.

Lowell's "philosophizing," as contemporary reviewers called it, gave his work a distinctive cast. Other writers of travel books, including Griffis, typically described their own personal experiences, various dimensions of Japanese life, and the evolution of events from past to present. For the most part they avoided the large generalizations for which Lowell had a penchant. His method was to identify the central concept governing the East Asian mind and to make it the organizing principle of his books. This concept then informed the essays he wrote on various Japanese institutions and practices. In effect, Lowell offered in *The Soul of the Far East* the first coherent picture of Japanese civilization in which the parts—language, family, art, and religion—were linked to each other through the expression of common values to form a cohesive whole. It was the systematic nature of Lowell's work that made it so attractive to contemporaries like Lafcadio Hearn and caused a discerning reviewer in *The Dial* to point out that Lowell's *The Soul of the Far East* was reminiscent of Emerson's *English Traits.* Both incorporated experiences and observations within a larger framework, offering an interpretation of the culture as a whole.[22]

Understanding the coherence of Japanese civilization was for Lowell, however, only a first step in his project of evaluating modernization. By identifying Japan with the traditional way of life, he laid the groundwork for comparisons between Japan and modern Western societies. The comparisons would yield a better understanding of the virtues and drawbacks of traditional and modern ways. By defending both Japanese art and Western science Lowell expressed his ambivalence about modernization. But the greater value he gave to scientific achievement suggested that, overall, and consistent with Spencer's scheme, the civilization of the West was superior to that of East Asia. This contrast theme was also calculated to reassure readers, since in their minds Western science was associated with power while Japanese art suggested feminine graces.[23]

The subject of Lowell's books—the significance of the opposition between West and East—was intended to provide a complete picture of the life of the globe. Juxtaposing the two civilizations would help to define the character of each by highlighting similarities and differences between them. Indeed, Lowell promoted the introduction of the Japanese language into the American college curriculum as a way of clarifying the distinctiveness of Western languages. However, Lowell departed from his plan to explain racial development on a worldwide scale as he came to realize that the very premises of Japanese civilization threatened the basic values of the West. "Can it be that the personal, progressive West is wrong, and the impersonal, impassive East right? . . . Is the other side of the world in advance of us in mind development, even as it precedes us in the time of day?"[24]

He answered these questions by offering a Spencerian interpretation of the differences between East and West. In his analysis, Lowell equated the differentiation of species in the organic realm with the achievement of individuality in the social order. Both would be evidence of the development of heterogeneity, which, in turn, could be equated with progress. Accordingly, he argued that "the degree of individualization of a people is the self-recorded measure of its place in the great march of mind." And although the process of evolution was fully operative in the West, it had been arrested in East Asia. "We behold here in the case of man the same spectacle that we see cosmically in the case of the moon, the spectacle of a world that has died of old age." This evolutionary stagnation was manifested, he said, in the arrested mental state of East Asians, whom Lowell equated with children. The experience of self-consciousness, a kind of second birth, though accessible to all people, was muted in East Asian culture. Accordingly, East Asians were unlikely to experience either the joys or the agonies of self-consciousness. Most important, they would never struggle with the hard choices that alone brought about change and thus progress. Evolutionary development, by contrast, had proceeded full tilt in the West, where societies were characterized by the prominence of individuality, increasing specialization of function, and mobility.[25]

Focusing on institutions and culture, Lowell proceeded systematically to demonstrate the truth of his thesis. He attended first to the family, which, as the basic unit in society, bore the responsibility for the undeveloped and undifferentiated character of Japanese society. "To-day a whole nation, resembling a nebula indeed in homogeneity, is swayed by a single patriarchal principle." Lowell offered as evidence of muted individuality the fact that birthdays were less celebrated than communal festivals like girls' and boys' days. He lamented arranged marriages, which reflected the "patriocentric pivot" of the family and neglected the preferences of the bride and groom. Under such a regime, the concept of romantic love was unimaginable, since it assumed and confirmed a powerful emotional attachment between self-conscious individuals. In the East, in short, the individual was given what identity he had by his membership in the family.[26]

From the family Lowell turned to a consideration of language, art, and religion, which, he believed, would mirror impersonality in the past, present, and future, respectively. Lowell argued that language provided a sensitive indicator of the values of any people, since it evolved to meet their needs. That being the case, he said, Japanese grammar reflected the impersonality of the people. He contrasted, for example, small talk in Japanese and Western languages and found it significant that Japanese small talk had no subject, while Westerners talked about people. He argued that the famous politeness of East Asians resulted from a lack of self-regard, which, put another way, meant greater regard for others. Such politeness was reflected in the complicated honorifics so central to the language. Impersonality was also evident in the absence of words to identify sex differences and the creation of sentences without subjects, thus suggesting that actions need no actors.[27]

Lowell's chapter on art dealt with images as a reflection of the Japanese mind. His appreciation of the Japanese aesthetic sense was already abundantly clear to his readers. Lowell, after all, supported the work of Fenollosa and Morse and was thus very much attuned to the aesthetic charm of Japanese life. His books are full of testimony to the elegance of Japanese gardens and works of art. Like Hearn, he also regarded the Japanese woman as one of the most remarkable specimens of such art. During his stay in Korea, he noted the "pretty coquetry" of a geisha named Fragrant Iris. "Winsome she was and as my eyes, wandering over her jet-black tresses, simple as the silver pin that bound them, fell at last upon her upturned face, in the smile I found there I forgot that I was foreign and my home so many thousand miles away." Lowell applauded the traits of "gentleness and delicacy" that characterized "far-Eastern" women. These were evident in the person of the Japanese maid who served him in Noto with "a grace that beggared appreciation. 'His Augustness is well come,' she said . . . and the voice was only too human for heaven. Unconsciously it made the better part of a caress." She poured tea as if it were "some beautiful rite." His conclusion that the "far-Eastern woman never seems to forget under any circumstances that she is a woman" suggests the enormous cultural gap that separated Griffis from Lowell and Hearn. Griffis rebuked the Japanese for the subservience of their women, while Lowell condemned American feminism.[28]

Lowell was equally responsive to other examples of beauty in Japanese life. He claimed that "from the tips of his fingers to the tips of his toes, in

The geisha "Fragrant Iris," northeast of Seoul, 1884 (courtesy Museum of Fine Arts, Boston)

which use he is surprisingly proficient, he [the Japanese person] is the artist all over. Admirable, however, as is his manual dexterity, his mental attitude is still more to be admired; for it is artistic to perfection." With this skill, the Japanese could endow even a mundane shopping area with an aura of beauty. Lowell remarked of the Ginza that "to stroll down the *Broadway* of Tokio of an evening is a liberal education in everyday art. As you enter it, there opens out in front of you a fairy-like vista of illumination . . . As your feet follow your eyes you find yourself in a veritable shoppers' paradise, the galaxy of twinkle resolving into worlds of delight." The effects of the Japanese artistic sense were also evident in the transformation of nature that they wrought in their gardens. Lowell spoke of a "quaint fairyland" where "diminutive rivers flow gracefully under tiny trees, past mole-hill mountains, till they fall at last into lilliputian lakes, almost smothered for the flowers that grow upon their banks."[29]

Taken together, these passages constitute a powerful endorsement of the Japanese capacity to create beauty in everyday life. This pervasive spectacle obviously made a deep and continuing impact on Lowell during his various trips to Japan—which makes it all the more curious that Lowell's chapter on art was cast in a defensive mode. He devoted two pages to establishing the failure of the Japanese to grasp the principles of modern science before enumerating the particular arts that gave evidence of Japanese prowess in that area. ("His conception of beauty is as keen as his comprehension of the cosmos is crude.") Lowell then informed his readers that the aesthetic sense in Japan could not be equated with individuality. The Japanese, as he explained, focused their art on the natural rather than the human world. Even when Japanese artists painted men, they treated them as a species in nature rather than as individuals. And whereas the Japanese idealized nature, they treated men prosaically by showing them in everyday dress. Furthermore, Japanese artists ignored the female figure, one of the favorite subjects of Western artists interested in the individual. In conclusion, Lowell insisted that the Japanese were more thoroughly artistic than Westerners were scientific. This judgment, however, served to confirm his contrast between Japanese impersonality and Western individualism and to reassure the author and his audience that artistic skill had a lower value than scientific imagination.[30]

For evidence of the Japanese view of the future, Lowell turned to religion, a subject rendered complex by the overlapping of Shinto, Buddhism, and Confucianism. However, Lowell excluded Shinto and Confucianism from consideration on the grounds that they were essentially extrinsic to the soul—Shinto because it only furthered the Japanese passion for nature, and Confucianism because it served only as a moral system. On this basis, Lowell claimed that Buddhism accurately represented the essence of the Japanese soul—a position he would later reverse. No doubt he insisted on Buddhism's priority in part because of his close association with Bigelow and Fenollosa, both of whom studied and converted to Buddhism.

Having affirmed the centrality of Buddhism, Lowell sought to understand Japanese religious views by comparing and contrasting Buddhism with Christianity. For Lowell, the two religions were similar in important respects. Both contained symbols, the aspect of religion that appealed to the masses, and spiritual truths accessible only to elites, and both inculcated love, but they differed in other respects. Buddhism emphasized contemplation rather than action. Moreover, and more important, the Buddhist road to salvation required a kind of annihilation of the soul, while Christianity affirmed the self. Thus, even in the domain of religion, Western individuality was a contrast to Eastern impersonality. For an evolutionist like Lowell, however, the road to truth lay not through religious devotion of any kind but through science.[31]

In his concluding chapter to *The Soul of the Far East,* Lowell clarified the contrast between East and West, emphasizing that the former was a stagnant civilization whose growth had been arrested during its adolescence. East Asia had achieved a kind of perfection on its own terms, but one in which change was no longer possible. The resulting stagnation had been brought on by the absence of imagination among the people. As Lowell recognized, this was a paradoxical assertion, since anyone would suppose that a nation of artists possessed imagination in abundance. Indeed, in *Chosön* Lowell claimed that imagination, the faculty requisite for originality, was equally present in poets and mathematicians. However, in the conclusion of *The Soul of the Far East,* Lowell, drawing on Spencer's analysis, argued that art required imitation and taste more than imagination. In point of fact, he insisted, most artists were severely limited in their ability to transcend everyday realities and perceive what is behind and beyond them. Imagination, by contrast, enabled the individual to conjure up an alternative world suggesting the possibility of change. In fact imagination and individuality were perfectly correlated, as were imitation and impersonality.[32]

The consequences of impersonality were serious, according to Lowell, since nature abhors the abandonment of self like a vacuum. In competition with the individualistic and imaginative West, the East would inevitably fare poorly. "If individuality be the natural measure of the height of civilization which a nation has reached, impersonality should betoken a relatively laggard position in the race." Imitation would yield only superficial change, and without real change, Asia would perish.[33]

In the scheme of things, ironically, Lowell's powerful testimony to the quality of Japanese art required an equally powerful denunciation of the incapacity of the Japanese to think scientifically. In Lowell's mind, the success of the Japanese in matters of art raised awkward questions about the primacy of the West. His discussion of science clarified the idea that only Westerners were possessed of imagination, and therefore identified a realm in which Easterners could never achieve to the West's level. Thus at one and the same time Lowell satisfied the complicated desire he shared with his public: to validate something truly valuable in East Asian culture that was, nonetheless, not the equal of Western achievement in science. And by establishing the relative poverty of imitation as compared to imagination, he reassured Americans that the Japanese would never be able to challenge Western hegemony.

Moreover, Lowell's conclusions about the imitative character of Asians were all the more persuasive because they reinforced the long-standing racial stereotypes inherited from the antebellum period and echoed the more systematic and up-to-date approach of Spencer and Fiske. In one important respect, however, Lowell deviated from Spencer's approach. Spencer had insisted on the value of cooperation at an early stage in a society's development to maximize its success in competing with nearby rivals. Later on, in Spencer's view, social progress would depend on balancing cooperation with individualism. We may speculate that Lowell chose to ignore the cooperative dimension of progress because it would have complicated his argument and violated his own commitment to individualism and the competitive ethos.[34]

Percival Lowell's emergence as an interpreter of the East Asian experience to the American reading public reflected significant changes both at home and abroad. The popularity of his work indicates the ambivalence many Americans felt about the industrial revolution. They could find an expression of this ambivalence in Lowell's systematic approach to "far-Eastern" civilization. Lowell, like Spencer, valued the economic development, specialization, and political liberalism that characterized Western countries. His endorsement of science and the dynamic ways of the West was a clear affirmation of the industrial revolution, and the contrast he drew between "impersonal" East Asians and individualistic Westerners was a thinly veiled formula for clarifying the superiority of the latter to the former.

At the same time, however, Lowell brought the aesthetic dimension of East Asian life into focus for an American audience already aware of the Japanese success at the Philadelphia World's Fair. In so doing, he and his audience abandoned their missionary concern for the "improvement" of Japan and focused attention on the continuing importance of traditional elements in contemporary Japanese life. Lowell's appreciation of East Asian artistic achievements clearly aligned him with the rising voice of critics of the industrial system, who denounced both its effects on the landscape and the monotony of mass-produced goods.

This critique of industrialism had its roots in Jeffersonian views but departed strongly from them. Jefferson spoke for rural Americans committed to the primacy of an agrarian society, while the new critics were themselves descendants of the founding fathers of industrialism, who strongly opposed Jefferson's views. Far from advocating a return to agrarian ways, Lowell and company viewed industrialism as the source of American power and sought merely to uphold the quality of its products and the environment. Indeed their subtly expressed fear of the potential of a modernizing Japan required the success of the American industrial system, to assure America's continued domination in the Pacific.

Lowell's work departed dramatically from Griffis's pathbreaking study and, in turn, prepared the way for Lafcadio Hearn. Hearn's initial enthusiasm for Lowell's work is understandable. Hearn was himself, like Lowell, interested in the traditional and aesthetic dimension of Japanese life and hostile to Christianity. He too was captivated by the charm of Japanese women, the beauty of the gardens, and the delicacy of Japanese art—all subjects that Griffis had left aside. And both men were ardent disciples of Herbert Spencer. Moreover, Lowell and Hearn, who both published their books at Houghton Mifflin and their articles in *Atlantic Monthly,* were addressing the same cosmopolitan audience.[35]

Lowell's judgments regarding Japanese impersonality, however, were problematic for his disciple. Although at first Hearn accepted them, the longer he lived in Japan, the more he bristled at Lowell's lack of sympathy for the Japanese. Hearn came to question the extent of Japanese impersonality and, more important, to view Western individualism, in its aggressive form, as distasteful and uncivilized. In this sense, he accepted in modified form Lowell's claim that the Japanese were impersonal, but attached a positive value to this condition.

The gap between the views of Hearn and Lowell, however, has been exaggerated because of the way they handled the question of "Westernization" in Japan. Hearn was quick to blame Westerners for the advent of any modern practices and defined Japanese life as if it were perfectly static and traditional. In his book *Kokoro,* for example, he maintained that the modernization then occurring in Japan had in no way tainted the deep commitment of the Japanese to traditional values. Lowell, of course, also viewed modernization as an essentially Western development. However, he insisted that imitation was a faculty deeply embedded in the Japanese psyche. In this way, the Japanese had to bear the responsibility for the modernization of their country. Of course it is true that Hearn's revulsion against modern life was more powerful than Lowell's; the latter's objections were directed, rather, at the awkward mixture of modern and traditional in Japanese life.[36]

In any event, in the 1890s Hearn's work began to displace Lowell's in the United States as the authoritative version of Japanese culture. We may surmise that the rising popularity of Hearn reflected a preference for a more thoroughgoing rejection of industrialism and a more appreciative view of tradition than Lowell's work presented. In the questioning mood of the decade, Hearn seems to have struck the right chord with the educated American public.

Lowell's contributions to the interpretation of the East Asian experience for an American audience are often hidden by certain offensive features of his work. His claim to scientific objectivity, based on the application of a Spencerian ranking of races, does not withstand scrutiny. Particularly offensive in this regard was Lowell's use of Spencerian language to give "scientific" sanction to the stereotype of the imitative Japanese. Further, the notions of arrested development and the incapacity of the Japanese people for science were clearly wrong. Finally, the reduction of a whole civilization to a single principle is not credible.

However, Lowell's effort to contrast the East Asian and Western psyches, if understood in terms of ideal types, does offer some interesting insights. Remove the assumption of racial causation, and what we have is one characterization of the mind-sets of traditional and modern peoples. In traditional societies, individual identity is conferred by family or class, while in modern societies individuals must discover their own identity. Lowell, it would appear, grasped this difference between East Asia and America—quite similar to the contrast that Tocqueville articulated between democratic and traditional societies—but wrongly assumed that it had a racial basis. Thus, however flawed in its details, Lowell's effort to show the interlocking character of different aspects of East Asian culture remains a signal contribution.

We may also credit Lowell for his effort to move beyond the limited role of a travel writer interested in the minutiae of everyday life in another culture, to that of a cultural analyst attempting to understand the implications of local phenomena in a global perspective. To be sure, Spencer's ranking system was inadequate as a device for comparing cultures, but the effort to understand one culture in a systematic fashion and then to view it as part of a dynamic world of cultures was quite laudable. No doubt the popularity of Lowell's work rested in good measure on its ambitious character. We need to remember this point even as we reject the conclusions he reached.

7

Psychical Researcher

In *The Soul of the Far East,* Percival Lowell preached the superiority of the Western scientist to the Japanese artist. In *Occult Japan,* Lowell practiced what he preached by joining the ranks of psychical researchers attempting to shed light on the phenomenon of trance states. In retrospect, it is clear that Lowell's decision to engage in psychical research was the first step in moving toward a career in science. His research contributed significantly to the West's increasing knowledge about East Asia and the occult, but it also revealed approaches that brought him into conflict with authorities in the field and anticipated problems he was to encounter in his astronomical work.

His brief foray into psychical research thus provides an important opportunity to examine the assumptions and values underlying Lowell's scientific work and gives us a preview of his mind at work on a scientific problem. Striking similarities become evident between Lowell the psychical researcher and Lowell the astronomer. Among those similarities are the fact that he was drawn to exotic topics—trances and extraterrestrial life—for which evidence is difficult to gather and interpret. In both fields his approach was driven by a prior and unyielding commitment to the truth of Spencerian evolutionary principles. Accordingly, for him scientific inquiry consisted of a search for evidence to confirm these principles. On no occasion would the evidence be permitted to falsify the theory. Furthermore, Lowell's research methods were regarded by important authorities in both fields as out of date or ineffective and contributed to doubts about the validity of his results. In the face of criticism, Lowell persisted in his ways and thus isolated himself from dialogue with mainstream scientists. In the end, he preferred to disseminate his findings to a larger lay audience that appreciated his lively approach to exotic subjects, rather than reporting exclusively to the experts. Lowell's marginal status as psychical researcher anticipated the position he was to occupy among astronomers.

One methodological issue separating Lowell from other psychical researchers as well as East Asian specialists was his insistence on distancing himself from his subjects, whom he saw only briefly, as a group, in a laboratory setting. By contrast, Lowell's counterparts advocated close contact with a small number of subjects so they could better understand the psychic dynamics that motivated their behavior. From a few carefully researched case studies, they then hoped to generalize more effectively about the true character of the individuals they were studying. These contrasting approaches developed into an informal argument between Lowell and his contemporaries—Williams James, Lafcadio Hearn, Erwin Baelz, and Basil Hall Chamberlain—about the value of psychical phenomena and the appropriate method for conducting a scientific inquiry into them.[1]

Psychical researchers were also disturbed by Lowell's insistence on using his investigations of occult practices to confirm Japan's ranking as a civilization below that of Western countries. In effect, Lowell denied that such practices were worthy of study in and of themselves. Furthermore, by assuming a link between occult practices and traditional groups, he predetermined the outcome of his study. Nonetheless, Lowell's pioneering study provided important evidence on the reality of trances that served other researchers in the field.

Occult Japan as a Scientific Study

On April 11, 1900, Percival Lowell and the psychologist and philosopher William James, both recovering from neurasthenia on the French Riviera, lunched together at the Hotel Costabello. Regarding that occasion, William wrote his brother Henry that "Lowell wears very well; and it has cheered me very much to have some grown-up talk." In fact, an intellectual exchange of sorts had been going on between the two men for the better part of a decade. Beginning in the summer of 1891, Lowell, accompanied by his friend George Agassiz (grandson of the naturalist Louis Agassiz), launched what would later become a project in psychical research. The two men climbed Mount Ontake in Japan, where they were surprised to encounter Japanese pilgrims near the summit who were putting each other into a trance. Since no other Westerners had witnessed these "possession" rituals,

Lowell thought they would make an excellent subject for a study—an idea that was surely buttressed by William James's *The Principles of Psychology*, in which Lowell encountered the following invitation: "I am persuaded that a serious study of these trance-phenomena is one of the greatest needs of psychology, and think that my personal confession may possibly draw a reader or two into a field which the *soi-disant* 'scientist' usually refuses to explore."[2]

James's challenge to enter a field avoided by more timid investigators must have resonated with the aspiring scientist in Lowell. Indeed, just ten days after ascending Ontake, Lowell was back in Tokyo reporting to William Lowell Putnam on his rapid progress with the new project. On August 16 he wrote that "the Shinto mesmeric show is as far as I can learn undescribed and unheard of. Bigelow and Buddhism are not in it!" He thus deemed the project worthy because it would enable him to move into new territory and, at the same time, exploit a truly exotic subject. Less than two weeks later Lowell had already made contact with the Ontake sect's high priest, whom he interviewed in Tokyo. "On one side of the table, the head of this particular Shinto sect; on the other side, Miyaoka [Lowell's Japanese secretary] and me. A most intelligent man the Kwancho [high priest] as he is called." On September 9 Lowell reported to Putnam that "my studies progress finely. The Good Kwancho Primate of the sect, explains as much of Shinto philosophy as his brain can manufacture . . . He says the god tells him. So you see I get it as near first hand as possible." By the end of September Lowell could report a "Shinto exhibition of god possession . . . taking place privately in this my dining room . . . I shall keep this at my elbow and jot down interesting items red hot."[3]

Fifteen months later, when Lowell returned to Tokyo to continue interviews and seances, he brought "James' two fat tomes," *The Principles of Psychology*, which he said were "among my best friends . . . I am giving them some real value by scribbled comments on this fair margin." Lowell spent the balance of the year completing his research and presenting it to members of the Asiatic Society of Japan.[4]

Lowell's investigations culminated in the 1894 publication of *Occult Japan*, which was immediately recognized as one of the first systematic studies of possession trances. The series of four lectures he had delivered in 1893 to the Asiatic Society of Japan in Tokyo was published in the society's journal, *Transactions*, and following his return to Boston, Lowell's father, the trustee of the Lowell Institute, asked him to give a series of six lectures titled "Japanese Occultism." The ensuing publication of *Occult Japan,* incorporating the material presented in *Transactions* with a long theoretical chapter, was greeted with considerable acclaim. William James himself, in his Lowell lectures of 1896 entitled "Exceptional Mental States," drew heavily from *Occult Japan* to provide evidence of the reality of possession trances. In addition, the *Journal of the Society for Psychical Research* published a review and an article praising the book for penetrating practices usually not accessible to foreigners' eyes and for contributing to the understanding of abnormal states of consciousness.[5]

The scientific community's interest in *Occult Japan* was precisely what Lowell had hoped for. From his own perspective, conducting a *scientific* investigation of trance phenomena was far more important than merely recording for Westerners the existence of practices already well known to the Japanese. The problem, in Lowell's view, was that no one yet understood the significance of the trances: "Japan is scientifically an undiscovered country even to the Japanese, as a study of these possessions will disclose." Accordingly, Lowell intended his investigation to clarify the nature and significance, as well as the existence, of Japanese possession trances.[6]

The possessions could be revelatory in two interrelated ways, according to Lowell: "archaeologic" and "psychic." "They are other-world manifestations in two senses and the one sense helps accentuate the other," he wrote. The trances shed light on the most deeply embedded cultural traits of the Japanese people, while also offering an unparalleled opportunity to explore the mental processes that occurred in alternate states of mind. In effect, Lowell could use psychic phenomena to continue his investigation of Japanese culture, and, by virtue of the fact that trances were quite commonplace in Japan, he would have access to data that would contribute to the continuing study of psychic matters.[7]

Lowell's complicated scientific agenda is reflected in his division of *Occult Japan* into two sections, phenomena and noumena. The former and larger section incorporates the data Lowell gathered on Shinto practices from his observation of trances on Mount Ontake and in his home, which he used as a laboratory. Meanwhile, he attended local performances of miracles, such as climbing sword ladders or walking on hot coals, and searched the *Nihongi* and the *Kojiki,* the two great texts of traditional Japan, for early reports of Shinto trance practices. Having collected his data, Lowell explained them in a conclusion linking trances to other psychic phenomena.[8]

Lowell's Quest for the "Savage" and the "Irrational"

Lowell brought to his psychical research project predispositions that shaped his efforts to apply science to the understanding of trances. The uneasiness he had expressed in *The Soul of the Far East* about the superiority of Japan in the aesthetic sphere had now grown into a pervasive sense that what he regarded as the rational modes of discourse promoted by male elites of Anglo-Saxon ancestry were under threat. These modes were threatened not only by the Japanese but also by women and the French. Japan, of course, was still in most ways a traditional society, as it had been in 1883 during Lowell's first visit, but rapid economic and political development was putting Japan on course to challenge Western nations both politically and culturally. Meanwhile, on the home front, the new women's movement threatened to displace male Anglo-Saxon leaders and, in the process, to privilege "irrational" modes of discourse. Witness the appearance in Boston of Mary Baker Eddy's new Christian Science Church, which, like Shinto, promised to cure disease through faith rather than modern medicine. In this respect women were similar to the French, who refused to restrain the expression of their feelings and were abandoning themselves to experiments with hypnotism. The persistence of France as a major world power thus also constituted a potential threat to Lowell's values.[9]

Almost as important as perceived threats from various irrational voices was the defection of members of the Anglo-Saxon elite from the industrialist ranks. Many of Lowell's fellow investigators were promoting the reality and value of alternate states of mind to counteract the effects of a scientific ideology they blamed for legitimating the excessive materialism and rationality of the industrial order. While some critics advocated abandoning science altogether, most believed that a modified methodology could verify and validate the spiritual realm as a source of transcendence in a material age. Such a program clearly threatened Lowell's assumption that trances were manifestations of the most primitive societies.[10]

This debate about the scientific study of "unscientific" subjects was by no means an abstraction. Just as Lowell was writing his book, vigorous discussion of these issues was taking place in two organizations with which he was closely associated, the Society for Psychical Research and the Asiatic Society of Japan. Both were attempting to use the paraphernalia of science, including research projects, meetings, papers, and journals, to determine the truth about subjects that appeared to be difficult to study using scientific methods of quantification and measurement. Inevitably the investigation of psychic and cultural phenomena, as opposed to matter and energy, would require adjustments in methodology to accommodate the radically different nature of the subject.[11]

The Society for Psychical Research (SPR) and its American counterpart paralleled in many ways the Asiatic Society. The former was founded in England in 1882 by a group of Cambridge graduates that included Frederic Myers, Henry Sidgwick, and Edmund Gurney, while the American branch, known as the American Society for Psychical Research (ASPR), came into being two years later under the leadership of William James. Both societies were part of a larger movement, centered in France, that was intent on verifying and exploring the existence of subliminal consciousness. Anglo-Americans were particularly interested in the work of Jean-Martin Charcot and Pierre Janet, who were exploring the subconscious mind through the use of hypnotism. The SPR and the ASPR, however, left hypnotism largely to the French and focused on a scientific approach to the existence of various psychic phenomena.[12]

The program of the psychical researchers was strained by tension between its traditional goal of verifying the existence of a spiritual dimension of life and its modern strategy of using the scientific method to accomplish this goal. Under these circumstances, it is not surprising that Lowell and others were divided over which elements of the program to embrace. Committed as he was to scientific rationalism, Lowell could not share the enthusiasm of many psychical researchers about the possible existence of an authentic and autonomous spiritual realm. Nor could he accept the program of researchers like the ASPR's first president, Simon Newcomb, who wished to use the organization to establish beyond a doubt the fraudulence of spiritual phenomena. Instead he would stake out a middle ground by affirming the reality of the trance state while, at the same time, portraying it as inferior to "normal," rational, consciousness.[13]

Despite the divisions among researchers, all agreed on the necessity of rescuing the study of psychical phenomena from the spiritualists by introducing a rigorous scientific approach to their investigations. They further agreed that it was important to provide solid empirical evidence of the phenomena under investigation. Accordingly they advocated such practices as testing results by repeating experiments, examining a single subject on a continuous basis, and using several witnesses, wherever possible, for the same case. In effect, they preferred making a thorough examination of a small number of cases to giving superficial attention to many of them. In appropriate situations they also advocated the use of such techniques as testing consciousness by sticking subjects with pins. They were adamant about exposing fraud and accounting for phenomena wherever possible with naturalistic rather than mystical explanations.[14]

Lowell's psychical research project thus grew out of the Victorian fascination with spiritualism, which in turn provided a vehicle for responding to the new world of science and industry. Lowell found some guidance for his project from the Society for Psychical Research and its American branch, with which he was in touch. He had an agenda of his own, however, and though he borrowed methods from the SPR, he was selective in doing so. The outcome of Lowell's research, as we shall see, was effectively predetermined.

Trances and the Japanese Spirit

The "archaeological" section of Lowell's project reassessed Japan's place among civilizations of the world in the light of dramatic new events and with the new evidence available from the study of trance sects. Lowell had argued that Japan could be faulted for the absence of individuality but must be praised for her artistic achievements. Like other Western analysts, he ranked Japan lower than Western countries, where the scientific spirit prevailed, but higher than aboriginal cultures lacking the aesthetic capacity of the Japanese. While in principle that ranking might change, in practice the likelihood of progress or regression was small in East Asian civilizations, where evolution had been arrested.

Nonetheless, new events had clearly taken Lowell by surprise and created a sense of urgency about reassessing the ranking. During Lowell's 1889 visit to Japan, the emperor promulgated the country's new Constitution, and the railroad from Tokyo to the Japan Sea was completed. Four years later, Lowell noted that "the Japanese upper classes had found a new faith; and Herbert Spencer was its prophet." In short, the Japanese were now absorbing not only Western technology and institutions but also the very ideology Lowell himself had embraced. These events suggested that Japan was rapidly becoming a serious rival of the West.[15]

The "true" character of the Japanese people was brought into question in a different way by another dramatic event. On the very day of the promulgation of the Constitution, an assassin took the life of Arinori Mori, the minister of education and a former minister to the United States. The Mori tragedy affected Lowell more deeply than any other issue of the time, as is evident in his gripping account of the event in an 1890 *Atlantic Monthly* article. Its significance in his mind stemmed from the fact that Mori was a strong Westernizer. He had become a convert to American ways and had even espoused the adoption of English as the national language of Japan during his term as minister to the United States. Such proposals provoked the wrath of Japanese nationalists, who were also angered by reports that Mori had defiled the mats of the sacred Ise Shrine by failing to remove his shoes and had viewed Shinto treasures only the emperor was allowed to see.[16]

The aftermath of the murder, in Lowell's view, was even more revealing than the murder itself. Mourners at the grave of his assassin far outnumbered those at Mori's funeral, which was attended by the Westernized elite. Since the assassin had been motivated by his strong objection to Mori's supposed defilement of the Ise Shrine—and therefore of traditional Japanese religion—Lowell was impelled toward the idea that underlying Japan's superficially Western behavior lay a powerful commitment to a primitive, indigenous, and anti-Western tradition.

These events, then, raised serious questions about Japan's place among world civilizations. On the one hand, modernization was occurring with such rapidity as to suggest that Japan should be ranked with the Western powers in the hierarchy of nations. On the other hand, the support that the Japanese people accorded to Mori's assassin, who had acted in behalf of an "aboriginal" religion, suggested that the country should be demoted to the bottom of the hierarchy with other indigenous groups. The only way to resolve the ranking question was to investigate the Shinto faith and assess its popularity among Japanese people. Such research would also enable Lowell to consider, by extension, the cases of women and the French, who, like the Japanese, rejected the scientific ethos.[17]

It is important, then, to recognize that Lowell's "scientific" project was based on certain predispositions very much affecting its outcome. The rising power of Japan and the challenge of American women to their subordinate position powerfully gripped his emotions and shaped his research orientation. To find decisive answers to troublesome contemporary issues, Lowell would need to discredit Japan's challenge to Western hegemony, as well as women's challenge to men, by reconsidering his initial ranking of

Japan as "half-civilized." If Japan could be demoted in rank to an aboriginal society, then the claims for her equality with Western countries would be deflated and Westerners could rest assured of their hegemony. At the same time, and by association, women and the French would also fall in the ranking system.

Lowell justified demoting Japan by acknowledging that he had wrongly classified the Japanese as a Buddhist people. Based on that classification, he had proclaimed in *The Soul of the Far East* that "the Japanese are not a savage tribe." Though he continued to have a high regard for Buddhism, Lowell argued in *Occult Japan* that Buddhism was a foreign import that had only a superficial relation to the essential elements of the Japanese character. "Shinto," by contrast, he wrote, "is so Japanese it will not down. It is the faith of these people's birthright, not of their adoption . . . Buddhist they are by virtue of belief; Shinto by virtue of being." Thus a fuller understanding of Japanese character, as well as an accurate assessment of Japan's place in the hierarchy of world civilizations, could be obtained only by coming to terms with the authentic religious tradition of Japan.[18]

In his research Lowell discovered overwhelming evidence that Shinto was the authentic, primitive faith of the Japanese people. The central elements of Shinto, namely the worship of ancestors and nature, had to be viewed as "aboriginal instincts. Next to the fear of natural phenomena, in point of primitiveness, comes the fear of one's father, as children and savages show." The primitive dimension of the faith was evident in its "skeleton-like" buildings and "scant" paraphernalia. To be sure, there was a certain beauty in this simplicity, but it was based on beliefs that were "childish conceptions embalmed in an exquisite etiquette." Take, for example, the Shinto practice of remembering the gods with food offerings placed on altars, which Lowell called "a divine dinner party" and an "aboriginal banquet." Similarly, Shinto pilgrimages were characterized as "peripatetic picnic parties, faintly flavored with piety." None of these practices required serious reflection or intense religiosity.[19]

The behavior of Shinto adherents also suggested to Lowell the undeveloped character of the faith. In a trance state, the possessed exhibited a startling loss of control at various points, but especially at the outset. Lowell noted the "semi-suppressed vehemence with which the [finger] knots are tied" and described the "spasmodic jerks" of those about to be possessed. In these practices spectators found a "compelling fascination," in spite of their "repellent uncouthness." A similar lack of control was exhibited by children who were put into trances and proceeded to somersault around the room. Less vulgar but more disturbing to Lowell was the conformity in actions and sounds of all practitioners and the fact that trances occurred only when the gods initiated them. Such behavior clearly violated the standards of Victorian decorum and the commitment to the autonomy of the individual that served Lowell and others as indexes of civilized behavior.[20]

Because Lowell believed that the conduct of trance subjects revealed their true character, he regarded trances as useful opportunities to compare the behavior of individuals from two different cultures. He tested this proposition in a thought experiment that, not surprisingly, yielded precisely the results he was expecting. By his logic, members of advanced races, when in a trance, would exhibit their individuality *unless* it were possible to "undo what evolution has done and return us a primeval savage in the body of an end-of-the-century man." As a confirmed evolutionist, Lowell denied the possibility of such reversion. Less evolved races, like the Japanese, in whom individual idiosyncrasies were far less marked, would also exhibit their true character by behaving in ways identical to their ancestors. "In Shinto god-possession we are viewing the actual incarnation of the ancestral spirit of the race. The man has temporarily become once more his indefinitely great great-grandfather. It is a veridic incarnation."[21]

Such evidence indicated that Lowell had been in error in ranking the Japanese with other semicivilized societies. By virtue of their aboriginal character, they deserved to be ranked with savage groups. Indeed, in a hierarchy running from simplest animals to the most advanced human beings, who could think for the sake of thinking, the Japanese, according to Lowell, ranked closer to the animals.[22]

Lowell could confirm the validity of the new ranking by comparing the Japanese with other primitive groups. Such comparisons revealed that Shinto priests were "civilized cousins of the medicine-men of North America, of the shamans of savage tribes the world over, and of Christian Scientists generally." No need to worry, then, that the Japanese would challenge the West or, by extension, that women gifted with the capacity to communicate with spirits would threaten male hegemony. A country like Japan, where trance-practicing was a central activity, could not have evolved beyond the stage of savagery. And, by extension, women, who had strong predilections to mediumship and Christian Science, were closely akin to primitives. They lacked the individuality that set apart the most developed societies and groups.[23]

In effect Lowell exploited the Spencerian ranking system to argue that challenges to Anglo-Saxon male hegemony were illusory. Readers could rest assured that the Japanese character was shaped by the primitive practices of Shinto, which was the true native faith. Buddhism could be dismissed as a foreign import. Taking this evidence into account, it would be necessary to demote Japan from the rank of semicivilized society to aboriginal, thus placing that nation in the company of women as well as African and Native American tribes.

Trances as Psychic Phenomena

In addition to revising Japan's rank among world civilizations, Lowell took up William James's invitation to study trances in a scientific manner as a way of demonstrating their relation to other alternate states of mind. In attempting to understand the psychical dimension of trances, however, Lowell was placed on the horns of a dilemma. He had been attracted to the project in part by his desire to contribute to the scientific quest for an understanding of such phenomena. For this purpose, he had read James's *The Principles of Psychology* as an introduction to the subject of alternate states of consciousness and had discussed these issues in Boston with Richard Hodgson, the salaried secretary of the American branch of the SPR. However, Lowell had already come to certain conclusions about the significance of trances in the Japanese context that he was not prepared to abandon. Given these conclusions, Lowell would not be able to conduct a freewheeling investigation and risk uncovering evidence that would falsify his hypothesis about either the "primitive" character of the Japanese mind or the subordination of mind to brain; he would have to work selectively with James's concepts—accepting those consistent with his own preconceptions and rejecting the others. Only in this fashion would he be able to reach the predetermined conclusion that trances were evidence of a primitive state of mind among the Japanese people and others.[24]

Lowell envisioned the project as an opportunity to confirm a body of European scientific theory regarding the nature of various forms of alternate consciousness, including dreams, hypnotism, and mediumistic experiences, by gathering data about the Ontake sect. The theory would suggest the meaning and significance of the data. Lowell thus looked to Japanese trance practitioners for evidence to support what "was scientifically known in Europe." Accordingly, he recognized the commonality between Japanese trance practices and those of James's favorite and much investigated mediumistic subject, Mrs. Piper, while at the same time suggesting how much more common the practices were in Japan than in the United States: "Think! where we make much of one Mrs. Piper they make little of many Mr. P's." In a similar fashion, Lowell could explain the fact that during the throes of possession, the *gohei* wand, used by Japanese trance practitioners, settled on the *zone hypnotique* at the top of the nose, between the eyes. In effect, he said, the Japanese had "discovered experimentally" and put into practice a principle only Western science could explain. This contrast between "science" and "experience" was also implicit in Lowell's reference to Japanese pilgrim clubs as "possession Salpêtrière"—Salpêtrière being the hospital where Charcot experimented with hypnosis. Lowell thus made the point that the Japanese were as effective as the French in inducing trances but, unlike the French, they did not have a laboratory for investigating the nature of these possessions.[25]

The experimental techniques Lowell used were drawn from European scientists who had worked with trance subjects. In addition to interviewing experts in Shinto practice and observing the behavior of the possessed, Lowell performed experiments on them to verify the reality of the trances. By sticking pins into his subjects and taking their pulse, he demonstrated to his satisfaction that the subjects—or "pin cushions"—experienced no pain and had low pulse rates during the peak of the trance. To assure the objectivity of these experiments, Lowell not only conducted a number of seances but also invited Western friends, including Erwin Baelz, the German expert on fox possession, and W. B. Mason, to witness the trances. Mason reported to Lafcadio Hearn that he had found the trance seances "pretty impressive and earnest. The idea of trickery or deceit could not be entertained for a moment by one present."[26]

These practices indicate that Lowell operated within the experimental norms of psychical research but preferred to use techniques for generating data without the active participation of his subjects. His method contrasted with that of James, who relied heavily on participants' accounts. Lowell's preference for hypnotism over possession trances also reflected his conviction that scientists ought to control their subjects. In both trance states, he claimed, the brain was reduced to "an inactive state." However, in the case of hypnosis, that result was "obtained by the application of a little science," while in the other the scientist was reduced to a passive observer of the process.[27]

Lowell also launched the project with strong preconceptions regarding the aptitude of the Japanese for entering trances. Given this view, he would reject James's claims for the dignity and autonomy of subliminal states of mind and the irrelevance of gender or race to the production of trances—points that were to prove problematic for Lowell, as we shall see.

To help explain the prevalence of trances, Lowell had to modify James's conception of the operation of the mind. Both men, along with Herbert Spencer, rejected the dualism between mind and brain that, they agreed, prevented a proper understanding of mental processes. Lowell, however, following Spencer, diverged from James when he argued that human beings are "part and parcel of the material world" and insisted that mind and matter are inseparable. Consistent with this point, Lowell described consciousness as "nerve glow" and claimed that ideas were "neural impulses." As such, they could be described as forces that were both "inevitable" and "impersonal." Clearly Lowell meant to suggest not only that human beings had little control over their ideas but also that the very same ideas occurred in every human being.[28]

James explicitly rejected this understanding of the mind, which he identified as a "producer" model. In a telling footnote in his essay "Human Immortality," James cited both Lowell's concept of consciousness as "nerve glow" and Spencer's claim that consciousness arises from "force existing as motion, heat or light" as being inadequate. Spencer and Lowell had erred, he argued, in trying to explain consciousness entirely in terms of material factors. Their conceptions ignored important cases, such as dreams and trances, in which preexisting ideas were transmitted to and received by the brain. In such cases, James maintained, ideas could not be understood as merely a product of force and matter.[29]

While Lowell borrowed from James the conception of the mind as the location of a chaotic stream of thoughts that could be managed only by the will, they strongly disagreed about the nature of will. For James, it was an instrument enabling the individual to control his destiny; by contrast, Lowell minimized the capacity of human beings to make independent choices: "All I can do is hold on to, or let go, what my stream of thought is kind enough to present me with." And even this apparent power was illusory. Despite the capacity for selective attention, individuals "selected" involuntarily as they were captivated by, and then acted on, certain ideas that never entered the thought process. In this sense, Lowell argued, the will was the captive of an idea, or vice versa, in effect ruling out the possibility of mental autonomy. It should be no surprise, then, that Lowell rejected one of James's central propositions: that consciousness is "a loader of dice in the game of life."[30]

Lowell's repudiation of mental autonomy served him well in explaining the ease with which many people could be put into a trance. And so long as he was discussing the Japanese, the French, or women, this argument was quite useful. However, it would not do to deny mental autonomy to the WASP, male elite. Instead, Lowell claimed that the process of evolution had propelled members of the elite to a more advanced stage in which they could think for themselves. Unhappily, the masses were stuck at the level of animals, engaged in "thinking for the sake of things."[31]

Lowell also deviated from James in refusing to recognize trances as a distinctive kind of consciousness having a status equal to normal, waking consciousness (James's position). The two men did, however, agree that the trance state was akin to sleep, since in both instances the mind only appeared to be inactive. Such appearances were belied by dreams, which constituted evidence of brain activity. However, Lowell maintained that during dreams and trances brain activity remained at a very low level. Only the existence of a few neural pathways associated either with long-standing habits or with recent events of a vivid character could be activated by stimuli from outside the individual. Thus, the brain in a trance state was a partially operating version of itself under normal waking conditions. In this sense, trances were inferior versions of normal consciousness, not true alternate states of mind.[32]

James, by contrast, argued for a complete distinction between the two states of mind. He pointed out that in a normal conscious state an individual could never recall the mental activity that had occurred during a trance, and vice versa. For James, this separate consciousness had to indicate the existence of two separate sets of pathways in the brain. He thus maintained that "each of the selves is due to a system of cerebral paths acting by itself." On this basis, James concluded that the brain "must be supposed capable of successively changing all its modes of action, and abandoning the use for the time being of whole sets of well organized association paths." Accordingly, James dignified the trance state by identifying it as "double-consciousness" or a "second personality."[33]

These differences between Lowell and James over the nature of the trance state prepared the way for their differing views on the kinds of human beings who were most susceptible to entering the trance state. Lowell's argument that trance states were inferior versions of normal consciousness was a useful prelude to his insistence that the characteristics of mind conducive to trances were to be found among less developed groups such as women, the Japanese, and the French. Since normal consciousness operated in these groups at a rather low level, he believed, fewer pathways were activated and fewer would need to be shut down in order to enter into the trance state. In this way, the Japanese in trances exhibited the same characteristics of mind as they would in a conscious state. The incarnate gods were mentally similar to the possessed.[34]

Lowell's claim that certain groups were predisposed to entering trances ran directly counter to James's categorical denial that age, race, or sex could bring about a predisposition to hypnotic states. "A certain amount of mental training, sufficient to aid concentration of the attention, seems a favorable condition," wrote James, "and so does a certain momentary indifference or passivity as to the result. Native strength or weakness of 'will' have absolutely nothing to do with the matter. Frequent trances enormously increase the susceptibility of a subject." In effect, James maintained that susceptibility to trances was a product of nurture rather than nature.[35]

Though he disagreed with Lowell, James's language may have encouraged Lowell's views. He used the term *savages,* for example, to identify individuals who had not learned to reason with abstract terms, although he rejected the idea that this deficiency had a racial or gender basis. James also spoke of individuals who possessed a strong will and personality, but denied any correlation between this condition and racial or national origin.[36]

Clearly Lowell came to his reading of *The Principles of Psychology* with a set of preconceptions drawn from Spencer that led him to reject many of James's most important ideas. Committed as he was to the notion that trances were a lower form of consciousness and that those who entered them easily were less evolved human beings, he necessarily rejected James's claims for mental autonomy, for a trance state equal in standing with normal consciousness, and, most clearly, for the equality of races and both sexes in their disposition to enter the trance state.

Despite these differences, James and Lowell borrowed from each other. Lowell preferred James's conception of the integral relationship between the brain as a body organ and the mind to the dualism of faculty psychologists. He was also drawn to James's picture of a mind besieged by many ideas that required a will to sort them out. James, in turn, welcomed *Occult Japan* as one of the first systematic studies of possession trances that offered convincing evidence of their reality. He could also learn from Lowell's observations about relations between hypnotic and possession trances that gestured toward a more comprehensive approach to alternate states of mind. Lowell noted, for example, similarities between the sleeplike appearance of the subjects and their lack of control over their actions; among differences, Lowell noted that subjects of hypnotic trances were listless and responded to the hypnotist, while trance subjects experienced violent physical contortions as they responded to the possessing god.[37]

Lowell played a significant role in the movement for psychical research. He adopted an aggressively scientific stance reflecting his commitment to the utility of the scientific method in assessing the nature of trance states. However, despite his promise to approach the study of trance sects with the "coldly critical eye of science," the preconceptions driving his study ultimately rendered his investigation unscientific. These preconceptions made it virtually impossible for Lowell to take seriously the practices he was investigating.

Lowell's stance reflected the sense of threat from various directions that he was experiencing as he wrote *Occult Japan*. He was at this time keenly aware of challenges to the hegemony of Anglo-Saxon male elites posed by non-Western countries, women, and immigrants. More important, however, Lowell was experiencing a real sense of isolation as he confronted the widespread questioning of modernization and science among WASP elites. In this regard, he was equally at odds with authorities in the Asiatic Society who doubted the adequacy of his scientific method and those in the SPR who expected that it would verify and validate the autonomy of the spiritual realm. Whether rejecting or embracing the scientific method, authorities on Asian life and psychical researchers were giving a careful and appreciative hearing to the subjects of their studies. It is curious, then, that we find no statements or documents in which Lowell recognizes explicitly how much at odds he was with other experts on psychical matters. He never once commented on the sections of James's *The Principles of Psychology* devoted to trance sects that contradicted his own position. In a similar vein, Basil Hall Chamberlain wrote Lafcadio Hearn that Lowell "jokes one off, and thinks one below the level of general ideas,—unscientific" whenever someone questioned his method, as when Chamberlain pointed out the limited usefulness of a deductive approach to Japan linking all phenomena to the single principle of impersonality. Lowell simply ignored Chamberlain's warning regarding "the limitations of science where things organic, and especially things human, are concerned." Our only clue to

Lowell's feeling of isolation was his emotional commitment to a rigidly "scientific" approach and his transparent satirical thrusts at "nonrational" groups that clearly reflected a deep uneasiness about their behavior.[38]

Of course, the willingness to recognize the limitations of science or to value nonrational consciousness was closely tied to feelings about alternative cultures. Those, like Lowell, who were strongly committed to the hegemony of Anglo-Saxon males would have little incentive to study sympathetically women, the Japanese, and non-Anglo-Saxon groups. Indeed, their best strategy was to clarify their intellectual superiority to these groups. By contrast, James, Hearn, Baelz, and Chamberlain expressed reservations about Western ways and sought models in cultures at the margins. In *The Varieties of Religious Experience,* for example, James listened carefully to the very same mind-cure exponents whom Lowell had dismissed in a sentence. Devoting twenty pages to the subject, James passed quickly over Mary Baker Eddy, the founder of Christian Science, to quote at length from the accounts of anonymous practitioners, including a number of women. The focus of James's book was the recorded experiences of the devotees of mind cure, from which he drew observations and impressions.[39]

Erwin Baelz and Basil Hall Chamberlain also understood the importance of a close involvement between investigator and subject. Baelz, who married a Japanese woman, noted the importance of learning the language, customs, and modes of thought of the Japanese. And in what could have been a repudiation of Lowell's satirical approach, he asserted "that those who find the lives and the ways of an alien people ridiculous are really themselves ridiculous, making fun of things they do not understand." In a similar vein Chamberlain advised Lafcadio Hearn that "an enormous deal may be found out about the Japanese by living among them, sympathising with them, noting down what they do and say and write." He must have known that he was preaching to the converted, since Hearn had already taught in Japanese schools, married a Japanese woman, and adopted a Japanese lifestyle.[40]

Nonetheless, despite Lowell's inability to develop an insider's view of trances, his *Occult Japan* deserves recognition as a pioneering work advancing the study of nonrational consciousness. Lowell answered James's call to investigate possessions and grappled with difficulties of language and access to remote geographical areas to assemble an impressive body of information about Shinto practitioners. He was ingenious in ferreting out the Shinshiu sect and in setting up seances in his own house. His measurement of pulse rates and bodily sensations was useful even if it was inadequate to fully explain the nature and value of trance experiences. He struggled to learn the complex theory of subliminal consciousness, and even if he disparaged that kind of thinking, he made a noble effort to synthesize his findings with those of other psychical researchers. In the end, however, Lowell became frustrated with investigating "unscientific" subjects like Japan and psychics and moved on to a more promising field: the search for evidence of life on Mars. There he discovered, to his obvious satisfaction, what he believed was evidence of an admirable race, more modern and scientific than any population on the face of the Earth.[41]

Even after taking up Mars, however, Lowell did not entirely abandon his interest in Japan and the occult. In 1895, he told a St. Botolph Club audience in Boston that he had thought of becoming an "impresario of a traveling circus" made up of the Shinto trance practitioners then performing in his Tokyo house. Thirteen years later, instead, Lowell hosted Dr. Suga, a Shinto priest, who performed the miracles of sword climbing and walking on hot coals before an audience of 200 Bostonians, including Harvard and MIT professors. The *Boston Evening Transcript* pointed out that Lowell took as much pleasure in disseminating what he had learned as he did in acquiring knowledge in the first place. In *Occult Japan* he had served as host to American audiences in a figurative fashion, for the performance of miracles and trances. On Huntington Avenue he resumed the role of host in a literal fashion. Once again he turned Japanese religion into a spectacle for the enjoyment and edification of Westerners.[42]

In important ways, Lowell's brief excursion into psychical research anticipated his approach to astronomy. In both cases, he selected an exotic research topic for which it would be difficult or impossible to collect conclusive evidence. Once engaged in research, Lowell collected and interpreted evidence in such a way that it would confirm the evolutionary theory to which he was wedded at the outset. He never entertained the possibility of falsifying that theory. In his research Lowell chose methods that would give him control over his subjects and lead to the answers he was seeking. When these methods generated criticism from authorities in the field, Lowell ignored them and thus isolated himself. In the end, he preferred to present his findings to an appreciative lay audience rather than to the experts who criticized his work.

8

◆ ◆ ◆

Cosmogonist

Lowell's work on Mars was just as theory driven as his studies of East Asia. In fact, Lowell embraced the theoretical justification for extraterrestrial life even before the Italian astronomer Giovanni Schiaparelli presented empirical evidence to support the idea. As early as his senior year at Harvard he had taken up Spencer's version of the nebular hypothesis, the first stage of cosmic evolution, to present as his graduation disquisition. By proposing a common origin of the various planets from a nebular cloud and their development through identical stages, the hypothesis rendered plausible the idea of life on those planets.

The importance of the nebular hypothesis as the foundation for Lowell's claims about life on Mars became particularly clear in 1894, when he launched his quest for evidence of life on Mars, and at frequent intervals over the next fifteen years. As it turned out, however, Lowell had embraced a theory that was just coming under attack on various grounds. The rising criticisms of the hypothesis over several decades led to a proposal for a new theory by two University of Chicago professors, Thomas Chrowder Chamberlin and Forest Ray Moulton, a version of which was published in 1905. The new theory accommodated geological and biological evolution within the period since Earth's origin, but provided no plausible grounds for believing that other planets would have life-supporting environments. Lowell was thus wary of endorsing the theory and turned instead to a revision of the nebular hypothesis in order to address its weaknesses while retaining its strong points.

The Chamberlin-Moulton theory was also troublesome to Lowell because it ignored Spencer's hierarchy of disciplines. Not only had a geologist initiated the project but also Chamberlin and Moulton had come together as specialists from two autonomous disciplines to design their hypothesis. This arrangement departed sharply from Spencer's notion that astronomers dealt with cosmic phenomena and geologists with local phenomena. Lowell was bound to protest the equal—or even superior—role that Chamberlin appropriated for geologists in creating a cosmogonic plan.

The resulting conflict between Lowell and his rivals was aired in scientific and popular sources by geologists and astronomers. In the end Lowell suffered an embarrassing defeat at the hands of his university-based rivals. On grounds of academic expertise, they denounced him for pronouncing on matters geological and for addressing the public rather than an academic audience. In this way they clearly hoped to assert the hegemony of university specialists in the domain of scientific research and thus anticipated the campaign they were to wage against Lowell's observational work on the canals of Mars. However, ironically, a recent renewed interest in the nebular hypothesis suggests that Lowell may have come closer than his rivals to explaining the origin of the solar system.[1]

The Nebular Hypothesis, 1876–1894

Eighteen years before the establishment of the Lowell Observatory, Percival Lowell was already a convert to the nebular hypothesis. In this respect, he was perfectly representative of the educated public, which was drawn to the nebular theory because it accounted for the origins of the inorganic world by natural law without dismissing the role of the deity. So benign and appealing was this explanation that the instructor of Lowell's freshman ethics course was an enthusiastic proponent, as was Lowell's mentor, Benjamin Peirce, who remarked that the nebular hypothesis "was the most magnificent conception that was ever formed in the mind of a philosopher."[2]

The French mathematician Pierre Laplace first formulated the hypothesis in 1796, and he updated it on no fewer than four occasions. In the first version of the hypothesis, published as a short chapter of his classic *Exposition du système du monde,* Laplace argued that the solar system had evolved from a hot, rotating nebular disk, which, in its initial condition, stretched beyond Neptune. As the surface of the disk cooled, the center contracted, heated, and rotated more rapidly. With the development of sufficient centrifugal force, the nebula spun off rings of nebular material from its equatorial regions. These rings, in turn, disintegrated, and the material from them was drawn by gravitational force into spherical masses, which then rotated as planets around the shrinking nebular disk. The same process was repeated in the formation of satellites for each of the planets.[3]

The educated public, however, rarely confronted Laplace's rather sketchy version of the nebular hypothesis directly. Instead, they read various interpreters, who presented the main themes of his work and frequently incorporated a provision for divine initiative, with which Laplace had dispensed, or an endorsement of the plausibility of extraterrestrial life, which he had supported. In researching his twenty-four-page disquisition for the 1876 Harvard graduation, Lowell read Laplace, but he relied almost exclusively on the version of the nebular hypothesis published by Herbert Spencer in 1858.[4]

Lowell was particularly interested in the impressive congruities of the system, including the fact that the planets lay in the sun's equatorial plane and, with the exception of Uranus and Neptune, rotated in the same direction. Lowell, through Spencer, attempted to account for the retrograde rotation of Uranus and Neptune by noting that the inner and outer portions of the rings farthest removed from the sun had different velocities. And he also focused on the curious arrangement of planets and satellites by size vis-à-vis their primaries. He noted that smaller planets and satellites were located near their primaries, larger ones at a middle distance, and medium-size bodies at the farthest remove. He also portrayed the outer and original planets as actors in the formation of the planets inside of them by virtue of their capacity to draw off greater or lesser rings from the primaries. Happily, the original ring formation of Saturn, because of its closeness to the planet, had survived to testify to the dynamic by which the system had been formed.[5]

While Lowell made no reference to extraterrestrial life in his paper, the connection between the nebular hypothesis and the plurality of worlds had already been clearly established by Laplace. As Michael Crowe has shown, virtually every late-nineteenth-century advocate of pluralism was also a believer in the hypothesis. In some cases the logic of the nebular hypothesis suggested to its advocates the probable existence of life on other planets, while in others, the hypothesis was a convenient rationale for an idea that was irresistible on other counts.[6]

Lowell's graduation disquisition provides an indispensable record of his approach to astronomical problems during the formative period of his life.

However, the value of this exercise for Lowell was fully evident only as he launched his Arizona expedition in 1894. In his first public account of the purpose of the Lowell Observatory to the Boston Scientific Society, Lowell announced that "with regard to the observatory's plan of work, its main object is the study of our own solar system. This may be put popularly as an investigation into the condition of life in other worlds, including last but not least, their habitability by beings like if unlike man." As for the probability of extraterrestrial life in the solar system, Lowell said, "If the nebular hypothesis be correct, and there is good reason at present for believing in its general truth, then to develop life more or less distantly resembling our own must be the destiny of every member of the solar family which is not prevented by purely physical considerations, size and so forth, from doing so. In La Place's [*sic*] day, so much was but the natural deduction from celestial mechanics."[7]

Lowell began his search for life on Mars with a conviction, derived in good measure from his belief in the nebular hypothesis, that evolution would lead to conditions on other planets favorable to the development of intelligent life. This theory undergirded his observational work in Flagstaff. Indeed, Lowell's first book, based on the study of Mars's surface during its near approach to the Earth in 1894, invoked the nebular hypothesis at the outset, as support for the likelihood of intelligent life on other planets.[8]

The Planetesimal Hypothesis, 1897–1905

A serious challenge to the central tenets of the nebular hypothesis was a century in coming. Moreover, even after it came almost a decade passed before the planetesimal hypothesis was fully articulated and accepted by an important group within the astronomical community. In 1905, T. C. Chamberlin and his astronomer colleague F. R. Moulton published the first full outline of their new hypothesis, which sometimes bore their names and at other times was known as the planetesimal hypothesis (Chamberlin's designation) or the spiral nebulae hypothesis (Moulton's term). In so doing, they seized the initiative in the field. Astronomers for whom cosmogony was an important issue would of necessity confront the arguments they advanced.[9]

The years between 1880 and 1905 may be regarded as an important transitional period when criticisms of the nebular hypothesis mounted and astronomers, including Lowell, began to take account of them. Critics argued that the angular momentum of the original nebula, judged by the current momentum of the sun, was insufficient to account for the spinning off of rings. They also maintained that the rings would be unlikely to break up and form themselves into planets. Norman Lockyer and George Darwin later suggested that the original nebula must have been composed of meteorites rather than gas, and Darwin in 1897 proposed that tidal pulls, rather than the centrifugal force of a rotating nebula, would account for the formation of satellites. Even so, none of these critics proposed abandoning the nebular hypothesis. Indeed, Darwin explicitly stated that his criticisms should be taken as a revision of the hypothesis.[10]

In undertaking to develop a new hypothesis, Chamberlin sought to refute Lord Kelvin's claim that the cooling of a gaseous mass after its separation from the nebula would occur too quickly to allow completion of the geological and biological processes necessary to account for the present condition of Earth. He also doubted that such processes progressed in linear fashion, as the cooling model required. Accordingly, in consultation with Moulton, Chamberlin investigated alternative origins of the solar system that would not involve the rapid cooling of a gaseous body. One possibility was an accretion model, according to which bodies in the solar system were formed by the coming together of particles at a relatively low temperature.[11]

Chamberlin's quest took a new direction following his examination in 1900 of James Keeler's remarkable photographs of nebulae. These photographs cast doubt on Laplace's claims by revealing only spiral nebulae, none of which were surrounded by rings. As a result, Chamberlin and astronomer colleagues considered the possible origin of the solar system from a spiral rather than a ring nebula. They proposed that an encounter between two stars might have liberated material out of which the solar system was formed. Given momentum by the encounter, particles would be attracted to each other and thus accrete to form the future planets and satellites in the solar system.[12]

But the development of the Chamberlin-Moulton hypothesis was far more than a story of theory building. It was also a product of the newly structured scientific establishment, involving collaboration of autonomous disciplines within a university setting and patronage from private philanthropy. And despite the fact that Moulton was a celestial mechanician, the hypothesis illustrated the rising importance of astrophysics within the astronomical profession. Chamberlin and Moulton procured essential photographs of spiral nebulae from James Keeler and W. W. Campbell of the Lick Observatory on Mount Hamilton, near San Jose, California. Further, they had continual encouragement from George Ellery Hale, their colleague at the University of Chicago and editor of *Astrophysical Journal,* in which some of the research was published. Hale, Campbell, and Keeler were leading astrophysicists. In addition, Chamberlin successfully approached the Carnegie Institution in 1903 for funding to support the project. The resulting grant freed Moulton from his teaching on a half-time basis and enabled a number of calculators to do part-time work. Clearly the new establishment in the profession, entrenched in the universities and supported by wealthy foundations, had aligned itself behind the planetesimal hypothesis.[13]

Lowell Engages the Planetesimal Hypothesis, 1903–1909

For Lowell, the rising criticism of the nebular hypothesis presented a potentially dangerous situation. His long-cherished theory was now regarded as beyond retrieval by respected scientists who had been critical of his work. It was being replaced by a hypothesis that took no account of extraterrestrial life and therefore might not provide theoretical support for the existence of life on Mars. Furthermore, there was no clear evidence that the planetesimal hypothesis could be understood as part of the grand theory of evolution to which Lowell was committed. His reaction to this situation was revealing.

Lowell addressed the criticisms of the nebular hypothesis in his characteristically energetic way, and, in contrast to the collaborative approach of his rivals, he set about to consider issues of cosmogony on his own. Furthermore, he relied entirely on his own resources to underwrite the costs, and he carried out his campaign exclusively in lectures delivered at MIT and the Lowell Institute and in articles and books based on the lectures—all addressed to the general public. In his work, Lowell took account of the criticisms of the nebular hypothesis and adjusted his own claims about the origins of the solar system accordingly.

Between 1900 and 1903, while Chamberlin and Moulton were finalizing their hypothesis, Lowell articulated his provisional revisions of the nebular hypothesis in *The Solar System.* Along lines suggested by George Darwin, Lowell argued that tidal pulls rather than centrifugal force explained the breakup of the nebula. Indeed, he referred to Laplace's centrifugal force explanation as a "fatal fallacy." He also considered the possibility of a collision of stars as the basis for generating the material out of which the solar system was formed and noted congruities Laplace had not explained. At the same time, Lowell continued to be impressed by the capacity of Laplace's hypothesis to account for other congruities as well as the family relationship of different bodies in the system, which suggested the likelihood of extraterrestrial life on various planets at various times.[14]

Six years after the publication of *The Solar System,* Lowell made his most sweeping effort to offer a cosmogonic synthesis. Of course his *Evolution of Worlds* and *Mars as the Abode of Life,* as we have seen, provided a comprehensive picture of cosmic evolution, but clearly an essential part of the story was their depiction of the origin of the solar system. Borrowing unmistakably from Chamberlin and Moulton, Lowell saw the origin of the solar system in a spiral nebula, at the center of which was a nova disturbed by tidal pulls of an approaching star rather than by a collision, as he had maintained in *The Solar System.* This star would draw off material from the nova while giving it and the separated material a rotation. Composed of fragments and knots, the material would become the arms of the nebula. In this dynamic condition, the fragments would gravitate toward the knots, thus forming planets and satellites rotating around the nova, now a sun.[15]

Still following Chamberlin and Moulton, Lowell described the development of different kinds of bodies from the accretion of different proportions of the same elements from the spiral nebula. Lowell claimed, for example, that the outer planets contained more of the lighter material from the crust of the disrupted body, such as hydrogen, while the inner planets consisted of larger amounts of the heavier material from the interior of the original body.[16]

The accretion of the planets and satellites from fragments was a process involving considerable interaction between them. Lowell argued that each planet and satellite swept clear a certain area and also influenced the orbits of those closer to the primary. Once a planet came into existence, its gravitational force would sweep away particles within a certain area. Only beyond that area would it be possible for particles to gather to form yet another planet, and so forth, as one moved toward the sun. In similar fashion the exterior satellite, by sweeping clean a certain area, determined the nature of the orbit of the satellite inside of it.[17]

Lowell again paid attention to the congruities of the system identified by Laplace and those discovered since his time. In doing this, he demonstrated his refusal to break fully with the nebular hypothesis. For example, while Chamberlin and Moulton argued that the size of all bodies in the solar system depended largely on the unpredictable way in which the catastrophe forming the nebula had scattered the knots from which planets and satellites developed, Lowell pointed to the perfect correlation of the sizes of these bodies with their distances from the primaries.[18]

Lowell departed more conclusively from Chamberlin and Moulton in his explanation of the retrograde rotation of Uranus, Neptune, and certain satellites. Chamberlin and Moulton were at pains to demonstrate that accreting particles could have either a direct or a retrograde rotation. Lowell instead accepted the more conventional notion that accreting particles would have a retrograde rotation in their early stages, after which tidal effects—or the force of gravity—would gradually turn the rotating bodies upside down and leave them with a direct rotation. Only in cases where tidal effects were weakened by distance would the bodies rotate in their original direction—which would explain why the most distant planets and satellites had retrograde motion. Thus Lowell opted for a theory in which regularity and predictability were central features.[19]

A subtle but important difference between Lowell and his opponents was Lowell's emphasis on the homogeneous character of the solar system and his opponents' emphasis on its heterogeneity. Lowell, for example, often began his popular books by pointing to the family resemblance between different bodies in the system, which he said reflected the fact that they all had a common origin. Of course he also recognized differences between them, including the size, temperature, and proportions of elements that made up each body. Chamberlin and Moulton, by contrast, though they acknowledged the common origin of bodies, emphasized the irregularities within the spiral nebulae, including nuclei of various sizes as well as finer material. This emphasis on irregularities suggested a more haphazard development of the solar system than that proposed by the nebular hypothesis and thus raised questions about the probability of intelligent life or of any discernible pattern of evolution in the planets and satellites. For Lowell, such drawbacks made the planetesimal hypothesis much less appealing.[20]

Lowell was similarly concerned about Chamberlin's insistence that the formation of the planets through the accretion of planetesimals had occurred at a slow rate, thus generating only small amounts of heat. In this way, there would have been little need for cooling to prepare the planet for life. For Chamberlin, such a scheme was advantageous because it would speed up geological and biological history. In the process, of course, he could do away with the neat progression of stages that geologists like James Dwight Dana, assuming the validity of the nebular hypothesis as a starting point, had envisioned. In his new textbook, Chamberlin argued that "the hypothetical stages arising from the Laplacian view have nowhere been stated more definitely and succinctly than by Professor Dana, and his synopsis is here introduced as a typical statement of the general conceptions that have been entertained by the majority of geologists during the past century." In their place Chamberlin erected a new set of stages that was far less linear. Among other things, he complicated the development process by doing away with not only cooling but also with the idea that life must have originated when the whole Earth, bathed in mist, supported a tropical climate.[21]

We can understand Lowell's reluctance to accept Chamberlin's account on three grounds. First, it lacked the elegance and apparent inevitability of cosmic evolution. Second, it seemed far less likely to result in the regular development of intelligent life in the solar system and elsewhere. Indeed, life on other planets was not a subject that was treated anywhere in Chamberlin's three-volume text. And last, Chamberlin's approach was theoretically backward from Lowell's point of view. Chamberlin had created a cosmogony explicitly responsive to the needs of geology, rather than proceeding, as Lowell pretended to do, from the top down. Lowell's cosmogony defined parameters within which the evolutionary course of Earth had proceeded.[22]

The Response of the Scientific Community

Nothing better illustrates the changes in scientific practice and the character of the scientific community than the responses of professional scientists to Lowell's efforts to develop a viable cosmogony. These responses, featured over a six-month period in the journal *Science,* clarified the way in which academic specialization had transformed the community and placed the generalist in a precarious position.

The exchanges in *Science* began with a withering attack by Eliot Blackwelder of the University of Wisconsin on Lowell's erroneous understanding of geology as exhibited in *Mars as the Abode of Life.* Blackwelder was later supported by Joseph Barrell, a geologist at Yale. By contrast, Lowell's position was defended by his former colleague T. J. J. See, who was an ardent supporter of the nebular hypothesis and had been serving in splendid isolation at the Naval Observatory's station on Mare Island, in California. Wrote Blackwelder, "It is not surprising that Mr. Lowell, an astronomer, should have only a layman's knowledge of geology; but that he should attempt to discuss critically the more difficult problems of that science, without, as his words show, any understanding of the great recent progress in geology, is astonishing and disastrous." More specifically, Lowell had erred in relying on the outdated work of James Dwight Dana—an error Blackwelder attributed to Lowell's "implicit" belief in the nebular hypothesis, a theory "now on the defensive." Here Blackwelder aligned himself with Chamberlin's views.[23]

Lowell's rather lame response to Blackwelder asserted merely that he had relied on the established authorities in the field and that Blackwelder's critique, motivated as it was by a desire to defend his colleague Chamberlin, could be dismissed on that account. But most of Lowell's letter was devoted to defending himself against an attack of a different sort by Moulton, who alleged that Lowell had adopted the planetesimal hypothesis without acknowledging that fact. Moulton's allegation is particularly startling in view of Blackwelder's previous assertion that Lowell was an implicit believer in the nebular hypothesis. Unfortunately, Lowell ignored the charge in the public exchange, though he later attempted to reckon with it in private. Instead, he identified mathematical errors in Moulton's calculations that raised questions about the validity of his theory. Moulton's rejoinder acknowledged the error but faulted Lowell's logic in suggesting that an error in calculation should invalidate an entire theory. In a more damning fashion, he quoted Lowell's recent writings to show his acceptance of the main elements of the planetesimal hypothesis and taunted him for not acknowledging his debt. "From these quotations it is clear that Dr. Lowell has a real affection for the main features of the planetesimal hypothesis, and if I had not been so unfortunate as to have utterly destroyed it (according to the new logic) by the blunder in my book 83 pages before I took the hypothesis up, he might almost have reconstructed it from his own recent writings." Moulton's allegations were telling and probably explain Lowell's failure to respond directly to the charges.[24]

Lowell returned to the issue of cosmogony in *The Evolution of Worlds.* While he made no direct reference to the previous year's spirited debate in *Science,* Lowell clearly intended to rebut his critics. He refused, however, to drop Dana's version of Earth's development. Instead, he sprinkled the text with numerous footnotes identifying his sources and frequent references to discoveries made at the Lowell Observatory, designed to enhance the credibility of his argument and the prestige of the observatory.

He acknowledged the unavoidable hazards of constructing cosmogonies that he and his fellow cosmogonists accepted as a matter of course. Lowell insisted that qualified men were strongly impelled to create cosmogonies, even though they were almost inevitably blind to their mistakes. "With the fatal fondness of a parent for his offspring it is rare for the defects to be so glaringly apparent to their perpetrator." And he acknowledged that equally qualified men stood ready to find their errors: "For there is almost as good a trade in exposing cosmogonies as in constructing them. But no special opprobrium attaches to failure, because everybody has failed, from Laplace down, or up, as you are pleased to consider it." Surely Lowell was thinking about his own recent bouts with the critics and placing himself, as the subject of criticism, in the good company of Laplace. And as if to anticipate the critics who had already accused him of borrowing, he acknowledged that his own cosmogony would not be "entirely homemade."[25]

Following the publication of *The Evolution of Worlds,* Lowell withdrew from further discussion of cosmogonic issues. As a result, it is difficult to sort out his views, but the evidence suggests that both Blackwelder and Moulton were right. Lowell did indeed borrow from Chamberlin and Moulton the concept of the origin of the solar system in a spiral nebula. Why he refused to explicitly acknowledge this borrowing is not entirely clear. But Blackwelder was also correct in claiming that Lowell continued to have a soft spot for the nebular hypothesis.

The evidence for these apparently contradictory points is abundant. First, by 1909 Lowell had come to speak of "nebular hypotheses." The text of *The Evolution of Worlds* suggests that the use of the plural was deliberate. There Lowell noted with admiration Laplace's identification of congruities that had subsequently been verified, while asserting that the discovery of new congruities in the solar system demanded "a theory different from his [Laplace's]." It seems evident, then, that Lowell had come to accept a much revised version of Laplace's original hypothesis, which he continued to label as *a* nebular hypothesis (my emphasis). This interpretation is justified by the prefatory remarks in *Mars as the Abode of Life,* where Lowell outlined his task as linking "the Nebular Hypothesis to the Darwinian theory, bridging the evolutionary gap between the two." Just a year later he spoke in glowing terms of the history of the nebular hypothesis as "the chronicle of the mind's most brilliant march, which had been more than human had it not been cumbered by mistakes."[26]

Lowell's claim that he was not borrowing from the planetesimal hypothesis may be understood as a semantic problem—one, indeed, that originated with Chamberlin rather than Lowell. On occasion Chamberlin identified his theory with the accretion of planetoids rather than the creation of spiral nebulae. Lowell himself insisted on this point in letters to friends. What Lowell wished to explain was not how the nebula had originated, but, assuming its existence, how the various bodies in the solar system were formed and set into motion. In this sense, Lowell could believe that although he had borrowed the concept of the spiral nebula, he had not borrowed the planetesimal hypothesis. This interpretation is borne out by Lowell's enthusiastic reception of V. M. Slipher's discovery of the high radial velocities of nebulae that indicated they were galaxies rather than proto–solar systems. Slipher's discovery, important in its own right, had the additional virtue of freeing Lowell from the only element in his theory of the origin of the solar system he had borrowed from Chamberlin.[27]

In conclusion, it is evident that Lowell embraced an ideology and a style of operation that set him apart from specialists in the new university science departments. He hoped to use his observational findings to buttress the evolutionary scheme articulated in the work of Laplace and Spencer. He would do so by drawing on the findings of other disciplines, and he would report his work to the general public as well as to astronomers. In this sense, he operated very much in the manner of grand theorizers such as Spencer and Fiske.

When the nebular hypothesis was challenged, Lowell demonstrated his concern. Not only did it undergird his work on extraterrestrial life, but it was also an essential part of the larger evolutionary scheme he strongly embraced. And, not surprisingly, he resisted elements of the Chamberlin-Moulton hypothesis that raised questions not only about the likelihood of planetary development, and therefore about extraterrestrial life, but also about the validity of his larger developmental hypothesis, which was to explain all of the important phenomena of the universe, organic and inorganic.

Lowell's continued adherence to Laplace and Dana, as Blackwelder claimed, stemmed from the close connection between their respective theories of cosmogony and geology. Dana's notion that the gradual contraction of Earth's crust from a hot, gaseous mass was the primary dynamic in planetary history took as its point of departure Laplace's similar claim regarding the composition and temperature of planets in the early stages of development. Under the Chamberlin-Moulton hypothesis, shrinkage would play a lesser role in view of Earth's supposed development from solid rather than gaseous material. And the evolution of Earth from an aqueous to a terrestrial stage, which Dana and Lowell regarded as a pivotal moment, had no place in Chamberlin's account. Moreover, Chamberlin's interest in testing multiple hypotheses ran counter to the unified approach to cosmogony and geology that Dana, in tandem with Laplace, was able to provide. Thus Lowell preferred the intellectual elegance of cosmic evolution to the more complex mechanisms of the new cosmogony.[28]

Lowell's project generated opposition from the supporters of Chamberlin and Moulton not only because of his claims for extraterrestrial life but also because his attempts to cross disciplinary lines violated the new professional norms. Thirty years earlier such attempts would have attracted less attention. At that time, the autonomy of disciplines was less clearly established, and generalists like Spencer had little hesitation in drawing from a variety of fields. Furthermore, the difficulty of mastering disciplines had increased with the proliferation of scientific research in large university departments. By 1900, Lowell would have had to familiarize himself with a body of highly specialized literature in order to successfully engage geologists on their own turf. And he would have had to address his colleagues in technical terms in their own disciplinary journals; he would not have been allowed to affront them with popular works written in a grand theoretical mode. In a book published in the same year as *The Evolution of Worlds*, George Ellery Hale acknowledged these changes. He could speak authoritatively as an astrophysicist about the "production of a planet like the Earth. It remains for the geologist to explain the changes which the surface of the Earth has undergone," he said.[29]

In the new era, Hale's colleagues Chamberlin and Moulton modeled the appropriate way to cross disciplinary lines. Instead of going it alone, as Lowell had done, Chamberlin, as a geologist, sought an alliance with an astronomer so that his project would achieve recognition in the astronomical community. Furthermore, he understood that communications about the new hypothesis would have to be conducted in academic journals and associations. These standards became the grounds for objecting not only to Lowell's nebular hypothesis but also to his claims for extraterrestrial life, as we shall see.

Lowell, like T. J. J. See and W. H. Pickering, operated in the mode of the strongly individualistic "gentleman amateurs" of the mid-Victorian era. All three men were drawn to cosmic evolution because of its sweeping account of the universe. They were sometimes too impulsive in finding "conclusive" empirical evidence to sustain theories that were difficult to prove. But even if they had been more cautious in their work, their relative isolation from the universities, scientific journals, and disciplinary meetings, as well as endowments for research, would have created serious obstacles to the reception of their work.

The future of astronomy would belong to those like Chamberlin and Moulton, who could take advantage of the new channels of communication and research funds. It would belong especially to those who were making astronomy into a laboratory science by applying the methods of physics to astronomy, and to those who respected disciplinary boundaries even while engaging in interdisciplinary projects. The new approach doomed the grand theory of Lowell and Spencer, whose work may be regarded as the last chapter of a story that began with the encyclopedists of the eighteenth century. Lowell's quest to provide a coherent explanation of the rise of the solar system ran headlong into the insistence on disciplinary autonomy. The conflict between him and his rivals illuminates the important changes that had already reshaped the interaction between disciplines at the time.

Lowell's philosophy of the cosmos might have fared better if he had followed his original plan to embark on the project in 1891, when the intellectual climate was more supportive of such endeavors. After all, Herbert Spencer's *Synthetic Philosophy* had been launched in 1862 and, except for the final volume, had been fully published by the 1880s. The best work of Ernst Haeckel and T. H. Huxley had long since appeared. By the time Lowell's work was published, Huxley had been dead for thirteen years and Spencer for seven, while Haeckel lingered for another decade, after retiring from the University of Jena.

Meanwhile, for various reasons, the cosmic approach was going out of fashion. Nineteenth-century individualism, which Lowell and Spencer so strongly embraced, was being partially replaced by the cooperative ethos of the Progressive Era. In addition, the materialistic assumptions of Huxley and company were being challenged by a new generation of intellectuals, including William James and Henri Bergson, who embraced vitalism, while the new physics—brought on by the discovery of radioactivity, the rise of quantum mechanics, and the conception of special relativity—subverted the very distinction between matter and energy. No longer would it be possible to think of matter as a tangible substance.[30]

The deterministic inclination of the cosmic philosophers was also under attack by intellectuals. James, in particular, endorsed the idea of the contingency of events. Under the circumstances, and with one interesting exception, Lowell's evolutionary scheme, which led with certainty from nebula to man, ran counter to the new currents of thought. The exception was the unpredictable and catastrophic event that had launched and would possibly end the evolutionary process. Only this chance encounter of two dark stars resonated with the idea of contingency. In this regard, the Chamberlin-Moulton thesis, with its element of unpredictability, was far more in keeping with the new intellectual environment than was the idea of cosmic evolution.

Despite these changes, however, it is perfectly evident that the evolutionary approach continued to exercise a powerful attraction on the large public that flocked to Lowell's lectures and read his books. The Spencerian method appealed intellectually to its adherents because it offered a comprehensive approach to understanding all facets of the universe, social and natural. Relatedly, Spencer's method appeared to bridge the widening gap between science and culture. Armed with evolution, intellectuals like Lowell were convinced that they could master the important truths about both nature and society.

The intellectual advantages of cosmic evolution were reinforced by practical ones. Spencer's theories offered a ranking system that placed Western societies above those in East Asia, thus confirming the superiority that Lowell and other Westerners assumed. The ranking of disciplines, another Spencerian idea, also served Lowell's purposes. He could assert the superiority of astronomy over geology and thus defend his own pretensions to speak with some authority on geological and biological matters. Most important, the evolutionary approach provided theoretical support for his claims regarding the existence of intelligent life on Mars. For all of these reasons, Lowell and his followers remained loyal to Spencer's system.

III

◆ ◆ ◆

Lowell as Astronomer

The Lowell Observatory was launched during a temporary expedition to Arizona to observe Mars during the 1894 opposition, or near approach, of Mars to Earth. For the next seven years the future of the observatory was uncertain. Not until the turn of the century did it assume the status of a permanent institution, and over the next decade and a half the observatory grew, in unanticipated ways, to become a far different institution than it was at the outset. A number of factors help to explain the evolution of the observatory over its first twenty-two years, but two stand out as most significant: the powerful imprint of Percival Lowell himself as founder, owner, and architect of many of its programs, and the equally powerful influence of the rapidly changing world of American astronomy. Within this context, Lowell struggled successfully to bring a permanent institution into being and to achieve recognition for it.

Lowell's forceful personality was a major factor in shaping his observatory. His rebellious inclinations had been clearly revealed in his decision to defy his father and leave behind the family investment business for a leisurely career as traveler and writer. In this manner he had secured his independence from many family and community obligations. In founding his own observatory, he would create an institution in his own image and tenaciously defend it against the imperial ambitions of larger and more powerful rivals. The Lowell Observatory, reflecting Lowell's prickly sensitivity to infringements on his autonomy, would have its own distinctive program, setting, instruments, and staff. Indeed, Lowell's outsider status increased his willingness to take up issues, such as extraterrestrial life, and employ methodologies, like planetology, that were considered far too risky by mainstream astronomers.

Lowell seems to have thrived on the conflict generated by his program and publicity. He—and some of his rivals—regarded the competition between observatories as a virtual war in which offensives on one side required defensive operations on the other. Indeed, during his lifetime the Lowell Observatory was almost continually embroiled in conflict over a variety of issues, including the Martian canals, the markings on Venus, and the knots in Saturn's rings. It is interesting to note that almost all of the controversy at the observatory revolved around the rather risky projects that Lowell had originated rather than those initiated by his assistants. As a result, during most of Lowell's twenty-two-year stint as director, he and his staff were isolated from the professional activities of mainstream astronomers.

Lowell brought to astronomy the same impatience for public recognition manifested in his other undertakings. This eagerness for recognition was a strong factor in determining the research program and location of the observatory. Lowell not only took up the most sensational astronomical issue of the time but also expected to achieve virtually instant confirmation of his theories by publishing his results in short order. Moreover, he defiantly published them in popular books and magazines that would generate recognition among the general public, to the dismay of his astronomer colleagues.

The astronomical world, which Lowell entered in 1894, was about to be reorganized in such a way as to confirm his outsider status. Astronomy, like other disciplines in the 1890s, was modernizing. Its leaders were establishing a new professional organization and a new journal; equally important, they were turning their observatories, most of which were affiliated with large universities, into research factories—an early version of big science. Navigating in such a highly organized environment required the skills of an insider—a team player. Astronomers who wished to get ahead would now have to play the game cautiously. They would be part of large data-gathering enterprises; they would publish their results in the newly established professional journals and at professional meetings. Such behavior did not preclude conflict between individuals and observatories, but it would be conflict tempered by a common agreement on standards for doing research and judging the quality of the work.[1]

Lowell's outsider status may be nicely captured by juxtaposing the creation of his observatory with that of George Ellery Hale's Kenwood facility, which was located near the University of Chicago campus and built just three years before Lowell's. Both Lowell and Hale were the scions of wealthy families who had provided their sons with the finest scientific education available. Their observatories were built with family money, as was quite common for wealthy amateurs in the late nineteenth century. (William Lassell in England, William Parsons, third earl of Rosse in Ireland, and Henry Draper and Lewis Rutherfurd in the United States come to mind immediately.) There, however, the similarity between Lowell and Hale ends. Lowell struggled to assure the independence of his observatory against the imperial ambitions of the Harvard College Observatory. Once survival was assured, Lowell proceeded with his Martian research, which was implemented by a staff of two or three astronomers. Lowell would initially prefer older research approaches, including using a smaller telescope and relying more on sketches than photographs or spectrograms. Only reluctantly did he turn to the spectrograph, camera, and reflecting telescope in his second decade of operation. Furthermore, in such a small, remote setting as Flagstaff, Lowell's insistence on the romantic aura of scientific discovery could be given full play.[2]

By contrast, Hale merged his Kenwood Physical Observatory with the University of Chicago, and, as a faculty member, exploited the name of the university to fund and build the Yerkes Observatory, one of the great research facilities in the country. There, Hale committed himself to a program based largely on pioneering methods of astrophysics that required a laboratory, spectrographs, cameras, and a large staff of highly specialized researchers. By its founding, in 1897, he was ready to exploit the largest refracting telescope in the world. In this setting Hale and his staff carried on the data collecting and reduction appropriate to a large scientific enterprise. Such an operation depended on impressive organizational skills to harmonize the work of dozens of astronomers and computers.[3]

Hale and Lowell represented two different directions in American astronomy. However, the profession's resources were not equally divided between them. As an outsider, Lowell could endow his own institution with his considerable personal resources. He could command the attention of the media through his facility for communication. And he could take advantage of a cadre of well-placed friends, some of whom were members of the National Academy of Sciences, to bring attention to his search for Martian canals. Nonetheless, he and his observatory staff would be condemned to virtual isolation from the mainstream of American astronomy. A letter from one of Lowell's astronomer collaborators, Frank Very, describing

Very's reception at the Astronomical and Astrophysical Society meetings, makes this point quite clearly. "I have the impression that with the exception of two or three personal friends, the American astronomers rejected me . . . Just now they have all jumped the fence with [Charles G.] Abbot and if it were not for you, I should be alone." Very despaired of finding an audience among astronomers unless his former director, Frank Schlesinger of Allegheny Observatory, should read a paper for him.[4]

Hale's situation was much more promising. He and his friends were the ones who refused to listen to Very. Hale was an organization man who helped to build the new world of American astronomy. At the center of this world were the directors of the leading observatories, who formed a kind of power elite: E. C. Pickering of Harvard, E. B. Frost of Yerkes, W. W. Campbell of Lick, and Simon Newcomb, superintendent of the Nautical Almanac Office. They and their associates effectively controlled the *Astrophysical Journal* and the Astronomical and Astrophysical Society of America. Lowell's opponents also had direct access to funding through their universities, the government, and eventually private foundations such as the Carnegie Institution of Washington. When important decisions were made on such matters as developing a new endowment for research, Newcomb recommended a committee composed of himself and the directors of the Lick, Yerkes, Harvard, and Wisconsin observatories to consider the matter. The power elite also controlled the selection process by which rewards were apportioned to members of the profession. The directors of the big observatories were themselves leading members of the National Academy of Sciences and, as such, informally designated those who would join them. In this regard, the grounds on which they campaigned to select E. B. Frost, Hale's replacement as director of Yerkes, are revealing. In addition to evidence of the high quality of Frost's research, Hale listed "broad interest in science, . . . [a] tendency to take part in the work of scientific societies, and . . . [a] willingness to co-operate with other investigators." Here was the perfect description of the team player prepared to do work in an organizational society. Based on these criteria, Percival Lowell, no matter how brilliant his research, would have been flatly rejected for membership.[5]

It might seem implausible that a maverick astronomer like Lowell, operating a small observatory in the Arizona Territory, should have been able to threaten the astronomical establishment of his era for more than a decade. However, the astronomical community was in the midst of transition and the establishment had only begun to organize itself. Three of the four leading observatories were themselves quite new. Mount Wilson and Yerkes, in fact, were newer than Lowell, and Lick had been founded just six years before Lowell. Only Harvard truly outdated the Lowell Observatory, and its modern stellar research program preceded the creation of Lowell Observatory by only twelve years. Furthermore, the key institutions of the new professional establishment, the *Astrophysical Journal* and the Astronomical and Astrophysical Society of America, had been created only in 1895 and 1897, respectively. In short, the astrophysicists, the dynamic element in the astronomical community, were just beginning to organize and establish lines of communication. In such a climate, an energetic individual like Lowell with a clear-cut program that appealed to the larger public could easily provoke a crisis.[6]

Percival Lowell sought recognition for himself and his observatory by staking out a distinctive program and developing strategies to defend it. In the process, he threatened the recently established norms of the astronomical profession and thus generated strong opposition from other astronomers and observatory directors. Those directors used the occasion to solidify the institutions they had already put in place and to clarify their role in settling an important issue of scientific truth. At the same time, the crisis forced establishment astronomers to divert attention from their work in stellar astronomy to challenge Lowell's claims about Mars.

The conflict informed a whole range of issues, beginning with the question of whether the canals of Mars existed and leading to concerns about telescope size, atmospheric conditions, and psychological factors that might affect seeing. The stakes were high inasmuch as the publicly aired conflicts raised the possibility that the disputants would question the very existence of standards for determining the truth of astronomical findings and thus threaten the scientific enterprise. Accordingly Lowell and his opponents engaged in sometimes bitter recriminations about the validity of their positions. In the end, the establishment won a decisive victory over Lowell, and the Lowell Observatory itself came to operate in many respects along the lines originally supported by his mainstream opponents. While Percival Lowell remained committed to his program of planetary research through mainly traditional approaches, his assistant V. M. Slipher contributed key evidence for an expanding universe. At the same time, Lowell assistants C. O. Lampland and E. C. Slipher were recognized by colleagues outside of Lowell for their excellent photographic work. In this way, Lowell's assistants prepared the way for a rapprochement between Lowell's observatory and mainstream astronomy.[7]

In the midst of the crisis, however, Lowell was well equipped to give his rivals a run for their money. He had, as Campbell recognized, a sensational issue that stirred the imagination of many people. He was equipped with energy, financial resources, and an extraordinary gift for communication. Furthermore, he was a wily combatant who used the media effectively to challenge his opponents. He was also shrewd in conceptualizing his views as interdisciplinary and innovative and presenting himself as an underdog who would stand up for truth against the bullying of the power elite.

The dynamic of the crisis can be seen clearly in retrospect. At its founding, the Lowell Observatory was viewed by Harvard astronomers as an opportunity for expansion rather than a threat. Lowell thus struggled initially to protect the observatory from such designs. Having survived this threat, Lowell then sought to determine the nature and location of the observatory. In its early days, the observatory's sole function was to collect data on Mars during periods of opposition at different sites. Following heated arguments over Lowell's first published claims, in 1895, for the existence of intelligent life on Mars, the conflict between Lowell and his opponents subsided. Not until Lowell's recovery from his nervous breakdown in 1901 was the observatory permanently established in Flagstaff and a new staff hired. The conflict between Lowell and his opponents then reemerged in a more substantial way in 1905 with the coming of more favorable oppositions. By then the power elite at mainstream observatories was in high gear, and the Lowell Observatory had clearly established its identity and was launching a new campaign to place beyond doubt the existence of life on Mars. The conflict came to a head between 1907 and 1910, by which time the establishment coalition prevailed over Lowell.

9

From "Astronomical Picnic" to Observatory

The founding of the Lowell Observatory was a daring enterprise. Percival Lowell risked his own funds on a speculative, high risk, sensational, and very idiosyncratic project. Only with the advantage of hindsight, however, can we see that his astronomical expedition of 1894 was the entering wedge for the establishment of a permanent observatory. From the perspective of 1894, the nature, duration, and location of Lowell's project were uncertain. There was, after all, no guarantee that a thirty-nine-year-old world traveler from Boston would actually take to an astronomical career in the Arizona desert. Equally uncertain were the viability of Flagstaff as a site and Lowell's program of research on the canals of Mars. Over a period of seven years, Lowell gradually increased his commitment from a single expedition to Flagstaff to a series of expeditions at various locations to gather evidence regarding life on Mars. Not until 1897 did he decide to keep the observatory operating between oppositions, and not until 1901 did the institution assume a permanent status. Central to this important change of status was Lowell's bout with neurasthenia, which, among other causes, grew out of doubts about the validity of his own observations and his future prospects in astronomy.[1]

The future of the observatory was called into question very early in its history by the designs of the Harvard College Observatory. Lowell's desire to create an autonomous expedition could not be instantly realized. He was an inexperienced amateur whose project required support from an experienced professional astronomer. In teaming up with William Henry Pickering, a member of the Harvard College Observatory staff, Lowell gained the needed expertise but risked control of his project. As strong individualists who resented any infringement on their autonomy, Pickering and Lowell were cut from the same cloth. They were also drawn to highly speculative projects, which they pursued with rather traditional methods. Despite these interesting parallels in personality and approaches to astronomy, Lowell could never be sure that Pickering's loyalty to the new observatory and its program would take precedence over his allegiance to Harvard, where his brother, E. C. Pickering, was director. The brief collaboration of these two men reveals much about their character, the process of creating a unique institution, and the context in which it developed.[2]

Having launched his observatory with the aid of Harvard, Lowell struggled against the imperial ambitions of the Pickerings to establish an independent identity for the Flagstaff enterprise. The founding is, in part, the story of this struggle; it is also an episode in the history of the Boston elite. Though the Lowell Observatory was located in Flagstaff, Arizona, its founder and his collaborators in the first year were Boston Brahmins. Their interactions took place within, and were shaped by, the network of clubs and institutions created by this elite.

The Boston Context

To launch an astronomical expedition in the Arizona Territory in the late nineteenth century, and as an amateur astronomer using one's own resources, was a formidable undertaking. Location notwithstanding, the founding cannot be understood without taking into account the Boston social network. Through his association with cultural and scientific organizations in his native city, Lowell made connections with scientific amateurs and their professional counterparts, with whom he would later collaborate.

Collaboration was furthered by geography and common educational backgrounds. Through much of the nineteenth century, Brahmin Boston was relatively intimate in geographical, as well as social, terms. The main actors in this drama grew up on or near Beacon Hill. The Pickering brothers were raised on Mount Vernon Street, where the Lowell family had lived for several years. Abbott Lawrence Rotch's home was around the corner, on Commonwealth Avenue. These future scientists gravitated either to MIT, from which Rotch (1884) and the younger Pickering (1879) graduated, or to Harvard, where Lowell (1876) and Edward Pickering (Lawrence Scientific School at Harvard, 1865) were educated. The cohesiveness of the Brahmin world, then, had a strong basis in urban geography as well as common educational institutions.[3]

Aside from educating the rising generation, Harvard and MIT helped to cement relationships between amateurs and professionals in several ways. Members of the Boston elite served on governing boards of these institutions. Others, as faculty members, effectively involved interested amateurs. Edward Pickering, for one, made the observatory facilities available to Brahmin friends and kept them abreast of its findings. Both Percival Lowell and his father, Augustus, were members of the observatory's visiting committee, which examined its research activities and recommended future directions. Augustus contributed financially to these projects as well. Father and son were similarly involved on visiting committees and boards at MIT.[4]

Three other local institutions in which the Lowells were active provided opportunities for communication between amateurs and professionals in the community and thus played a role in the origin of the Lowell Observatory: the Lowell Institute, the American Academy of Arts and Sciences, and the Appalachian Mountain Club. Edward Pickering was one of many scholars invited to lecture at the Lowell Institute by Augustus Lowell and was also consulted on the qualifications of other prospective speakers. Since many of the lecturers dined with the Lowells before they took the podium, the institute cemented relationships between the family and leading scholars.[5]

Equally important as an arena for such encounters was the American Academy of Arts and Sciences, the Boston counterpart of Philadelphia's American Philosophical Society. Election to the academy honored achievement in science and the arts, and provided a forum for members to discuss each others' findings. At the academy, the leading scholars of the region met on the same turf with amateurs like Augustus and Percival Lowell. The four principals in the founding of the observatory were all members.[6]

They were also on the rolls of the Appalachian Mountain Club, founded by Boston Brahmins in 1876 to promote their interest in vigorous exercise. Such recreation could easily be combined with scientific purposes. By the late 1870s, as astronomers came to value mountain observatories for enhancing the quality of seeing, they looked to mountain climbing both to satisfy a passion for adventure and to test the atmosphere at high altitudes. Science and sport would advance hand in hand. Accordingly, Edward and William Pickering were charter members and Edward was the first president of the Appalachian Mountain Club, while Abbott Lawrence Rotch and Percival Lowell were regular participants.[7]

These upper-class Boston institutions, reflecting the commitment of the elite to cultural and scientific advancement, created a setting in which Lowell, Rotch, and the Pickering brothers regularly interacted. The members shared a passion for intellectual pursuits and, in the cases of Lowell and Rotch, an eagerness to use inherited wealth to pursue them.

Origins of the Lowell-Pickering Collaboration

In retrospect, it is not surprising that Lowell and Pickering collaborated on the expedition to collect data about Mars. They were joined not only by the Boston cultural network but also by their imaginative approach to astronomy. While other astronomers of their generation were gravitating toward stellar research requiring expertise in photography and spectroscopy, Pickering and Lowell focused on a highly speculative reading of the surface details on Mars that could be captured in their drawings. Their outsider status made them natural allies against their mainstream colleagues. Indeed, as Seth Chandler, editor of the *Astronomical Journal,* astutely remarked, their collaboration encouraged the speculative tendencies to which they were both drawn. Chandler noted that "Lowell . . . [had] not selected the right kind of companions for his astronomical picnic . . . What he needed was some young well-equipped astro-physicist . . . sound and conservative, to sit on his coat-tails and keep him down to business and prevent wild flights of fancy."[8]

The origins of the Lowell Observatory may be traced directly to the interaction that occurred at the Harvard College Observatory between its staff and the Boston amateurs, beginning in 1890. Edward Pickering was the central figure in this network. As director of the observatory, he encouraged interested Boston amateurs to use its facilities and consult with him on astronomical projects. Percival Lowell's friends and fellow travelers in Japan, George Agassiz and William Sturgis Bigelow, were among those who took advantage of this opportunity.[9]

As Lowell became interested in pursuing astronomical research in a more systematic way, he turned to the observatory. He discussed Mars with William Pickering in 1890. Two years later he asked Edward Pickering if he could see the charts and maps of Mars that his younger brother had drawn. Apparently a cordial relationship had developed between William Pickering and Lowell. The former wrote to his brother: "I have always found him [Lowell] very agreeable."[10]

In the fall of 1892 Lowell was once again in touch with the observatory on the subject of Mars. He wrote Edward Pickering to request copies of the maps of the Martian surface drawn by the Italian astronomer Giovanni Schiaparelli. In November he asked to see the photographic equipment at the observatory. Probably in recognition of his increased interest in its activities, Lowell was appointed in that same month to the committee to visit the observatory. Edward Pickering was clearly responsible for Lowell's selection.[11]

Another close collaborator of Edward Pickering was Percival Lowell's cousin Abbott Lawrence Rotch. Lowell and Rotch were both grandsons of Abbott Lawrence, who in 1846 had endowed the Lawrence Scientific School at Harvard. Rotch, after graduating from MIT, founded a meteorological station on the outskirts of Boston. The Blue Hill Meteorological Observatory was built with the founder's money and opened in 1885. Percival Lowell and Edward Pickering were among the early visitors to the observatory. So successful was the new facility that Harvard quickly closed its own weather station. Rotch, in turn, published his findings in the observatory *Annals*.[12]

Rotch took advantage of his Harvard College Observatory connection to join two of its mountain expeditions. In point of fact, meteorologists and astronomers had a common interest in mountain sites. The meteorologists sought to observe a wide range of phenomena in the upper atmosphere, while astronomers were seeking the steady air in high desert country. Rotch accompanied William Pickering to Mount Wilson in California in 1889 and four years later established a meteorological station as part of the mountain observatory that Pickering was setting up in Arequipa, Peru.[13]

The astronomical work at Arequipa contributed in important ways to shaping the Lowell Observatory. Harvard's Arequipa station was part of Edward Pickering's ambitious plan to collect and analyze stellar spectra. He hoped to maintain two high-altitude stations, one in the southern hemisphere and the other in California, thus making Harvard the first observatory to cover the whole sky. The stations would be Harvard's eyes, while Cambridge would become a computing center where the data would be analyzed and published. Though the California station never materialized, Pickering's successful fund-raising ventures enabled him to support what was essentially an early version of big science.[14]

Edward Pickering chose his brother to continue the work begun by Solon I. Bailey in collecting stellar spectrograms at Arequipa. William

Pickering, however, preferred viewing Mars through the telescope to taking spectrograms of the stars, and he rejected the camera in favor of old-fashioned drawings. With the excellent seeing at Arequipa, he discovered mountain ranges and lakes on Mars that he reported to the *New York Herald.* Such sensational findings appealed to the general public, but not to Edward Pickering, who in 1893 relieved his brother of the Arequipa post. In retrospect, it is clear that William Pickering was unsuited to a subordinate role in a large data-collecting enterprise. He wished to pursue his own projects without interference from others.[15]

The younger Pickering returned to Boston—after stopping in Chile with Abbott Rotch to view the solar eclipse of April 16, 1893—and immediately began planning for a new Martian expedition in time to exploit the October 1894 opposition. Martian oppositions occurred every twenty-six months, but it would be another fifteen years before the planet would be so close to Earth and so favorably positioned for viewers in the Northern Hemisphere. Communication with the U.S. Weather Bureau indicated that the Arizona Territory offered desert conditions similar to those in Peru. Accordingly, William Pickering suggested to his brother a Harvard expedition to Arizona to compare the seeing with that in Peru as well as to investigate the surface of Mars. Edward Pickering, whose budget had been depleted by his brother's excesses in Peru, rejected the proposal. Over the next three months William sought support from various other sources.[16]

The surviving documents tell us little about how Percival Lowell reached his decision to collaborate with William Pickering. It is evident, however, that his plan to visit the Arequipa station in Peru in 1893 fell victim to his research on esoteric religion in Japan. His 1891–92 communications with the Pickerings reveal a strong interest in investigating the surface of Mars; moreover, on his way to Japan via Chicago and San Francisco, he had requested and received from S. W. Burnham, the Chicago double-star observer, a letter of introduction to E. E. Barnard of Lick; and Lowell took with him a 6-inch Clark refractor through which he viewed Saturn at his Tokyo residence. However, Lowell had apparently not decided on his future course of action by November of 1893, when he wrote Ralph Curtis to urge an Easter jaunt to Seville in the following spring.[17]

A Christmas gift of Camille Flammarion's *La planète mars et ses conditions d'habitabilité* was apparently the catalyst for Lowell's decision to launch an astronomical expedition. Inside the cover of the book, Lowell wrote "Hurry" in his own hand, apparently in recognition of the fact that the Martian opposition was approaching. From the Pickering and Rotch records, it is possible to piece together the likely course of events leading to the initial informal understanding between William Pickering and Lowell. Pickering had already developed a clear plan for an Arizona expedition, and among his several proposals to attract funding, one was for a joint expedition. Lowell shared Pickering's interest in the surface of Mars and good seeing, though he had had virtually no time to think about where such an expedition might locate and how it should be organized and equipped. These matters would now have to be addressed.[18]

Collaboration between Lowell and Pickering, arranged through the Boston social network, removed the obstacles that both men faced. Rotch was in a good position to play the role of middleman, since he had already collaborated with the Pickerings and A. E. Douglass, the young assistant who had accompanied William Pickering to Arequipa, and could advise Lowell about their reliability. He had also built a research station with his own funds. Accordingly, Lowell sought counsel from Rotch on January 16, the day before William Pickering responded to Lowell's inquiry about an expedition, and on seven subsequent occasions.[19]

The Struggle for an Independent Expedition

Shortly before January 17, 1894, Percival Lowell sought out William Pickering at the Harvard College Observatory. Less than a week after their lunch at the observatory, the two men reached agreement on a plan for an Arizona expedition. While Lowell left no record of their conversations, Pickering's January 24 letter to the Pittsburgh instrument maker John Brashear offers evidence of an understanding between the two men—and of Pickering's proprietary feelings about the expedition: "I am happy to be able to inform you that Mr. Percival Lowell, a wealthy Boston gentleman, has decided to go out to Arizona with me to observe Mars at this opposition. Mr. Lowell belongs to the prominent Boston family from which the city of Lowell was named. He is much interested in large telescopes and particularly interested in Mars. I have asked him to write you and in future all financial arrangements will be made through him."[20]

While Pickering and Lowell agreed on the purpose, equipment, and personnel for the expedition, Lowell was uneasy about who would receive credit for the expedition's findings and where they would be published. After thanking Lowell for supporting the project, Edward Pickering offered the Harvard Observatory *Annals* as the medium of publication. On the related matter of crediting discoveries, William Pickering proposed to Lowell that most reports would be jointly written, though each man, contingent on his partner's agreement, would be able to publish his own work independently.[21]

Lowell remained uneasy about these proposals and especially about his right to independent publication of the expedition's findings. In a January 27 letter to Edward Pickering, he insisted on the right to publish findings "both in the Annals of the Observatory and in any account I determined to publish." By way of conclusion he noted that "I am thus specific because a corporation, Harvard College, is the other party and needs more formality than another man."[22]

Lowell's concerns were particularly legitimate in view of Harvard's expanding operations around the globe and Pickering's desire to acquire a Harvard station in the West. After all, William Pickering had just returned from one Harvard station and might very well look upon Flagstaff as another one in embryo. Indeed, William Pickering's proprietary sentiments in the letter to Brashear justified such suspicions. Under these circumstances, it was surely appropriate for Lowell to deal cautiously with Harvard. In attempting to exercise control over the publication of research findings, Lowell was effectively protecting the independence of the expedition.

Lowell's reservations about working with Harvard were intensified by a February 13, 1894, article in the *Boston Herald*. Both the title of the article and its contents were worded in such a way as to confirm Lowell's worst fears: "Another Harvard Observatory. Funds for the Work Provided by Mr. Percival Lowell, Who Will also Take Part in the Expedition." The article portrayed Lowell's role as largely financial, suggesting that the expertise lay with the Pickerings and Harvard. William Pickering was identified as "the head of the explorers," and his brother was quoted as saying that "our Arizona station will be just right for the opposition."

Lowell was understandably put off by the notion that the Arizona project would be just another station of the Harvard Observatory. It conjured up visions of Arequipa. The fact that William Pickering was identified as "head" of the expedition suggested that the Pickerings regarded the new observatory as subservient to Harvard. Of course, Lowell's objections to the article must also have been provoked by the public airing of an obvious truth: that his principal contribution to the expedition in its early stages would be funding rather than expertise.

In his response to Lowell's complaint, William Pickering pointed out that the article gave "public recognition of your interest in the investigation," but acknowledged errors in the text. Even a more deeply felt apology, however, would probably not have saved the Harvard-Lowell project. On February 14 Lowell had already written Edward Pickering to protest the "misstatements" to the press that would deny him the "recognition" he was due and threatened to terminate the arrangement with Harvard, which he did three days later.[23]

From the record and the context, the basis of the misunderstandings on both sides emerges clearly. As professional astronomers, the Pickerings assumed that Lowell would defer to their expertise. In this way, the new observatory would become part of Harvard's developing network of stations. From Lowell's point of view, the *Herald* article justified his suspicion that the Arizona project would be presented to the public as an exclusively Harvard expedition.

Percival Lowell's approach to the negotiations illustrates clearly his determination to maintain his own and his project's independence from Harvard, in part as a way of achieving public recognition for its results. In this respect, he followed very much in the footsteps of his Lowell ancestors, who were also wary of collaboration, as the provisions for governing the Lowell Institute indicate. The habit of running affairs without the interference of collaborators was presumably ingrained in the Lowell household, where the son had observed his father's handling of the Lowell Institute.[24]

The conversations that took place in the week following the publication of the *Boston Herald* article laid the foundation for a fully independent observatory. That the conflict between Lowell and the Pickerings could be resolved so quickly is testimony to the urgency of reaching an agreement before the rapidly approaching Martian opposition. In all likelihood Rotch, who spoke with his cousin on February 18 about "troubles with the Pickerings *re* the Arizona expedition," again played a mediating role. According to the new plan, negotiated by Edward Pickering and Lowell, A. E. Douglass and W. H. Pickering would receive a leave of absence from the Harvard Observatory and would look to Percival Lowell for support after March 1. Meanwhile Edward Pickering, at Lowell's insistence, wrote to the editor of the *Herald* to announce the termination of the original agreement.[25]

The new agreement for the "Arizona Astronomical Expedition," announced to the press on March 1, was, except for finances, similar to both the original Harvard-Lowell arrangement and W. H. Pickering's October proposal. The twofold purpose was to examine Mars, the moon, and the planets, and find the best atmospheric conditions for viewing. In return for gaining exclusive control over the expedition, Lowell would pay the salaries of Douglass and William Pickering. Nonetheless he was, in reality, very dependent on Harvard for the leaves of absence granted to his two assistants, as well as for the loan of a 12-inch Clark refractor.[26]

It is also clear that the denouement of this story was an embarrassment to the Harvard Observatory director. As Seth Chandler explained: "E. C. Pickering tried to capture it [the Lowell expedition], but found he had caught a Tartar in Lowell to play as a sucker. E. C. P. found his match . . . and he had to make a rather ignominious retreat." According to Chandler, Pickering's "absorbing and cooperating propensities" failed to work on Lowell and the episode "created a good deal of hilarity among the clubs and scientific 'coteries' in Boston and Cambridge."[27]

The Emergence of Lowell as Observatory Director

Once the Arizona expedition had been established as an exclusively Lowell-funded enterprise, the question of leadership came naturally to the fore. Amateur status notwithstanding, Percival Lowell had already signaled that he intended to control the expedition. Given time pressures, however, Lowell would have to accede to his collaborator's initiatives in selecting equipment and sites for the Arizona observatory. Nonetheless, within a few weeks Percival Lowell had assumed the directorship of "his" institution.

As early as February 6, Pickering, obviously with Lowell's approval, took charge of arrangements for setting up the observatory. He wrote *W. R. Burrows Monthly Weather Magazine* in Tucson, Arizona, to order maps of Arizona and to inquire about weather patterns in the Phoenix, Tucson, and Prescott areas. Pickering was also designing the observatory dome, which was larger than Harvard's but weighed only one-seventh as much. And well before the agreement with Lowell, he had negotiated with John Brashear for the use of an 18-inch refractor to go with the 12-inch one he would borrow from Harvard.[28]

Meanwhile, Pickering was preparing to send his collaborator, A. E. Douglass, to the Arizona Territory to choose a site. The instructions he provided clarify that Pickering hoped to duplicate the conditions at Arequipa in terms of altitude, steady air, and proximity to a city. To avoid cold air currents, a protected site was particularly important. Furthermore, the location had to be "suitable for ladies . . . We do not wish to be attacked by . . . indians or desperadoes." The instructions to Douglass also provided an itinerary and a way of comparing systematically the seeing at different sites.[29]

The fact that Pickering took charge of technical matters associated with the expedition made good sense in view of his expertise and experience. As the planning phase of the project came to an end in early March, however, Pickering's influence diminished. Symbolic of this decline was the naming of the enterprise. When Arizona newspapers repeatedly identified the expedition with Harvard, Lowell became increasingly irritated. Finally, on March 15, he wired Douglass: "Simply call it the Lowell Observatory." He also announced testily that Pickering and Douglass had been released "from all connection" with Harvard in order to join the Lowell Observatory, "of which I need hardly add that I shall myself be director." By mid-May the name was broadcast to outsiders on stationery that Lowell had printed in Boston.[30]

Increasingly Lowell involved himself in the critical decisions of the observatory. He had early on instructed Pickering to introduce him to Brashear by letter so that he could establish direct contact with the instrument maker. He also took over the task of communicating with Douglass about site selection. In this process, Douglass carried out Pickering's instructions and used his seeing scale but reported his findings by letter and telegram to Lowell. In short order Lowell contradicted Pickering's original plan by sending Douglass north, to Prescott and then Flagstaff, without having tested a second site at Tombstone, Tucson, or Phoenix. Later, in Flagstaff, Douglass asked for time to try out various sites after Lowell had already approved one. Permission was not forthcoming. The plan for a systematic exploration of sites was abandoned, probably because of time pressures. Further testing would have delayed preparation of the site for the May 1 arrival of the Brashear refractor.[31]

Lowell's determination to assume the leadership of the expedition helps to explain the curious fact that the observatory's chief architect, Pickering, played only a very brief role in its history, a role that was quickly obscured. Exactly when and how the break between Lowell and Pickering occurred remains a mystery. Certainly their collaboration in Flagstaff got off to a good start. Pickering reported to his mother on June 4 that "I like Mr. Lowell very much indeed . . . and we get on nicely together." Six weeks later, on July 14, William wrote to his brother in a euphoric mood to announce his enthusiasm for the purposes and equipment of the expedition. Only Arequipa, he said, could match the fine atmosphere of Arizona.[32]

Though amicable relations between Pickering and Lowell continued through the summer, by November 1894 Pickering had decided to leave the Lowell Observatory. In response to a letter from his brother encouraging him to stay on at Flagstaff, William Pickering announced that a short stint of another month or so in Arizona was all that he would consider in view of his strong desire to maintain his connection with the Harvard College Observatory.[33]

No doubt Pickering's decision reflected, in part, the fact that the expedition had originally been conceived as a temporary project to collect data about Mars. And as he said, he was reluctant to sever his ties with one of the most prestigious observatories in the world. Beyond those considerations, however, tensions between the two men were surely on the rise. Pickering, after all, never accepted Lowell's theory that the canals were of artificial origin and thus sustained intelligent life on Mars; nor would he have encouraged the swift pace of publication that brought Lowell into the public eye at Pickering's expense. Seth Chandler had made prescient observations in this regard: Lowell "is very much struck on William H.—or was at one time—but it would not be surprising if there were a rift in that lute, impending in the future."[34]

Moreover, Pickering's decision may very well have been encouraged by Lowell. After all, his continued involvement with the expedition reinforced the public's impression that it was a Harvard enterprise. And Lowell was certainly eager to bury his own and others' recollections of this early period when he had funded a project largely planned and implemented by William Pickering. No doubt both men also realized that, in view of their stubborn independence, the prospects for long-term collaboration were dim. Lowell's "unbounded confidence in his ability to keep W. H. P. under his thumb, as he put it" (Seth Chandler's report) must have been tested during the summer of 1894. Finally, Lowell could contemplate the departure of Pickering with equanimity because A. E. Douglass planned to stay on and supply Lowell with the needed expertise.[35]

This conjecture about the relationship between Lowell and Pickering gains credence from Lowell's hostile reaction to an account that Edward Pickering proposed to publish in the Harvard College Observatory's annual report for 1894. It credited William Pickering with the idea for the Arizona expedition while recognizing Lowell's willingness to assume the entire cost of the operation. Lowell upbraided Pickering for wishing to publish an account of the origins of an observatory that had "no connection" to Harvard.[36]

After this exchange with Lowell, E. C. Pickering appealed to Harvard president Charles Eliot for permission to credit his brother's role but averred that "I do not wish to antagonize Mr. Lowell and his family." We may also assume that he had to consider the feelings of other Brahmin patrons. Eliot, in turn, after denouncing Lowell's character, left the matter in the hands of his prudent director, who capitulated to Lowell and published an account without reference to his brother's initiative.[37]

Lowell, meanwhile, had second thoughts about burying Pickering's role as he prepared the introduction to volume 1 of the Lowell Observatory *Annals* in 1898. In a draft, which only came to light thirty-seven years later, Lowell acknowledged that his collaborator had pioneered in recognizing the importance of the atmosphere in advancing astronomical work during the Arequipa expedition, and that that work, in turn, had laid the foundation for the expedition to Flagstaff.[38]

Apparently, however, Lowell reversed himself before the *Annals* appeared, because the published introduction omitted any reference to Pickering's work at Arequipa, and Lowell took credit for "realizing the importance of research" on the subject of atmosphere. The impression Lowell conveyed was that he was the moving force behind the search for good seeing. In a curious turn of events, A. Lawrence Lowell rectified his brother's omission and restored proper credit to William Pickering by publishing the first draft of the introduction in his biography of Percival.[39]

As this analysis suggests, Percival Lowell's contribution to the establishment of his observatory was both financial and conceptual. In creating the observatory, Lowell had to manage a delicate situation at the outset. Given his inexperience, he desperately needed the expertise of Harvard. However, by accepting the Pickerings' assistance, Lowell risked succumbing to the agenda of the Harvard College Observatory. Indeed, Lowell could never be entirely certain that the younger Pickering would not be a kind of Trojan horse for his brother. Lowell's determination, as well as his resources, kept Harvard at bay and put the observatory on an independent course.

In responding to the danger of control from Harvard, however, Lowell overreacted. Acutely sensitive to the matter of personal recognition and perhaps as well to his amateur standing in a time of growing professionalism in the discipline, he attempted to sever all links between his observatory and Harvard and then to obscure any recognition of William Pickering's—and hence Harvard's—role in establishing the observatory. Nonetheless, it was precisely this determination that enabled the observatory to survive as an independent enterprise. Here, the fate of Abbott Rotch's Blue Hill Observatory is instructive. Rotch's cooperative posture toward Harvard from the opening of his station led quite naturally to his decision to have its findings published in the observatory's *Annals* and eventually to his leaving Blue Hill to Harvard's Department of Geology and Geography. By contrast, Lowell's insistence on autonomy from the outset laid the foundation for creating an independent observatory that could set its own research agenda.

Searching for Observatory Sites

Lowell's struggle for the autonomy of his project was only one of several issues he had to face in the early years. Equally important was determining the nature of his involvement in astronomy. By launching the expedition, Lowell had committed himself to the collection of data for the 1894 opposition and nothing more. In the aftermath of that operation, he would be free to choose from a number of options, the first and most obvious of which was to terminate his astronomical career with the publication of data on the opposition of 1894 as implied by the designation "temporary expedition." However, the publication of those results did not preclude the continuation of the project in one form or another. For example, Lowell could establish an itinerant observatory to collect data on Mars for an indefinite number of oppositions. Another option would be to create a permanent observatory operating year-round at Flagstaff or some other site. Such an observatory, in turn, would open up a range of research possibilities, from planetary to stellar investigations, in various combinations. Indeed, one reason for Lowell's confusion about his course over the next few years was that some of these options could be pursued simultaneously over a short period of time, without necessarily precluding others.

The evidence makes clear that Lowell did not regard the 1894 expedition as the first step in establishing a permanent observatory. Aside from William Pickering's proposals for a temporary expedition, Lowell telegraphed the *Astronomische Nachrichten* on July 31, 1894, more than two months into the observing season, and announced, "I have put up a temporary observatory, known as the Lowell Observatory, at Flagstaff, Arizona Territory, U.S.A. . . . chiefly for the study of the planet Mars during the present year." Indeed, all his arrangements were temporary at that point. He had secured the services of Pickering and Douglass for a year only and had borrowed two telescopes. Of course, he had endowed the operation with the name Lowell Observatory, which certainly had a greater ring of permanence than "Arizona expedition," but, as the telegram to Germany indicated, the designation Lowell Observatory did not in any way imply permanent status for the project.[40]

The question that arises, then, is: at what point and under what circumstances did Percival Lowell decide to make the observatory more than an expedition or a series of expeditions to investigate conditions on Mars? Unhappily, no firm answer can be given to this question, because Lowell has left no record of his thoughts on the subject. A. Lawrence Lowell, in his biography, tantalizingly entitled one of his chapters "The Permanent Observatory," but neglected to say exactly when the Lowell Observatory actually became a permanent institution. To shed light on Lowell's intentions, we need to follow closely the decisions he made about staffing, instruments, locations, and research program for clues to his evolving plans for the future of the expedition between 1894 and 1901.[41]

On the basis of such decisions, it would appear that Lowell's sense of his role in astronomy grew gradually over that seven-year period. What began as a single expedition at Flagstaff was extended to a series of expeditions at various locations to collect data about Mars. The momentum established by these expeditions, as well as Lowell's nervous prostration, forced him to consider keeping the observatory open between oppositions and eventually to establish it as a permanent institution. But this decision was not reached until 1901, seven years after the first expedition was launched. In the meantime, he moved toward expanding his research program beyond Mars, to stellar investigations, before settling on planetary research as the main focus.

Once the opposition of Mars had passed in October 1894, Lowell had to determine whether he would maintain a presence in Flagstaff. As Pickering departed, he decided to keep Douglass at work and suffered his first uncertainty about the location of the project in Flagstaff. Seeing grew worse at the end of the year, and through the winter and spring Douglass became virtually inactive as a result of poor weather conditions. This turn of events cast doubt on the desirability of Flagstaff as an observatory location and raised questions about the rather hurried process by which it had been chosen in the previous year. In consultation with Douglass, therefore, Lowell decided to close the observatory in the spring of 1895 and consider other locations. Indeed, had it not been for Douglass's insistence on maintaining the dome on Mars Hill in order not to disappoint the local citizens, Lowell might very well have abandoned Flagstaff completely at this point.[42]

However, it is equally clear that while Lowell was prepared to relocate his observatory, he was developing a strong commitment to the Mars project as well as to the profession of astronomy. Indeed, two important moves in the spring of 1895 suggest that he was seriously considering a long-run, if not permanent, involvement with astronomy. After returning the borrowed telescopes to Harvard and Brashear, respectively, Lowell ordered a 24-inch refractor from Alvan Clark and Sons, the most reputable telescope maker in America. Investing $20,000 in a new instrument—the fourth largest refractor in the United States at the time—suggests that Lowell planned to continue his work for the indefinite future. Moreover, while he was engaged in ordering the telescope, he was also delivering a set of lectures to the Lowell Institute on the observations of the preceding year and preparing those findings for publication. Inasmuch as this presentation was a kind of advocacy for life on Mars, based on only one season of observations, it seemed to invite a search for more evidence to firm up discoveries made in the first year. Indeed, in his conclusion to *Mars,* Lowell acknowledged that, while the existence of intelligent life on Mars was

Lowell at the 24-inch Clark Refractor on Mars Hill (Lowell Observatory photograph)

probable, the form it took was still uncertain. Thus, he wrote, "for answers to such problems we must look to the future." This formulation also implied continuing research on Mars. Moreover, the appendix referred to the volume of annals that was then in preparation and would appear in 1898. That project, too, suggested a longer-term commitment to the study of Mars.[43]

By the following year, Lowell's thinking on the subject of his future had apparently advanced. He was making plans for the opposition of Mars that would occur at the end of the year. And he had spoken to T. J. J. See of the University of Chicago about those plans during negotiations with him to join the staff of the observatory. In his very favorable *Dial* review of *Mars* in the summer of 1896, See spoke of Lowell's plan of attack as follows: "A movable observatory, equipped with an exceptionally fine telescope, to travel from point to point on the earth at successive oppositions of Mars, so as always to secure the best possible locality and atmospheric conditions for observations; and persistent devotion of the observatory to a definite limited line of work."[44]

See's report of Lowell's intentions for the future is borne out by the events of 1895 and 1896. Following disappointment with the atmospheric conditions at Flagstaff in early 1895, Lowell and Douglass developed a plan to explore options for a better observatory site in Mexico, South America, and Algeria. In 1895 Douglass had already proposed a test of "meteorological latitudes," beginning in Flagstaff and moving south to the equator. Because Douglass was under obligation to Harvard, however, his exploratory trip, supported by Lowell, became a hurried and abbreviated version of the original plan. He tested sites only in Mexico, at intervals between Chihuahua and Mexico City. These tests, in his judgment, revealed no site that would match summer viewing conditions at Flagstaff, though their prospects for winter viewing appeared more promising.[45]

In the wake of these disappointing results, Lowell set off to France in the fall of 1895. Through the Société Astronomique de France, he secured the assistance of C. Trépied, director of the Algiers Observatory, to explore astronomical sites in Algeria. Together with Trépied and Ralph Curtis, Lowell visited Boghari and Biskara and made a brief solo excursion along the northern fringe of the Sahara. On the basis of this investigation, Lowell ruled out the Sahara as a possible observatory site for the next opposition. By March of 1896 he was back in the United States, where the new Clark lens was being prepared for the Martian opposition of December 1896.[46]

The rejection of Algeria and Mexico owing to their inferior atmospheric conditions, the completion of the Clark refractor, and the onset of the Martian opposition, in turn, forced Lowell to make decisions that brought him nearer to establishing a permanent observatory. He had decided at this point to mount regular investigations of Mars during oppositions. The need for a home for the new Clark refractor and the desirability of using that expensive instrument for an expanded research program during the intervals between oppositions drove Lowell toward keeping the observatory open on a year-round basis. Such a program would require both an increase in the size of the staff and the range of research projects. We shall see how these factors intersected over the next four years to push Lowell toward establishing a permanent observatory in Flagstaff.

To try out the new 24-inch refractor, Clark and Lowell journeyed to Flagstaff, where they were joined by an expanded staff, including Douglass, T. J. J. See, and two graduate students, Daniel Drew and Wilbur Cogshall. In the late summer of 1896 Lowell conducted observations of Venus and Mercury (about which more will be said later) while See launched his investigation of double stars. Meanwhile, Douglass was dispatched once again to Mexico, where he located a site for viewing Mars near the Mexican National Observatory in Tacubaya.[47]

As it turned out, however, Tacubaya proved to be atmospherically inferior to Flagstaff and, following winter observations, the telescope was returned to Flagstaff. It might appear, in view of this decision, that Lowell, Douglass, and company had overcome their previous doubts about Flagstaff and that they were opting for a permanent observatory. But it is far more likely that the decision to return was based on a desire to exploit the upcoming summer season, when seeing would be good, as well as the need to find a home for the telescope and the staff; the only other option available to Lowell was to send the telescope back to Boston.[48]

This conjecture is borne out by a letter from T. J. J. See to President William Rainey Harper of the University of Chicago, discussing the itinerary of the observatory. According to See, the observatory was to be temporarily returned to Flagstaff and then shipped to Peru. See's report is consistent with what we know about the thinking of Douglass, Pickering, and Lowell on the matter of desirable sites. After all, W. H. Pickering had selected Arizona to duplicate the conditions at Arequipa in a North American setting. And at the conclusion of Douglass's Mexican site tests in the spring of 1895, he had written Lowell to say that "from all I can learn theoretically the west coast of South America is the most promising region in the world." Lowell must have agreed, because just four months later he asked to see Douglass "about South America—for it ought to be begun soon." Unhappily, however, Lowell and Douglass were so preoccupied with other matters that they were unable to explore and exploit a site in Peru for the 1896 opposition.[49]

Lowell's nervous breakdown put a temporary end to consideration of a South American site. In January 1898 William Lowell Putnam, who had been made temporary director of the observatory, wrote Douglass that it was inappropriate for him (Putnam) to "make any radical change," though he acknowledged that "it is extremely probable that [Lowell] would have moved [the observatory] there [to South America] this winter." As Lowell's recovery proceeded, however, he discussed with Putnam the possibility of taking a new assistant, Samuel Boothroyd, a former science student at the University of Chicago, to South America with him. That idea never materialized, but eight years later Lowell did mount a South American expedition under the leadership of David Todd, and he considered another one during the opposition of 1909. Obviously South America remained—for reasons of prior experience and proximity—the preferred location for good seeing. It was also a desirable site at summer oppositions of Mars, such as the one in 1907, when Mars is far south in the sky and can be seen better from southern latitudes.[50]

Meanwhile, other sites for observing the Martian oppositions had been under consideration. In 1895 Douglass, based on information published in *Science,* noted the ideal atmospheric conditions on mountains in the Hawaiian Islands, where the peaks rose above the trade winds. Immediately after the 1896 opposition he received a letter from Lowell, who said, "I think your idea of going to Hawaii excellent," though Lowell had already told Douglass that Hawaii would not be appropriate for astronomical purposes—no doubt because of its distance and the expense of relocating there. On a more realistic note, Douglass stopped at Santa Fe, New Mexico, and Denver, Colorado, in the fall of 1898, to test those sites. All of this activity and the unimplemented plans rule out the possibility that Lowell had decided to establish a permanent observatory in Flagstaff any time before 1900.[51]

Research and Staffing

The decision to carry on a year-round—though not a permanent—operation seems to have been made prior to the 1896 opposition just as the new telescope was completed. The extension of the observing period was closely linked, moreover, to the broadening of Lowell's research agenda. In point of fact, Lowell had considered from the beginning a research program on the solar system, of which the study of Mars was to be a part, as his 1894 pronouncements in the *Boston Commonwealth* indicate.[52]

The observations of Mercury and Venus, launched in the summer of 1896, broadened Lowell's research agenda and provided work during the intervals between Martian oppositions. They also indicate his continuing devotion to his astronomical master; in this, his second project, Lowell would attempt to confirm Schiaparelli's findings for the rotation periods of Mercury and Venus as, in his debut in astronomy, he sought evidence for Schiaparelli's canals of Mars.[53]

The second indication that Lowell intended to expand the observatory's research was his hiring of See, Cogshall, and Drew in advance of the 1896 opposition. See and Cogshall were double-star observers, whose work bore no direct relation to Lowell's original plan to explore the solar system. No doubt Lowell hoped to take advantage of See's still unclouded reputation to enhance the observatory's standing in the world of astronomy. But he had also apparently decided to broaden the research agenda from solar system to stellar research. In this way, he could compete with the stellar research programs of larger observatories. Indeed, See maintained that he had received "from Mr. Lowell an invitation to undertake a survey of the southern heavens for the discovery and measurement of new double stars and nebulas."[54]

That Lowell for a time fully intended to open his observatory to stellar research is evident from later correspondence with Wilbur Cogshall, who worked with See on double stars at the observatory. Seeking to rejoin the staff in 1903, Cogshall asked Lowell whether he had dropped "all the star work from the observatory." Lowell provided a revealing response. "As you know, work here I mean to make planetary as I think it better to concentrate than to scatter. Of course in the future I may add to it but I have no purpose to do so now." It would appear, then, that Lowell arrived only gradually at a definition of the observatory's research program. Not before 1900 did he settle on planetary research as the exclusive focus for the Flagstaff enterprise.[55]

No doubt the elimination of stellar research was not entirely Lowell's decision. The firing of See, following massive opposition to him from the rest of the staff, helped settle that particular issue. In a similar fashion, unanticipated factors, including the decline of his health and the reactions to the research on Venus and Mercury helped to precipitate the decision for a permanent observatory. In the midst of what amounted to a personal and institutional crisis, Lowell abandoned the program of stellar research but affirmed the permanence of the observatory and its location in Flagstaff.[56]

The crisis began with some startling results from the research on Mercury and Venus. Lowell saw spoke-like markings on Venus, which he said were quite different from the canals of Mars, and a pattern of intricate lines on Mercury; none of these markings could be seen from other observatories. And while his assistants also saw the Venus markings, what they saw had nothing like the distinctness of the lines Lowell saw. In the short run, this network of lines was quite useful in permitting Lowell to make determinations for the rotation periods of Mercury and Venus that confirmed the results of Schiaparelli. In the long run, however, they provoked disbelief at other observatories, as well as at his own. The skepticism of astronomers at the Lick and Yerkes observatories about the canals on Mars was now reinforced by serious doubts about Lowell's claims for markings on the surfaces of Venus and Mercury.[57]

These doubts were apparently shared by Lowell himself, in private, as evidenced by his orders to Douglass in 1900 to conduct optical tests on artificial disks, and by his recanting in 1902 of his claim to have seen markings on Venus. Such doubts, moreover, soon extended to the existence of the double canals on Mars. Both Lowell and Douglass had been able to see double canals on the Martian surface from Mexico, under poor weather conditions; however, when conditions had improved the double canals had disappeared. To make matters worse, Douglass, working in Mexico in 1897, detected surface markings on all four of Jupiter's satellites. In short, Lowell's entire astronomical project seems to have come crashing down upon him in early 1897. Serious doubts had been cast on the viability of Flagstaff as an environment for good seeing, and questions had been raised about the validity of the Mars work and other planetary observations. To what extent such doubts may have contributed to Lowell's depression is not clear. But it is certainly plausible that his own hesitations about the validity of the work contributed to the exhaustion brought on by overwork.[58]

And, without question, the coincidence of a crisis in observatory affairs and Lowell's own personal crisis impelled the decision to create a permanent institution. In his own absence, and in view of doubts raised both within and outside of the observatory, Lowell sought to assure the future by ordering his brother-in-law, William Lowell Putnam, to keep the observatory open. Putnam, in turn, assessing the implications of Lowell's illness, decided that a relatively large staff would be maintained, even though Lowell had decreased its size after the 1894 opposition. As Putnam remarked to Douglass, "The staff seems to be larger than it usually was while Mr. Lowell was there, but this is not necessarily a mistake." To do the same work in Lowell's absence would, as Putnam suggested, require extra hands. Furthermore, in a time of crisis, a larger staff would provide a certain stability to the operation.[59]

Lowell's breakdown also helped to assure that Flagstaff would be the permanent location of the observatory. To enhance the stability of the observatory and lower costs, it was advantageous to discontinue the search for a better site. Lowell had given instructions to Putnam to keep expenses below $6,000 per year. Under these circumstances, there was a clear tradeoff between mounting a costly expedition to South America and paying the salaries of two or three assistants at Flagstaff as well as publication costs of the *Annals.* Clearly, by opting for the larger staff, Putnam, with Lowell's approval, was effectively putting on hold the search for better seeing.[60]

Thus, by the time of Lowell's recovery Flagstaff had been operating as a year-round observatory with at least one—and usually more—staff member(s) in residence for the entire period of his four-year illness. The search for alternate sites had been abandoned. Meanwhile, some kind of middle ground seems to have been reached with respect to the research agenda: stellar research would be abandoned, while the solar system, including Mars, would become the focus of research activities. In effect, Lowell Observatory had become institutionalized through a series of decisions and unanticipated developments. It would now assume its place in the American astronomical world as a unique institution in terms of its location, program, instruments, and staff.

But the observatory would also bear the burden of addressing the skepticism of its critics. During the four years of Lowell's convalescence, Douglass and Lowell responded to questions raised about the reality of surface markings on planets and satellites in the solar system. Douglass especially wished to determine, by means of experiments pioneered in Arequipa, whether there were psycho-physical causes for the appearance of surface markings on the planets. Accordingly, artificial planets were set up at a distance from the observatory and viewed through the 24-inch refractor.[61]

The skepticism about the Venus markings clearly changed the direction of research at Lowell. Prior to 1900 the observatory had relied on such traditional methods as drawings by astronomers and use of a diaphragmed refractor, which effectively decreased the size of the lens. However, so distraught was Lowell about the controversy over the markings on Venus that he ordered a new spectrograph from Brashear in 1900 and, within a year, hired an assistant to operate it. In making this move, Lowell hoped to prove that his claims about Venus's rotation were correct.[62]

While Putnam managed the affairs of the observatory in 1897 and 1898, Lowell played an increasingly important role after that. And as he regained enthusiasm for astronomy, he resumed the combative role he had adopted earlier. Doubts about his observatory's discoveries were put aside as he prepared to defend the quality of the seeing in Flagstaff and to make it the home of a permanent observatory. A telltale letter to his sister Elizabeth indicates his own deep attachment to the place: "I am so much at home here of course and yet no one I know knows it."[63]

Percival Lowell jumped into astronomy as a full-time, temporary occupation during the Martian opposition of 1894. His first observations of Mars were made with borrowed equipment and borrowed staff. Only slowly, over a period of seven years, did he build on the observations of Mars to create a permanent observatory in Flagstaff, Arizona, devoted to solar system astronomy. His decision to make Flagstaff his home and planetary astronomy his field of research came about incrementally as he purchased a telescope, tested sites around the world, and expanded his observatory's research program. The personal and professional crisis he experienced in 1897 finally led him to establish a permanent institution.

Lowell's success in creating a new observatory was partly the result of a collaboration with the Pickerings that grew out of the Boston network of scientific and cultural organizations. In considering the Lowell-Pickering relationship as a case of amateur and professional collaboration, it seems to have worked out in strikingly different fashion from other such ventures of the period. After all, wealthy men had funded the Lick and Yerkes observatories but had taken no part in their operation. And Edward Pickering had organized amateur astronomers to play a subordinate role in stellar research. No doubt the Pickering brothers mistakenly assumed that Lowell would be an absentee owner or an occasional observer. If this was the case, it helps to explain why the plan for a joint operation was scrapped and why William Pickering's connection with the observatory was so quickly severed. Clearly, in Percival Lowell, the Pickerings found a collaborator who intended to run the show. As we shall see, Pickering was only the first among elite astronomers who would discover Lowell's aversion to cooperating with their projects. Over the years, Lowell's collaborators would be, like A. E. Douglass, young professionals who had been trained at less prestigious schools and were willing to follow his lead. In this sense, the relationship between amateur and professional at the Lowell Observatory was distinctive by contemporary standards.

The relationship between the two founders was also a product of personalities. In many respects, as Seth Chandler observed, Lowell and Pickering were cut from the same cloth. They shared a strong interest in the search for good seeing and in sensational projects, such as the exploration of Mars's surface, that would bring them public recognition and satisfy their wanderlust. Even after Pickering left the observatory, both men pursued these projects. While Lowell tried out Mexico, Algeria, and Chile, Pickering moved on to Jamaica. Meanwhile, both Pickering and Lowell took up the search for a ninth planet. Given their similarities, they were perhaps a bad match for each other in the first place. In any event, Pickering served Lowell well in providing technical guidance to get the Lowell Observatory under way. But the projects themselves achieved visibility largely through the funding and literary skill of Percival Lowell. He succeeded in presenting astronomical research to the public as an adventure story. And his stubborn insistence on the independence of the Lowell Observatory was a necessary strategy if the new observatory was to survive and gain recognition in competition with its larger, better established rivals.[64]

10

◆ ◆ ◆

Lowell's Campaign for the Canals of Mars

Once recovered from neurasthenia, Lowell launched a campaign on two different tracks to achieve scientific proof for the canals of Mars. On the one hand, he and his assistants conducted a series of empirical tests implemented with conventional astronomical techniques and instrumentation quite familiar to professional astronomers at other observatories. Through spectroscopy, photography, and visual telescopic investigation, the observatory staff hoped to answer the doubts of astronomers at rival observatories regarding the sufficiency of the Martian atmosphere to support life and the adequacy of instruments and atmospheric conditions in Flagstaff to detect the canals if, in fact, they existed.

On the other hand, Lowell, with other scientists, launched an imaginative, interdisciplinary investigation of the evolution of planetary bodies that he dubbed planetology. After acquiring data on the surface features of at least two planets, Lowell intended to compare their development over time and thus to provide insights into the possible existence of extraterrestrial life. Such investigations would also highlight the differences between what Lowell believed to be his daring enterprise and the data-gathering routines of professional astrophysicists.

While the first track required the employment of astronomers trained in using increasingly complicated instruments, the latter would draw on earth and social scientists who could extrapolate from their knowledge of the inorganic and organic development of Earth to the possibilities of life on another planet. At the same time, these scientists, all of whom were personal friends of Lowell, participated in a well-orchestrated campaign to support his claims for the existence of the canals. The prospect of collaboration between Lowell and scientists in other disciplines, however, was clouded by the staunch individualism of both parties as well as Lowell's insistence on the supremacy of astronomy in the hierarchy of sciences. For these reasons, "collaboration" came in practice to mean the collaborators' support and publicity for Lowell's views.

It is hardly surprising that Lowell should have articulated the fighting creed woven into his proposal for planetology in the spring of 1905. He was, at that moment, in Flagstaff preparing for the most promising Martian opposition since the founding of the observatory in 1894. He was fully aware of his opponents' skepticism, which he had encountered after the publication of results from previous oppositions. He could look with renewed optimism to current prospects, now that his assistant Carl Otto Lampland was on the verge of a major breakthrough in photographing the canals of Mars. And he was hopeful that his other assistant, Vesto Melvin Slipher, with the aid of the spectrograph, would be able to confirm the presence of water vapor on Mars. Furthermore, he now believed that the setting of his own observatory in the Arizona desert might prove to be more than fortuitous. The resemblance of the Arizonan and Martian deserts seemed to provide an important clue to unlocking the secrets to life on the red planet.[1]

Empirical Science to the Rescue

Through various experiments, Lowell sought empirical support for his deeply held conviction that what he saw through the telescope was a close approximation of the surface of Mars and that the atmosphere of Mars contained sufficient water vapor to support intelligent life. Spectrographic studies revealed to Lowell's satisfaction the presence of water vapor on Mars. Experiments in seeing convinced him of the reality of the surface detail that appeared through the telescope; photographs recorded the canals, while comparative tests of a single region of the sky at different observatories persuaded Lowell that his claims for the superiority of the atmosphere at Flagstaff were not exaggerated.

Lowell decided reluctantly to conduct experiments in seeing because of the widespread skepticism among astronomers about the likelihood of perceiving surface detail accurately at such a great distance from Earth. Aside from the disturbing effects of Earth's atmosphere, the skeptics noted, lenses in eyes and instruments, as well as misinterpretation of visual detail as it registers on the brain, might introduce significant distortions of any object under examination. As early as 1895 Lowell and Douglass had discussed the possibility of optico-psychic effects that might be tested by examining artificial planets through the telescope. Such examinations had already been launched by Douglass and Pickering in Arequipa and Flagstaff and resumed by Douglass while Lowell was suffering from neurasthenia.[2]

Lowell soon launched a series of his own tests and instructed Douglass to undertake others. On the basis of these tests, Lowell convinced himself that optical effects were responsible for the markings he had seen on Venus. Indeed, he wrote Douglass that "my drawings at Amherst dishearteningly suggest optical results as the rays show by measurement a convergence to the centre of the illuminated disk." And he instructed Douglass to "please make some experiments on artificial disks, unmarked and marked, as nearly the size and brilliancy of Venus as possible." A month later Lowell asked Douglass to observe Venus-like artificial disks with faint and then darker markings under brilliant illumination to determine whether the markings could be seen at a distance. No doubt these tests were the basis for Lowell's admission in *Astronomische Nachrichten* that optical effects were responsible for the markings he had seen on Venus.[3]

The experiments in optical effects continued even after Lowell returned to Flagstaff in 1901. At that time, he drew a double canal where only shading had appeared on the artificial planet. Despite these results and his capitulation on the Venus markings, Lowell's belief in the reality of the canals seems not to have been shaken. With Douglass's departure in 1901, the tests were abandoned.[4]

But not for long. The issue of optico-psychic effects was revived in 1903 in the wake of E. W. Maunder's British schoolboy experiment in which his subjects had drawn a network of lines after viewing at a distance artificial planets inscribed with a series of dots. These results had been challenged in turn when Camille Flammarion, performing the same experiment, showed that different results could be achieved with different subjects. Nonetheless, the issue was sufficiently important that in the same year Lowell deliberately launched a new series of tests in which he viewed a long thin wire at a great distance. In examining the wire, Lowell saw not one, but several lines alongside it; he was reassured, however, by his capacity to distinguish between illusory lines and the wire itself. Furthermore, the ability to see the continuity of the wire at a great distance convinced him that the canals were not at the limits of visibility, as his critics had claimed. In this way,

Lowell satisfied himself that optico-psychic effects did not play a significant role in the perception of the canals and that what he saw through the telescope was a close approximation of the actual surface detail of Mars.[5]

Lowell maintained that good seeing was a matter of the subtle interplay between the eye, the instrument, and the atmosphere. Relying on the work of Douglass, who had given much thought to these matters since 1890, Lowell argued that good seeing in planetary astronomy required a calm atmosphere, an acute eye, and a small, refracting telescope. At the root of his argument was a notion that certain eyes, like certain telescopes, were particularly adapted to light gathering, while others were better at detecting shapes. The acute eye, which Lowell claimed to possess, matched with a good, small refracting telescope, was perfectly designed for distinguishing planetary detail, while the sensitive eye and the large reflecting telescope with which, for example, Edward E. Barnard at Yerkes was equipped, would enable the observer to perceive faint stars and nebula. Whether engaged in planetary or stellar astronomy, however, good seeing was based on a calm atmosphere such as the one that prevailed in Flagstaff.[6]

To convince other astronomers and the public of the existence of the Martian canals, Lowell sought evidence that the atmosphere and instruments at his observatory were superior to those of his rivals. One kind of test that could be persuasive was a comparison between observatories examining the same section of the sky to see which one could detect the most stars. An initial search of the Delta Ophiuchi region conducted with the new 26-inch refractor at the Naval Observatory in the 1880s had recorded 63 stars. R. H. Tucker had then observed this same region with the Lick 36-inch refractor and charted 161 stars in the same area.[7]

Ever the opportunist, Percival Lowell recognized an opening for the claims he was already making about the superiority of the atmosphere and the 24-inch refractor on Mars Hill. The ensuing study, completed jointly with C. O. Lampland in late July of 1905 under what Lowell claimed to be less than ideal conditions, revealed 172 stars. The results of this study were soon published and republished on six occasions in the next two years in both popular and scientific journals. They appeared, however, with no information about the atmospheric conditions under which they had been made. Nonetheless the results buttressed Lowell's claim that the seeing at his observatory was superior to that at any other site, and they also provided, in Lowell's view, an effective rebuttal to Simon Newcomb's claim that the Martian canals were a product of optico-psychic effects. Indeed, in an article in *Outlook,* Lowell specifically rebuked Newcomb ("an eminent astronomer") for doubting the linearity of the canals and presuming to recommend what Lowell had done for years: namely, correct for chromatic aberration of his telescope.[8]

Lowell sought to allay his rivals' suspicions by recording the canals photographically as well as in drawings. This was not a simple task inasmuch as, until 1900, photography had been more useful in revealing points of light than surface details of solar-system objects. Enter Lowell's assistant C. O. Lampland, who made three important innovations that dramatically improved the capacity of the camera to record surface detail on distant objects. He introduced a new optical device to lengthen the focus of the 24-inch telescope; he developed a film holder that enabled the astronomer to rapidly record a number of images while they were clear; and he experimented with different plates, filters, and developing solutions to improve the quality of the photographs. After testing these innovations by photographing other planets, Lampland turned the camera on Mars in 1905 and produced a number of prints on which he and Lowell were able to discern as many as six or eight lines in places where they had drawn canals. The results were quickly announced to the astronomical world and the general public. Unfortunately, no canals appeared in the photographs when they were reproduced, so the public had to depend on the word of scientists to whom Lowell had shown the original photographs of the canals or on retouched versions of the prints. Nonetheless, Lowell was determined to take advantage of the recording power of the camera, and in 1907, during the Lowell expedition to Chile, he had E. C. Slipher take more than 7,000 photographs of the surface of Mars.[9]

Driven in part by the desire to clinch the argument for the reality of the canals by establishing that they could be photographed, Lowell, despite his preference for refractors, was "reflecting on a reflector." Indeed, he was considering the purchase of the world's largest reflector (84 inches) from its most innovative telescope maker, George Ritchey. The logic of using the large reflector, which, according to Lowell, was better adapted for gathering light than clarifying surface detail, was never explained. We may surmise, however, that Lowell's motivation was more complicated than he acknowledged publicly. He must have been eager to upstage George Ellery Hale, whose reflector was being highly touted in the press. Buying a larger reflector for the Lowell observatory would turn attention to Lowell and convince the public that his Observatory was at the forefront of astronomy. Later negotiations with Ritchey ran into difficulties over the matter of price, and Lowell settled instead for a 40-inch Clark reflector. This episode suggests that Lowell was sometimes prepared to beat his rivals by joining them.[10]

Absolutely essential for Lowell in his struggle to convince the public and astronomers of the reality of life on Mars was to show the existence of an atmosphere to support it. Despite Lowell's claims that the Martian polar caps contained significant amounts of moisture, doubts had been cast on this proposition by Campbell's spectroscopic examination in 1894. However, the difficulty in obtaining accurate measurements rendered his results inconclusive, leaving the matter unsettled as the first favorable oppositions of the twentieth century approached. Accordingly, Lowell set V. M. Slipher to work on the new Brashear spectrograph to determine the quantity of water vapor in the planet's atmosphere and to investigate the possible presence of chlorophyll on Mars's surface. Campbell, as it turned out, was fully abreast of Slipher's efforts as a result of correspondence they had been carrying on for several years.[11]

Early in 1908, Slipher got what he considered significant results showing a difference in the critical small *a* band on comparative spectrograms of Mars and the moon, thus indicating the presence of more moisture on the former than the latter. The spectrograms were shown to Slipher's brother, E. C. Slipher, to Lampland, and to Lowell, all of whom agreed that the Mars *a* band was stronger than its counterpart from the moon. However, in view of V. M. Slipher's hesitations about the results, Lowell encouraged him to state the conclusions in a firmer fashion than he was initially prepared to do. Lowell then spread the word of this latest evidence of Mars's habitability to the press, and on April 17, 1908, V. M. Slipher sent copies of his spectrograms to Campbell, Simon Newcomb, and other mainstream astronomers. Newcomb refused to comment on the results, given his own inexperience with the examination of spectrograms, while Campbell could not see the difference in the *a* band on the moon and Mars spectra that was visible to the Lowell observers. Disappointed with these reactions, Slipher climbed to the summit of the San Francisco Mountains near Flagstaff in the summer of 1908 to secure spectrograms above much of Earth's atmosphere, but the weather conditions were less favorable than those under which he had obtained his original results. And after hearing of Campbell's Mount Whitney expedition to challenge his own conclusions, he planned another trip to the mountains for the summer of 1909, but it was never carried out.[12]

Thus, using conventional instruments and techniques of the period, Lowell and his assistants found what appeared to them to be significant evidence supporting the reality of the canals and the habitability of Mars. They had established to their own satisfaction the existence of the canals by conducting a variety of experiments confirming the quality of the Flagstaff atmosphere and the capacity of the eye and the telescope to accurately perceive details at great distances; the canals had been photographed, and evidence of water vapor on Mars had been secured. Of course these results would not go uncontested, but certainly in 1907, many journalists and some astronomers were persuaded of their validity.

Lowell's Collaborators

Despite these successes, however, Lowell's situation, even at the peak of the canal mania in 1907, was perilous, largely because the leading astronomers at the leading observatories remained skeptical of his claims and thus constituted an effective obstacle to persuading the public of the reality of the canals. In the absence of support from other observatories or astronomical organizations, Lowell desperately needed to have authoritative testimony from individuals outside of the observatory staff to corroborate his claims regarding the existence of the canals. For this reason, he turned to maverick astronomers and to friends in other scientific fields. These individuals were all part of the New England intellectual elite, all affiliated with academic or scientific institutions, and well known to each other as well as Lowell. They were drawn to his cause by personal friendship and a longstanding interest in astronomy or in the possibility of extraterrestrial life. Moreover, it is not entirely coincidental that four of the seven colleagues he called upon had traveled in and written about Japan. Their predisposition to support the idea of life on Mars, combined with a strong affection and respect for Lowell, made them quite willing to accept his authority on the subject of intelligent life on the planet, a position that was confirmed by their own experience at the telescope.[13]

These friends served a variety of purposes. At least two of them were trained in allied disciplines—zoology and sociology—that would provide support for Lowell's interdisciplinary approach to the study of extraterrestrial life. Others would agree to testify to the reality of the canals to counteract widespread doubts about them. And all of his friends would be valuable allies in the public debate over the canals. They would trade on their names and prestige to give enthusiastic written and spoken evidence for Lowell's version of life on Mars.

Taken collectively, this group had considerable stature within the larger scientific, if not the astronomical, world. Three of Lowell's friends, Edward S. Morse, Elihu Thomson, and William E. Story, were members of the prestigious National Academy of Sciences, while George Agassiz was the grandson of one of its founders. Morse was a zoologist, director of the Peabody Museum in Salem, Massachusetts, and a longtime collaborator with Lowell on the preservation of traditional Japanese culture; Thomson was an electrical engineer who rivaled Thomas Edison for inventions in this new field, and he served with Lowell as nonresident professor and fellow member of the visiting committee to the department of electrical engineering at MIT; William Story was a mathematics professor at Clark University after having taught first at Johns Hopkins. By contrast, Agassiz was a businessman, philanthropist, and amateur astronomer as well as a member of the faculty at the Harvard Museum of Comparative Zoology, founded by his grandfather, and later of the Harvard Board of Overseers. All four of these individuals were part of the Boston cultural and social scene and members of the American Academy of Arts and Sciences; Agassiz and Morse were regulars at St. Botolph Club.[14]

Lowell also had on board Henry Fairfield Osborn, the president of the American Museum of Natural History in New York, who had been trained as a vertebrate paleontologist and served on the faculty at Princeton. He and Lowell were both members of the Half Moon Club, which met annually in New York with such prominent financial and cultural leaders as J. P. Morgan and Madison Grant. Another Lowell supporter, Lester Frank Ward, was a well-respected sociologist at Brown University. The only professional astronomer in the group was David Peck Todd of Amherst College. His wife, Mabel Loomis Todd, notorious for her affair with Emily Dickinson's brother and known for her editing of Emily Dickinson's poems, was also a popular author and an amateur astronomer.[15]

David Peck Todd, the professional astronomer outside of the Lowell Observatory who was most closely associated with Lowell's defense of the canals, was a genuine maverick in the Lowellian mold. Todd had been interested in the Lowell Observatory from its founding and had planned to visit Flagstaff in 1897, just as Lowell suffered his breakdown; instead, he and his wife hosted Lowell in Amherst in 1899 during his recuperation. Early in 1900 Lowell financed Todd's eclipse expedition to Tripoli and joined the Todds as a witness to this event. Based on this experience, Lowell chose Todd to head the Lowell Observatory expedition to Chile in 1907.[16]

Todd had already been carving out for himself a rather unusual career. His virtual obsession with solar eclipses was manifested in twelve expeditions that he organized and took part in during his life. Moreover, the two Todd eclipse expeditions to Japan, one in 1887 and the other in 1896, had been conducted in the company of his wife and provided the occasion for her to write about both Japan and the eclipses. So it could hardly have surprised Lowell that Mabel Todd once again accompanied her husband on the expedition to Chile or that she was active in disseminating her views of that event in the press and in public talks. Lowell, in fact, attended her lecture at the prestigious Tech Club at MIT, while the Todds were present at Lowell's talk at the home of Mrs. Henry Draper (the widow of the stellar photographer). On November 20, 1907, David Todd himself gave a presentation at the National Academy of Sciences, the audience for which included E. S. Morse and E. C. Pickering. Todd's interest in Mars persisted at least through the next opposition, for in 1909, two years after the Andean journey, *Scientific American* reported that Todd planned a balloon trip to take him above Earth's atmosphere to capture Hertzian wave signals from Mars. Consistent with their enthusiasm for matters extraterrestrial, Todd and Lowell joined one of the early aero-clubs.[17]

While the observation of Mars was a natural extension of David Todd's work as a professional astronomer, Lowell's other friends participated in viewing Mars as a kind of avocation. Morse took time off from his work as director of the Peabody Museum to spend five consecutive weeks in Flagstaff during the 1905 opposition, after which he devoted countless hours to writing his book. And, following the publication of *Mars and Its Mystery,* he was called on by Lowell to deliver public lectures on Mars. The two men, now in their second bout of collaboration, were, in Lowell's words, "fireflies flashing in unison." Similarly, George Agassiz put aside his work as an investor to spend several weeks in Flagstaff in 1907 and 1909 during two favorable Martian oppositions. Agassiz was also a frequent visitor at Lowell's Boston office, where he gathered material for writing and publishing his observations, and despite a reputation for shyness, he delivered a talk on Mars in Santa Barbara, California. His devotion to Lowell, however, may be measured by his willingness to undertake the "retouching" of photographs in order to "clarify" the image of the canals. He had but one condition for performing this dubious task: "It must be thoroughly understood that the results are not direct reproductions from the negatives." *The Century*, however, had scruples about publishing retouched photographs, so Agassiz was let off the hook.[18]

During an earlier favorable opposition (1894), William Story had launched the program of amateur visitations with a two-week stay at the observatory. Story had then represented Lowell's views in an 1896 lecture at the Brooklyn Institute on "Recent Progress in Astronomy" and defended them in writing in a 1904 article for *Knowledge and Scientific News.*[19]

Henry Fairfield Osborn performed a rather different service in supporting Lowell's cause. Although he may never have observed Mars himself, he wrote Lowell: "Your greeting and the wonderful photographs of October

E. S. Morse (left) and Lowell in Lowell's study in the Baronial Mansion, Mars Hill, 1905 (Lowell Observatory photograph)

20th rejoice me greatly. I am taking on a new birth in my interest in astronomy." In 1907 and 1915, Osborn gave over space in his museum for exhibitions of drawings and photographs from the observatory; the 1915 exhibit drew more than 21,000 visitors, and in connection with it, the *American Museum Journal* published articles by Lowell and E. C. Slipher to publicize the work on Mars.

By contrast, Ward and Thomson were drawn to Lowell by a genuine fascination with the idea of extraterrestrial life. Ward visited the observatory for a single evening in the aftermath of a southwestern botanical collecting trip, but the drawing he made on that occasion became the basis for a passionate defense of Lowell's Martian theory; according to Ward's own report, he "saw as much in those few hours as Morse saw in the first week." Thomson, who designed and tested his own telescopic lenses, used a 10-inch refractor at his home in Swampscott, Massachusetts, to observe Mars and then communicate his findings privately, through Morse, to Lowell.[20]

Lowell's eagerness for the support of his friends is evident in the extensive assistance he gave them in viewing Mars and publishing their observations. The bill for the Todd Andean expedition alone came to $3,500, in addition to which Lowell supplied the services of E. C. Slipher from his own observatory as expedition photographer. Slipher, in turn, had to be trained by C. O. Lampland. In comparison, support for E. S. Morse, consisting of his round-trip fare to Flagstaff and five weeks of room and board, was relatively modest. Lowell also hosted Story and Agassiz during their stays at the observatory.[21]

Lowell encouraged his friends to publish their findings on Mars, both to spread his views of the planet and to buttress the case for the canals. To be sure, Lowell threatened legal action against David Todd, whose *Cosmopolitan* article failed at first to credit E. C. Slipher for the photographs of Mars taken on the expedition. As a result, publication of the article, which was quite supportive of Lowell's views, was delayed for two months.[22]

But Lowell went to considerable lengths to orchestrate other publications by his friends. He provided the drawings, photographs, and miscellaneous information Morse used in *Mars and Its Mystery.* When Lester Ward suggested the possibility of writing more than the very supportive article published in the *Brown Alumni Review,* Lowell had Ward's 1901 drawing of Mars traced and sent to him. He also copied Ward's letter endorsing the existence of life on Mars to send to Morse. In that letter, Ward informed Lowell that as early as 1895 he and Morse were already considering the likelihood of life on Mars. While Thomson never declared himself publicly on the canals of Mars, it is interesting to note that Morse shared Thomson's 1907 letter affirming the reality of the canals with Lowell, who copied it word for word in his own handwriting on his personal stationery. In their way, Thomson, Lowell, Morse, and Ward served as a kind of mutual support network in affirming the canals of Mars.[23]

The range of responses to Lowell's work among his friends was quite broad, but a common theme, implicitly or explicitly rebutting Lowell's critics, was the easy visibility of the canals. William Story, in rejecting E. W. Maunder's optico-psychic theory, maintained that in the excellent atmosphere at Flagstaff, the canals "are so easily and certainly seen that there is no reasonable doubt about them." Accordingly, he castigated critics for raising doubts about the validity of scientific observations. "The scientific value of facts would then be liable to complete emasculation by the ignorance, carelessness, or malevolence of an observer." Rejoinders to Story's article were offered by the amateur astronomer W. F. Denning and by Maunder himself.[24]

If the canals were easily seen, as Story claimed, they could not be attributed to Lowell's idiosyncracies. Agassiz made this point explicitly: "Anyone of good eyesight who is accustomed to the use of the telescope" could see the canals. Without question, Agassiz was speaking from personal experience. Indeed, Lowell reported to his brother-in-law that Agassiz was "seeing canals to break the records." Todd, who, according to E. C. Slipher, had poor eyesight and could not detect new or double canals, nonetheless made precisely the same point in his *Cosmopolitan* article: "Nearly everybody who went to the eyepiece saw the canals and once I fancied I heard even the bats as they winged their flight down the pampa, crying 'canali, canali, canali.'" Todd's wife, not to be outdone, agreed about the canals, if not the bats. She spoke of her first glimpse of Mars as "epoch-making" and "an overwhelming revelation" in which a "strange labyrinth of crossing and recrossing lines" was "an unmistakable reality."[25]

On a somewhat less dramatic note, Elihu Thomson argued almost as forcefully about the canals' reality but offered a different sort of evidence. He pointed out that the canals could not have been illusory because they appeared most clearly, both individually and as a network, when the air was steadiest. Had they been illusions, they might have appeared in the telescope even at moments of bad seeing. Apparently he was less sure of the double canals, which nonetheless provided him an opportunity to express his own fears of future gender, racial, and generational arrangements in the United States. Thomson inquired of Morse: "Can it be that the Martians have put their niggers on a reservation canal off to one side? Can it be that the population is getting too dense and the men and women beginning to live on separate canals by edict of the central executive? Can it be that the parallel canal is a training area of youthful irrigators?" In Thomson's view, life on Mars consisted of vegetation that supported some animals but no intelligent beings. While only Todd among these defenders was an astronomer, the prestige of Lowell's advocates must have impressed the public. And their unanimity must have dissuaded more than a few doubters.[26]

A second, less persistent theme stressed the viability of life even in the harshest atmospheric conditions. The Todds, during their Andean expedition, were particularly well positioned to address this issue. Acknowledging that Mars must be inhospitable to life, Mabel Todd argued that it was nonetheless no more so than the Andes at elevations of 14,000 feet, where she had discovered various hardy species. David Todd remarked in his turn that "the more I visit arid areas of the earth and observe the devices of desert dwellers to coax the growth of even the sparcest vegetation, the more the truth of Lowell's theory of the Martian canals impresses itself upon me."[27]

Lowell's friends not only publicized his views of Mars but also secured academic honors in his behalf. The value of such recognition was not lost on Lowell. After all, many of his most vocal critics were attached to institutions like the ones that were now recognizing him. The honors bestowed on him would help to disarm his critics and improve his standing with the public. Accordingly, he had negotiated with MIT for the position of Nonresident Professor of Astronomy. Now his friends would do their part.

It is not entirely coincidental that Lowell should have been awarded an honorary degree by Amherst College in the summer of 1907 while David Todd was managing the Lowell expedition to the Andes. Fifteen years later, Todd explained to his daughter that he had lobbied the Amherst board of trustees in his friend's behalf as a way of fulfilling his obligation to Lowell for supporting the Andean expedition. He conveniently neglected to mention—now that Lowell's theory had been discredited—that he himself had been was an enthusiastic advocate for the canals of Mars at the time the degree was awarded.[28]

No motive was given by William Story for his support of Lowell for an honorary degree at Clark in 1909. However, that event should be placed in the context of the Story-Lowell friendship. The two men clearly admired each other's work and were committed to advancing each other's careers. Toward that end, Lowell helped to finance Story's "Mathematical Journal" in 1904 and supported him for an honorary degree at Harvard in 1902, and, on different occasions, for posts in mathematics at both Harvard and Yale. As secretary of the faculty, Story used his influence to persuade the Clark University authorities that Lowell was a worthy candidate to be awarded an honorary degree during the twentieth anniversary commemoration. On that occasion, Lowell joined fellow degree recipients Sigmund Freud and Carl Jung at the ceremonies.[29]

The testimonials of Lowell's friends to the reality of the canals constituted a minimal form of collaboration. Lowell provided the ideas while his friends became the vehicles for their dissemination to the public. They borrowed his telescope or used their own to record what they saw on Mars without soliciting his views. Of course, they had read his books and seen the drawings reproduced in them. In the end, they were most effective in putting their names and prestige on the line to advance his cause. With one exception, they worked in Lowell's behalf without attempting to alter his concept of life on Mars.

Planetology

The implementation of Lowell's planetology project required collaboration of a much higher order. In this project the issue was not simply the reality of the canals but, more broadly, the geological, biological, and sociological character of Mars. For such a project, Lowell would in principle draw on the work of specialists in those fields like Morse and Ward. However, two factors limited the actual collaboration. First, all of the men in question were strong individualists who resisted collective endeavors. And second, Lowell's hierarchical view of science gave a more prominent place to astronomy than to other sciences; accordingly, and consistent with his practice in cosmogonic work, he borrowed readily from other disciplines without bothering to consult the experts.

Lowell first coined the term *planetology* in 1905, first tested this concept in collaboration with Morse during the Martian opposition that same year, and popularized it in his books on planetary evolution. He defined planetology as "the study of the planets of our solar system considered, not with regard to their gravitational action upon each other, but in respect to their own nature, constitution and history." They were, thus, "evolving worlds" like Earth and could be studied by applying the methods of Earth sciences. In short, planetology was the comparative and interdisciplinary study of planetary history.

While the contemporary historian will be struck by the innovative ring of this formulation—and, indeed, Lowell has recently been recognized as a pioneer in planetology—it was clearly fashioned from familiar elements. Indeed the concept itself was an amalgam of evolution and the application of the interdisciplinary approach. Lowell, as we have seen, was drawn to astronomy in the first place by his attachment to an evolutionary understanding of the solar system.[30]

He was also well acquainted with the interdisciplinary approach as practiced by astrophysicists. Astronomers, with the aid of concepts from physics, had already shifted much of their attention from positional astronomy to a focus on the physical composition of the planets. What Lowell proposed, in effect, was to combine astronomical observations with the insights of scientific disciplines other than physics. In this regard, geology and biology, which focused on the issue of development, were particularly pertinent to research on the evolution of the planets. And sociology provided concepts to explain the nature of society on any planet possessed of extraterrestrial life.

Planetology, as conceived by Lowell, was a pioneering approach to the study of the planets. However, even a cursory review of the 1905 paper in which he first proposed the new field clarifies Lowell's continuing preference for small refractors as the most effective instruments for studying planetary detail. This old-fashioned view was interwoven with the proposal for a new discipline. As we shall see, both strands of the scheme to implement planetology were developed by Lowell to support his own approach to Mars against the relentless efforts of professional astronomers to raise doubts about the legitimacy of his results. He clearly hoped to justify the continued use of smaller telescopes and the focus on planetary, as opposed to stellar, astronomy. The claims for a new discipline would also offer Lowell a way of repudiating the authority of the professionals. In the field of planetology, planetologists, rather than astronomers or astrophysicists, would be the experts. Thus Lowell's version of the surface details of Mars could be authoritatively challenged only by someone well versed in the practices of the new discipline.[31]

Planetology, as an interdisciplinary enterprise, introduced a collaborative dimension into the study of solar system objects. The astronomer, in partnership with scientists from other disciplines, would study surface changes on the planets. Such a proposal, as Lowell pointed out, made eminent sense, but it also raised a number of important questions that he never addressed. Of particular concern would be the nature of the collaborative arrangements and the domain of responsibility assigned to each collaborator. A number of different arrangements were possible. The astronomer could supply information to his colleagues and allow them to evaluate that information; the colleagues could instruct the astronomer in the methodologies of their disciplines, which he could then apply to the data; or the astronomer might encourage his collaborator to collect the observational data for himself. In point of fact, Lowell never specified precisely how this interdisciplinary approach was supposed to work, nor did he ever systematically implement it. We are left to examine the ways and extent to which Lowell, Morse, and Ward actually shared information and theories during their study of Mars.

The brief experiment in attempting to implement the new discipline had several facets—all of them designed to exploit an interdisciplinary approach to the question of life on Mars. One facet was Lowell's effort to gain access to the literature of geology and biology. As Lowell was formulating the discipline of planetology, he wrote to Morse requesting "the best book,—pictures if possible,—on the fauna of this part of the country." In response, Morse recommended Dana's *Manual of Geology* (1875) and C. Hart Merriam's more recent *Results of a Biological Survey of the San Francisco Mountain Region and the Desert of the Little Colorado, Arizona* (1890). They contributed, respectively, to understanding the drying out of Earth and the capacity of the Arizona environment to sustain life. Lowell drew on these concepts to demonstrate the similarities between desert environments on Earth and Mars that were then articulated in *Mars and Its Canals.*[32]

Morse also influenced the direction of Lowell's research on water vapor on Mars. Of course, V. M. Slipher was already engaged in his spectrographic efforts to detect such moisture. Morse suggested, in addition, the possibility of a spectrographic search for chlorophyll on Mars, which Slipher carried out in 1905. Inconclusive results notwithstanding, the experiment indicated the kind of fruitful suggestions an earth scientist could make to support research in planetology.[33]

Morse also assisted Lowell in framing in a more persuasive way for specialists and the general public evidence alleged to prove the existence of life on Mars. Lowell wished to show that what he claimed to see on Mars—deserts and canals—was actually there. One way to do that was to simulate the extraterrestrial perspective by making Earth, rather than Mars, the subject of study: "to see ourselves as others see us." The observer would concentrate on how to distinguish natural from artificial markings, since the case for extraterrestrial life must be based on evidence that the planetary surface had been altered by intelligent beings. The only certain case in which this had happened was, of course, Earth. The issue, then, was how to take advantage of our knowledge of the way intelligent beings had marked Earth's surface in order to interpret the evidence now available on the nature of the markings on Mars.

For this experiment to be effective, of course, it would be necessary for the experimenters to distance themselves from Earth's surface. Otherwise comparing the appearance of the two surfaces would not be meaningful. In the prerocket age, two ways to develop an extraterrestrial perspective of Earth suggested themselves. One was to imagine the experience of looking down on Earth's features from high above. The second was actually to assume such a perspective by taking advantage of high mountains—or, in an unusual case, of a balloon ascent, as Lowell arranged in 1908—to examine the terrain below. Neither course was really satisfactory. Imagination at this level was hardly trustworthy, and even a tall mountain was millions of miles closer to Earth's surface than Earth was to Mars.[34]

Nevertheless the two men demonstrated to their satisfaction the usefulness of this perspective. Lowell believed strongly that the desert location of his observatory would not only promote better viewing through a steady atmosphere but also provide a clearer sense of what a desert environment like that presumed to exist on Mars would actually look like to an outsider. The San Francisco Mountains adjacent to the observatory, standing in the midst of the desert, provided the perfect setting for simulating a perspective on Mars. After undertaking this experiment, Lowell was impressed by "the resemblance of its [the desert's] lambent saffron to the telescopic tints of the Martian globe." As with Mars through the telescope, deserts and forests were reduced at high altitude to "washes of color."[35]

Lowell's exercise in simulating an extraterrestrial perspective was duplicated in Morse's *Mars and Its Mystery.* Morse, however, chose the Verde Mountains near Phoenix as the point of observation and hoped to understand a different Martian phenomenon, namely the effect of changing seasons on the appearance of the irrigation ditches. "The marvels of irrigation are impossible to conceive of without first seeing a parched land before the water channels are dug and the exuberant vegetation springing with the water's advent." The green of the irrigated areas contrasted strikingly with the gray-brown tints of the desert. In this way, the extraterrestrial perspective obtained by both Lowell and Morse suggested that Mars was essentially an irrigated desert like Arizona.[36]

Both men also used the extraterrestrial perspective to establish the clear distinction between markings made by men and those of natural origin. The appearance of railroads and canals as seen from above was strongly emphasized. In the most ingenious section of his book, Morse analyzed the difference between the configuration of lines resulting from natural cracks in various surfaces and those produced by construction projects. In the former category, Morse considered cracks in mud and in Satsuma vases of the kind he had collected in Japan. The irregularity of those cracks was shown on one page, while the articulation of artificial lines formed by railroad, canal, and street systems was laid out on the opposite page. The Martian canals drawn by Lowell and Schiaparelli, resembling other artificial systems, were straight, of uniform width, and converged on a common center. Morse not only mapped these lines for his readers but also actually verified the appearance of the transcontinentals and canal systems by examining them from the mountaintop near Phoenix. Both men also buttressed their argument that the double canals on Mars were products of intelligent life by comparing them to the appearance of transcontinental tracks viewed from above. In this fashion, they satisfied themselves that the surface markings of Earth provided a way to help interpret the lines and spots visible on Mars.[37]

While Lowell and Morse emphasized the detection of surface markings to determine whether there was life on Mars, they only occasionally moved beyond this strategy to consider the nature of such life. Lester Ward, by contrast, hoped to "read" the marks of intelligent life so as to determine the nature and organization of the Martian society. To justify such a step, he reminded Lowell of his own words in *Mars:* "That Mars seems to be inhabited is not the last but the first word on the subject." For a sociologist like Ward, an investigation into this topic was far more interesting than any discoveries that might be made about the terrain or weather on the planet. Accordingly, Ward considered writing a book entitled *The Sociology of*

Mars. And, while he never did so, he could not restrain himself from speculating to Lowell. Essentially, Ward's assessment confirmed Lowell's claims that the Martians must be a superior race who had created a planetwide organization devoted to the construction of the canal network as a matter of self-preservation. In this way, Mars anticipated the future course of Earth and provided "a grand 'parallax' for the study of the philosophy of history." Ward thus hoped to use Mars to gain a new perspective on the history of Earth. His views were, of course, welcomed by Lowell, who encouraged Ward's further research in this area: "It is most gratifying that you should be so interested in Mars."[38]

Planetology was, however, less useful as a concept that could be applied to the study of planets—since so little reliable data was available—than as a platform from which Lowell could attack his adversaries. In this campaign, Morse would be his bulldog, as Huxley had been Darwin's. In defending Lowell, moreover, Morse also became an advocate for the preservation of an older tradition in astronomy. Among the related ideas he endorsed on Lowell's behalf were the desirability of believing in, and the evidence for, extraterrestrial life, the error of modern professional astronomers in resisting this notion and directing their efforts toward the less productive area of stellar research, and the necessity of reporting and interpreting the results of astronomical research to the larger public rather than to specialists.[39]

The resistance to this program was certainly stronger among astronomers than it was in the general public, according to Morse. Indeed, scientists as a rule had been reluctant to take up new ideas, as was evident in their opposition over the last few centuries to every great scientific advance. In this sense, Lowell was in good company. Morse pointed out that Darwin and Kepler had suffered, like Lowell, for their audacity in suggesting new ideas, but that, in the end, they had been proven right. Considered in this light, the attacks on Lowell were evidence not of his mistakes, but of his adversaries' bigotry.[40]

This being the case, Morse berated professional astronomers for failing to take up the issue of extraterrestrial life and for refusing to believe the evidence carefully gathered by Lowell and others. He blamed their skepticism on the conservatism of the discipline, which had moved only recently from positional astronomy to astrophysics and its focus on the composition of heavenly bodies. Planetology now offered interesting possibilities if astronomers would only abandon their obsession with stellar research. "As to the question whether it is more important to add another to the thousands of variable stars and binaries and hundreds of asteroids, already determined, or to consider whether we are alone in the universe and, if so, the significance of it, I think with the intelligent public there can be no doubt."[41]

Like Lowell, Morse also regretted the obsession of scientists with data at the expense of theory. He protested "the way in which the gatherers of the raw material arrogate to themselves the science of astronomy, relegating the thinkers and generalizers to the limbo of speculation." Such an appropriation he regarded as equivalent to letting bookkeepers rather than entrepreneurs run corporations. Data gathering was, of course, necessary, but the data had to be interpreted. This in turn required the exercise of the imagination, although that imagination, as Morse, quoting the British physicist John Tyndall, argued, must be "bounded and conditioned by cooperant reason . . . Newton's passage from a falling apple to a falling Moon was a leap of the imagination."[42]

While astrophysicists presented their findings to specialists, planetology would address a new audience of generalists, including scientists, like Morse, from other disciplines. In this way, according to Morse, planetology would bring astronomy back to issues of great consequence to the educated public and present them in accessible language. Traditionally, scientists popularized out of a sense of obligation to explain, in terms accessible to the educated public, the complex discoveries of their disciplines. The scientist, in this way, talked down to the public. Morse and Lowell, however, maintained that nontechnical science was superior to the work of specialists because it responded to truly important questions and communicated ideas clearly. This process was, in fact, much more demanding than the use of obfuscating formulas. More was required of the scientists and more was gained by the audience when scientific findings were communicated to laymen. A formula could serve as "shorthand," Lowell admitted, but frequently, he believed, it was evidence of "fictitious understanding." Morse was in perfect agreement on this point. He denounced a scientific memoir "accompanied by pages of algebraic formulae which the layman turns over in despair, the only illumination consisting of a few words in English which render the gloom still more apparent."[43]

Morse's attack against the practices of the new specialists was, of course, a way for both men to legitimate their own credentials to enter the debate over extraterrestrial life. They had framed the issue so that astrophysicists and their stellar programs were put on the defensive. If, indeed, the necessary preparation for exploring extraterrestrial life was some combination of observational expertise and knowledge of earth sciences, then stellar astronomers and astrophysicists would be less qualified than Morse and Lowell to undertake this work. So long as the two men collaborated, they were effectively more expert than their adversaries. Thus were the tables turned. Ironically, Morse and Lowell, who had begun by decrying the narrowness of disciplinary boundaries, had drawn some of their own—though they were arguably broader than the traditional boundaries.

The proposal for planetology provided an occasion for Lowell and Morse to make the case for a nontechnical science aimed at illuminating larger issues. They were perfectly convinced that astronomers hid behind the technicalities of their discipline, thus preventing the uninitiated from playing a role. Too frequently, moreover, scientists dismissed the applicability of knowledge from other disciplines in reckoning with questions such as extraterrestrial life. Both men seemed to embrace the notion that science had turned into an obscurantist enterprise that served the interests only of those who were paid to produce results.

While he was preparing a public defense of this position in *Mars and Its Mystery*, Morse was also making the case in private to his distinguished astronomer friend Simon Newcomb: "Without the slightest experience as an observer, I am convinced that a fairly intelligent man familiar with the surface features of the Earth . . . is quite able to decide upon their [Martian surface features'] character and to interpret their meaning whether he be an astronomer or not. It is for this reason only that I am wrestling with the subject." Morse's letter makes a persuasive case for introducing methodologies from other disciplines and emphasizing interpretation as well as data, but his eagerness to set aside experience in observation as irrelevant is startling. On this matter, he went beyond Lowell's claims and surely confirmed Newcomb's suspicions that there were optical and psychological problems associated with the observation of the canals.[44]

The refusal of Lowell and Morse to recognize the barriers to entering an established discipline like astronomy remains something of an enigma. This is especially true for an experienced scientist like Morse, who had worked for years as a specialized researcher in zoology. It would seem that despite his (and Lowell's) professed admiration for Huxley, neither man had taken to heart Huxley's dictum that the ultimate court of appeal in science is "observation and experiment." In this regard, Morse's public admission that the Lowell Observatory janitor had seen the Martian canals more clearly than he is revealing. More surprising is that Percival Lowell, who frequently insisted that criticism of his Martian views should be dismissed because the critics were not specialists in planetary astronomy, allowed—indeed encouraged—neophytes like Morse and Ward to speak for his position on the basis of their very limited viewing of Mars. That practice seems to make a mockery of Lowell's claim that authority in disciplinary matters should be accorded to specialists. Furthermore, the strong emphasis by both men on imaginative interpretation of observations, even at the limits of seeing, was surely a violation of Huxley's principle. Perhaps the best explanation for their conduct is that an intense curiosity about extraterrestrial life and an eagerness to defend themselves against numerous critics overwhelmed their respect for disciplinary expertise.[45]

Lowell and Morse also set a poor example in implementing their proposal for interdisciplinary work. Indeed, there seems to have been virtually no sharing of the expertise that each man possessed in his own field. To be sure, Lowell sought recommendations from Morse regarding appropriate texts in the earth sciences, but Morse, as a zoologist, was surely the wrong person to ask about a geology text. Moreover, Morse's book on Mars rehashed issues already aired in Lowell's writings. In this respect, Morse seems to have duplicated Lowell's work instead of using his knowledge of zoology to interpret the surface features of Mars.

The success of planetology clearly rested on collaboration of scientists from different disciplines. Ironically, however, its chief proponents were strong individualists for whom collective endeavors were difficult, if not distasteful. This barrier to collaboration was reinforced by Lowell's commitment to a Spencerian version of disciplines that made collaboration between equals a virtual impossibility. The astronomer, as the master scientist, worked down from his grand truths to the more specific findings of other disciplines, without necessarily bothering to exchange views with other specialists. Such a conception created an obvious barrier to collaborating with scientists from other disciplines, especially those of National Academy stature.

Nonetheless, while recognizing Lowell's self-interested motives for the creation of planetology, and its flawed debut at Flagstaff in 1905, we must acknowledge that the idea itself was pathbreaking in its formulation. Lowell was surely correct in identifying planetary evolution as a subject worthy of study, and shrewd in sensing that for this kind of subject matter the interdisciplinary approach was appropriate. It may seem paradoxical that, while pioneering a concept like planetology, Lowell should at the same time have promoted the restoration of the gentleman amateur's role in astronomy as an antidote to the modern research observatory. From his perspective, however, the development of large observatories was a step backward. Their grandiose programs, based on narrow specialization and the collection of vast quantities of data, were more likely to constrain than to encourage innovation. On this score, he may have had a point. Seen from our perspective, however, planetology was a good idea whose time had not yet come. What Lowell desperately needed was something like the harvest of reliable data that followed the Viking and Mariner expeditions. Had it been available, however, it would at once have refuted his claims about life on Mars and affirmed the usefulness of large research programs administered by narrowly specialized data collectors.

Nonetheless, even if planetology posed no threat to mainstream astronomers, Lowell's campaign to provide scientific sustenance to the idea of the canals of Mars had scored some victories. The photographing of the canals, the growing evidence of good seeing at Flagstaff, and the spectrographic evidence of water vapor on the planet caught the attention of elite astronomers as well as the general public. Furthermore, the endorsement by major scientists of Lowell's theory gave it an apparent scientific credibility. Professional astronomers at other observatories would now have to address the issues that Lowell had placed in their court.

11

The Establishment Responds

Lowell's apparent success in photographing the canals, providing evidence of water vapor on Mars, and attracting the support of prestigious intellectuals like Lester Frank Ward and Edward S. Morse galvanized the American astronomical establishment to launch a counterattack to refute Lowell's claims for the canals. Leading mainstream astronomers used their positions as power brokers within the discipline to put to rest what they regarded as a serious threat to the public's faith in science. Among Lowell's opponents were current or future members of the National Academy of Sciences, including the directors of the country's most prestigious observatories and their assistants. Their collaboration took the form of experiments and expeditions designed to discredit claims that Lowell had made, efforts to close avenues of publication to him and his assistants, domination of key posts in honorary organizations, and systematic campaigns in the media to undermine faith in Lowell's work. By 1910 there was a concerted effort to isolate the Lowell Observatory and turn Lowell into a kind of pariah among astronomers. Indeed, Lowell's former assistant T. J. J. See, himself already a pariah, viewed the network of mainstream astronomers as the scientific equivalent of the infamous and corrupt Pennsylvania political ring, an alliance of party bosses and railroad officials who controlled state politics at the turn of the century.[1]

At the center of this operation was the highly combative and overbearing director of the Lick Observatory, W. W. Campbell. However, he had the willing and sometimes eager collaboration of other members of the establishment. In the campaign against Lowell, Campbell rarely made a move without first consulting George E. Hale, and in every case, Hale seems to have encouraged Campbell's efforts to discredit Lowell. In addition, C. G.

Abbot of the Smithsonian Astrophysical Observatory, who spent much of his time observing at Mount Wilson, collaborated with Campbell on his Mount Whitney expedition. Also part of this network were E. B. Frost, the rather ineffectual director of Yerkes, and E. E. Barnard, a member of Frost's staff who had been hired at Yerkes by Hale. Frost was uncomfortable with the public confrontation of Lowell and sometimes counseled Campbell to observe caution in his efforts, but he certainly hoped that Lowell's influence could be curbed. Barnard, quite naive in navigating institutional conflicts, managed to maintain good relations with Lowell but was strongly critical of his unscientific approach to the Mars work. On the East Coast, E. C. Pickering, a consummate fund raiser for Harvard Observatory's stellar research projects, and Simon Newcomb, an old-line celestial mechanician, actively resisted Lowell, encouraged their assistants or colleagues in projects designed to challenge his findings, and communicated their views of Lowell to West Coast colleagues. Thus, ironically, one of Lowell's most important, if unintentional, effects was to cement relationships between individuals and observatories who were on many other occasions rivals. The regular communications and planning that arose out of the conflict with Lowell clarified the makeup and the role of what was rapidly becoming a power elite in astronomy.[2]

The Problem Identified

Opposition to Lowell first mounted after his entry into astronomy in 1894. However, his meteoric rise coincided with the end, for a time, of favorable Martian oppositions and the onset of his extended nervous breakdown. Gradually, as Lowell recovered his health and exploited the increasingly favorable oppositions of the 1900s, establishment astronomers took notice once again of the insistent claims for the reality of intelligent life on Mars. In this regard, the 1907 opposition was a turning point. The Lowell expedition to Chile under David Todd's leadership, and the ensuing media blitz, convinced the establishment that Lowell's claims for the canals were based on unscientific methods that had to be challenged to protect the good name of astronomy, in particular, and science in general. Accordingly, they decided to take collective action.

In May 1908, Campbell and Hale had an exchange of letters that clarified their determination to confront Lowell. On May 11 Campbell wrote: "When you visit me I want to talk with you about the Lowell situation.

Perhaps . . . you could put your views in black and white. I think Lowell and Todd are going to be a trial to sane astronomers shortly. My question is just how far they should be allowed to go before somebody steps on their rope." On May 20, Hale responded as follows: "I fully share your opinion that Lowell is doing a great deal of injury to astronomy . . . What I particularly dislike is his absolutely unscientific method of dealing with his material and of stating his case. This is bound to do much harm with the public. If you should decide to express your views, I think you would do an important service to science."[3]

In suggesting that Lowell might do "much harm with the public," Hale was almost certainly concerned about the impact of a heated, public controversy between astronomers on the future funding of establishment projects. After all, Lowell's insistence on the greater significance of his projects might deter donors from supporting the establishment agenda. No doubt Hale, who depended heavily on grants from the Carnegie Institution, also resented Lowell's financial independence, which was reflected, for example, in the lavish publication of the *Annals of the Lowell Observatory.*

The Campbell-Hale letters were written with full knowledge that leading astronomers in the rest of the country were also ready to take on Lowell. E. B. Frost, for example, had written to Hale after Lowell's visit to one of the assistants at Yerkes: "I will tell you later about Lowell's recent visit here . . . Things were kept smooth—but what a specimen he is!" A year later, F. R. Moulton of the University of Chicago wrote Campbell to congratulate him on his recent Mount Whitney expedition to determine water vapor content in the Martian atmosphere. In a facetious tone, Moulton applauded Campbell's efforts to refute Lowell's position: "Your occasional mild reprimands to Lowell have been much appreciated in many places." He added: "In exposing this frenzied science I believe you are doing a real service." Campbell responded by praising Moulton's attack on T. J. J. See in *Science* and encouraging his forthcoming letter in the same journal denouncing Lowell's work. In Campbell's view, it was impossible to argue "along honest scientific lines" with Lowell.[4]

The establishment's campaign to put Lowell in his place and clarify its role as guardian of scientific truth can best be understood by examining six specific cases in which mainstream scientists challenged various practices, results, and publicizing efforts of Lowell and his colleagues. Viewed in isolation, each of these episodes might be regarded as a simple expression of the combative instincts of individuals within the power elite. However, in each case there is strong evidence of planning and collaboration among several members of that group. They clearly intended to orchestrate a coherent campaign against a perceived threat to the emerging world of modern astronomy. Newcomb and E. C. Pickering, for example, took issue with Lowell's claims for the canals on optico-psychic grounds, while Campbell and Frost, in quite different ways, challenged Lowell's claims for the superiority of the Flagstaff atmosphere and instruments. Hale and Barnard used the 60-inch reflector at Mount Wilson to obtain photographs of Mars that showed no canals, at almost the same time as Campbell and Abbot climbed Mount Whitney to obtain spectrographic evidence of the upper limit of water vapor on Mars. Meanwhile, in the East, Pickering, with Campbell's support, attempted to close the telegraphic service that informed subscribing observatories of astronomical discoveries so they could track them to reports from the Lowell Observatory of new details on the surface of Mars. Shortly thereafter Pickering went head to head with Lowell by backing Campbell for election to Simon Newcomb's seat in the American Academy of Arts and Sciences, against C. O. Lampland of Lowell's staff.

Experiments in Visual Interpretation

As we have seen, both the Lowell Observatory and its critics conducted experiments to test for optico-psychic distortions in the appearance of the canals on Mars. One of the most persistent advocates of the reality of such distortions was Simon Newcomb, who early on questioned the existence of the canals. To substantiate his position, Newcomb contacted E. C. Pickering, and together they sought to engage Lowell and Edward Morse as subjects in an experiment designed to cast doubt on the reality of the canals. In this way, they hoped to have Morse and—much more important—Lowell offer testimony contradicting their own claims regarding the existence of the canals.

Newcomb had long been interested in the optico-psychic dimension of astronomy and had interviewed Barnard and W. H. Pickering in 1896 about the reality of the canals. He had then stated in a letter to Barnard that "the only theory I can form to explain these discrepancies [in Pickering's views] with your views is that there are certain arrangements or features of the markings which, with the imperfect means at the command of Schiaparelli and Pickering, are sufficiently near to the appearances of long, dark lines to make the eye judge such to be the effect." Barnard was apparently in full agreement with Newcomb. On a trip through Flagstaff in 1905, he visited with A. E. Douglass, who was then serving as probate judge in Prescott, Arizona. Douglass told him that in visual experiments undertaken while he was assistant under Lowell, contiguous dots drawn on paper and viewed at a distance through the telescope appeared as straight lines. On the basis of such experiments (the results of which Douglass had urged Lowell to publish), Barnard insisted in a letter to Hale that the reports of the canals on Mars, as well as the markings on Venus, Mercury, and the satellites of Jupiter, should be "wiped out from the literature of astronomy." However, a public rejection of the reality of the canals was put on hold awaiting the further advance of Lowell's campaign.[5]

It was not until 1907 that Newcomb decided to pursue this issue further. He agreed with Lowell that tests on schoolboys, such as Maunder had conducted, were inadequate because of the age of the subjects, but he was convinced that the psychology of seeing was a complex business and that the human eye and the telescope could distort markings at a distance. In his correspondence with Lowell he acknowledged that canal-like lines had appeared in the lens of the telescope, but he insisted that even if the canals on Mars were in reality long, straight lines, they would not appear in that form to viewers at great distances from the planet. Lowell, in turn, assured Newcomb that the lines he was seeing were quite clear and unmistakable, and offered his standard defense of the uniquely excellent atmosphere at Flagstaff, which, in combination with a reduced lens and an acute eye, equipped him for successfully viewing the surface markings on Mars. Newcomb's frustration with these claims was reinforced by correspondence with his fellow academician Edward Morse, who had just observed Mars at Flagstaff and published his defense of Lowell's position. Indeed in early 1906, when Lowell and Morse were preparing their books, Newcomb discussed the canals of Mars over lunch in Boston with the two of them.[6]

Newcomb was spurred to devise an experiment in which he hoped to involve Lowell and Morse, as well as other establishment astronomers. To help him with the project, he recruited E. C. Pickering and arranged with him to use the Harvard College Observatory during the March 1907 meeting of the Harvard Board of Overseers, of which Newcomb was a member. The Newcomb-Pickering correspondence reveals that the two men hoped to gather with Morse, Lowell, and other members of the observatory staff to view a series of eight or nine scale drawings of Mars inscribed with some of its surface markings. These drawings would be placed at 100 feet from the subjects and viewed through transmitted light using an opera glass with a magnifying power of two to three. The drawings would be of three kinds: one for the study of lines, three others of Syrtis Major, and the last one to represent the surface of Mars as it appeared in Lowell's drawings. To duplicate the actual condition of the Martian surface, the paper on which these drawings appeared would be mottled. In this way, the experimenters hoped to resolve the question of how a rough-textured surface might affect the appearance and interpretation of lines on it. Each of the subjects would then be expected to determine which of the drawings came closest to representing what he had actually seen through the telescope.[7]

Newcomb and Pickering gave almost as much attention to how they would recruit the participants as they did to designing the experiment. Among other things, Newcomb thought that Morse and Lowell ought to be tested separately "in view of Lowell's sensitiveness and Morse's loyalty to him." Pickering responded that, based on long-standing friendship, he would have no trouble inviting Morse to the observatory. Lowell, however, might be unwilling to accept a Pickering invitation because he was currently embroiled in a dispute with Harvard. Newcomb had his own reasons for not inviting Lowell, including the fact that Pickering was hosting the event. Moreover, Newcomb explained, "altho we are perfectly friendly, he [Lowell] considers, not without some justice, that I am not a warm supporter of his views of the canals of Mars, and might possibly look upon the invitation from me as having a little the air of a challenge." Alphonse and Gaston could not have done any better than Pickering and Newcomb.[8]

Unhappily, neither Lowell nor Morse was present at the experiment. Lowell had to leave Boston for Flagstaff the day before the experiment, while Morse stayed home with a bad cold (perhaps diplomatic flu). Newcomb conducted the experiment using as subjects Harvard Observatory staff members Solon Bailey and the Pickering brothers. Later Newcomb took his experiment to Yerkes, where he tested E. E. Barnard and Philip Fox. The results of the research as reported by Newcomb in the *Astrophysical Journal* confirmed his previously held view that visual inference leads to distortion of surface detail. Broken lines looked continuous, while straight black lines appeared blurred. Furthermore, Newcomb calculated that the canals as Lowell reported them would actually occupy more than half the surface of Mars.[9]

The *Astrophysical Journal* gave space to Lowell for a rebuttal in which he criticized Newcomb for using transmitted rather than reflected light and insisted that he (Lowell) had not used black wire in the experiments he had conducted to test the visibility of lines at various distances. In short, Newcomb's experiment changed no one's mind. Lowell and his supporters continued to assert the accuracy of their image of the Martian surface, while the establishment, already doubtful about observations at the limits of seeing, was increasingly persuaded that surface features on Mars were distorted by distance and other factors. Nonetheless, the experiment offered an occasion for the astronomical establishment to demonstrate its solidarity at the expense of Lowell.[10]

The *Outlook* Controversy

At the time of Newcomb's experiment, W. W. Campbell was already privately embroiled in conflict with the Lowell Observatory. He was incensed by Lowell's repeated assertions in scientific and popular media of the superiority of seeing at Flagstaff as compared to Lick. Moreover, some of these articles cast aspersions on Yerkes Observatory, with which Lick had a close working relationship. Inasmuch as Lowell's comparisons had appeared in several journals on both sides of the Atlantic, Campbell felt that the time was ripe to defend his turf. Nonetheless, he was cautioned by E. B. Frost that efforts to combat Lowell's claims in the media would be fraught with difficulties. "Lowell is a rather agile party to deal with, and I think he will enjoy entering into any kind of scrap that you may (in his opinion) open up." Frost agreed, however, that an "authoritative correction" of Lowell's views would be of service to the less thoughtful reading public.[11]

Frost tried to settle his own differences with Lowell behind the scenes. In late 1907 or early 1908 he wrote Lowell to ask what was the basis for his claim regarding the superior power of the 24-inch refractor at Flagstaff as compared with Frost's own 40-inch refractor at Williams Bay. Lowell indicated that he had reached this conclusion by comparing the times required for spectrographic exposures and by drawing on George Ritchey's comparison of the relative visibility of test stars at Lowell and Yerkes. After checking with Hale to see if Ritchey could verify the comparison, Frost decided not to contest Lowell's claims in the media. And Campbell agreed that he would leave aside the charges Lowell had made against Yerkes as a matter between Frost and Lowell. Meanwhile, though, he assured Frost of his continuing belief in the quality of the Yerkes instrument.[12]

In early September 1908, Campbell prepared rebuttals to send to four journals that had published Lowell's claims. In each of these letters, Campbell focused on what he claimed was Lowell's unprofessional conduct in publicly insisting on the superiority of the seeing at Flagstaff as compared to Lick. Campbell also denounced as unscientific Lowell's failure to report the phase of the moon and the actual weather conditions during his observations. With this critique, Campbell incorporated information about a new star survey conducted by two of his observers who were much more experienced than his original surveyor, Richard Tucker, with the 36-inch refractor. Their success in turning up new stars was greeted enthusiastically by George E. Hale. "I am glad to hear that [Charles D.] Perrine and [Robert G.] Aitken have observed so many stars in the region mapped by Lowell. It was, of course, a foregone conclusion that the 36-inch would show more than his instrument, but it is a good thing to be able to prove it."[13]

The above arguments were repeated in four of the six periodicals in which Lowell had published his claims. In addition, Campbell wrote a letter to *Popular Astronomy* that contained an interesting proposal to settle the issue between the observatories. He suggested that Lowell and Lick invite a disinterested astronomer to view stars in an unphotographed region of the sky, after which the results would be compared. No response to this proposal came from Lowell, however.[14]

When *Outlook* (October 3, 1908) published a similar letter from Campbell, though omitting the proposal for a comparative test of seeing, it provoked a furious response from the normally very private V. M. Slipher. Already smarting from Campbell's rather negative reactions to his study of water vapor on Mars, Slipher was certain that Campbell had not given Lowell his due in the *Outlook*. He was accordingly prepared to respond in print, but sought Lowell's approval to do so. He also wanted to counter Campbell's proposal to test the seeing conditions at the two observatories by proposing spectrograms of "the same stars on the same kind of plates and developed at exactly the same time and with the same developer." Lowell, however, rejected that idea on the grounds that Campbell would have good excuses for refusing to participate in the contest and would present the invitation as evidence of Lowell Observatory's desperation. Lowell was, in any case, convinced that Campbell's rebuttal was actually a form of flattery for him, because it called attention to his keen-sightedness as compared to Tucker's—"a fact which astronomers generally will bear in mind when the next martian details are criticised."[15]

In his reply to the *Outlook*, Slipher acknowledged that the Lick astronomers had mapped more stars than Lowell had, but pointed out that Lowell was handicapped in the survey of star fields by virtue of being a planetary astronomer with no experience in charting stars. In addition, Slipher changed the grounds on which the case was to be contested. Instead of arguing that more stars could be seen at Lowell than at Lick in a particular region of the sky, he claimed that was the case *proportional to the size of the two telescopes* (my emphasis).[16]

These claims infuriated Campbell, who was wrongly convinced that Slipher's letter had actually been written by Lowell. Indeed, he later referred to Slipher as "Lowell's man Friday." Campbell also hastily drafted a rebuttal to Slipher, copies of which he sent to Yerkes Observatory director Frost, to President Benjamin Wheeler of the University of California, and to the *Outlook.* This letter has disappeared, but the revised version of it, dated May 29, 1909, almost certainly reflects the tone of the initial draft. In this later version, Campbell denounced Slipher for changing the basis of the argument and for maintaining that the 36-inch refractor, on the basis of its large size, should be able to locate stars that might not even exist; he then concluded: "The gratuitous publication, in purely popular journals, of such claims as these of Mr. Lowell and his assistant (clearly intended to promote his affairs in the popular mind at the expense of another institution) constitutes an extreme case of what physicians and others call 'unprofessional conduct,'—even if the claims were true and they are false."[17]

When Campbell's letter of rebuttal was rejected by the *Outlook,* he wrote the editor an accusatory letter claiming that the magazine was prepared to mislead ninety-nine readers rather than do justice to one injured person. Furthermore, he alleged that the editor had deliberately halted an exchange of letters at a point that would support claims made in Lowell's original article. Meanwhile, Frost weighed in with a warning that Campbell's second rebuttal would perpetuate a futile exchange unlikely to settle the issue at hand. President Wheeler, while also concerned that Campbell's response might backfire by generating boredom in the public, agreed to forward the letter if Campbell omitted his attack on the *Outlook* for publishing "unprofessional" charges. To Wheeler, Campbell wrote: "I do not like to lie down in the presence of the enemy . . . Lowell is an abnormal man, who will continue his unprofessional publications concerning his own greatness and the weaknesses of others, if he is not dealt with." President Wheeler apparently agreed with Campbell: "The only question can be whether it is now time to break up Lowell and slip the pieces out of sight. He is evidently a misguided soul misguiding others."[18]

The *Outlook*, however, rejected the revised letter, and the controversy terminated with an interesting exchange of letters between Slipher and Campbell in which Campbell, still smarting over the *Outlook* exchange, took advantage of Slipher's request for information about the Mount Whitney expedition to raise the former issue. In his letter, Campbell expressed particular irritation with Slipher for misleading the public into believing "that telescopes are efficient in proportion to size." He also announced that Lick had conducted a photographic survey of Delta Ophiuchi that showed that "the stars of the magnitude you describe, to the number you describe, do not exist."[19]

Slipher, in response, explained that he had written to the *Outlook* to protest what he felt was the unfair treatment of Lowell meted out by Campbell. In an apologetic mode, Slipher confessed, "I deplore controversy" and said that immediately after sending the letter to the *Outlook*, he had "wished it back." Further, Slipher acknowledged that his own claims had oversimplified the problem of viewing, which included many factors besides the quality of the air.[20]

Campbell, insisting that he was not looking for the last word, responded to Slipher's response by adopting a military metaphor to describe the Lowell-Lick conflict: "Expressed in military language, he [Lowell] in this took the offensive attitude . . . It was my plain duty to make a *defensive* reply." After reviewing the conflict, Campbell came back to the real issue, which had too frequently been obscured. Lowell had launched the claims in the first place to convince the public that he could see delicate details on Mars. For that reason, neither Lowell nor Slipher had a right to complain if others expected them to be able to see faint stars as well.[21]

The military analogy was clearly appropriate. The effect of this war of words on the public is uncertain, but it seems clear that it served as a nice rallying point for Campbell and other mainstream astronomers, who certainly agreed—as even Slipher did—that astronomical issues should not be aired in popular magazines and that no observatory should tell another one what it should or should not see.

Challenging the Existence of the Canals

Given the previous work of such well regarded professionals as E. E. Barnard and James Keeler, the astronomical establishment was not inclined to take Lowell's drawings of the canals or even Lampland's new photographs at face value. After all, comparable drawings by Barnard and Keeler featured a plethora of detail, but the canals were conspicuously absent. As the Mars debate heated up again, observers hoped to challenge Lowell, Todd, Morse, Agassiz, and Ward by offering their own drawings of the Martian surface, supplemented by their own photographs. The establishment would also seek testimony from former and present Lowell observers and renowned scientists from other disciplines, and use their own viewing of Mars and interpretations of the Lowell negatives to cast doubt on the reality of the canals.[22]

By the opposition of 1909, the question of the reality of the canals of Mars was of such importance that Hale himself set aside time from his solar observations on Mount Wilson to view and photograph the planet with his 60-inch reflector, then the largest in the world. He reported to Campbell: "I observed Mars under fine conditions with the 60-inch, not long ago, and failed to see his [Lowell's] narrow linear canals. As a matter of fact, I was able to analyze one of those so-called canals into its constituent parts and found it totally unlike his drawings."[23]

During the 1909 opposition, Hale not only viewed Mars but also made "a few photographs . . . with the 60-inch," which he promised to send to Campbell. "You may be surprised to know that they fail to show any trace of the canals!" Three weeks and numerous photographs later, Hale had still not obtained any images of canals. "Apparently the magic atmosphere of Flagstaff or the imagination of Lowell are the things we really need." Hale also reported the same news to Frost and added, for good measure, that the seeing was good enough to enable him to obtain photographs of the moon showing "smaller details than any previous photographs."[24]

The appearance of the canals in Lampland's photographs had been disputed as early as 1905 by a number of astronomers who saw the original negatives. Campbell himself was quite skeptical, as was Barnard, who, during a visit to Flagstaff, inspected Lampland's negatives on which the canals were supposedly recorded and spoke with A. E. Douglass. Barnard, noting the defects in the focus of the photographs, then reported to Hale: "I am perfectly sure you would have agreed with me that they did not show the canals as claimed." Hale replied, "I would rather have your opinion of the planet than anyone else," and said he regretted that "the facts are not generally known, since everyone will now be convinced of the reality of the canals on the supposition that photographs cannot lie."[25]

It is interesting to note that neither Hale nor Campbell took an interest in the reproduction of the canal photographs in the *Proceedings of the*

Royal Society of London until October of 1909. When Hale finally examined the reproductions, he wrote Campbell to question whether these photographs were "genuine." He asked rhetorically: "How could such experienced observers as you and Barnard fail to see these conspicuous markings?" Campbell's response reflected his own doubts about the Royal Society reproductions as well as Lampland's original Flagstaff photographs, on which he could see no evidence of the canals. Indeed, the grain of the plate was "several times larger in diameter than the width of the drawn canals." Campbell concluded: "I still believe that your photographs with the 60-inch are greatly superior to any results made with other instruments."[26]

Meanwhile, outsiders were helping to confirm Hale and Campbell in their skepticism about the canals. It was, no doubt, especially satisfying to have Lowell's former assistant Douglass on Mount Wilson confirm that no canals were visible through the 60-inch. Meanwhile, Godfrey Sykes, an English mechanic and the builder of the dome for Lowell's 24-inch refractor, told Hale that he had seen the canals through the same 24-inch, but found no evidence of them in Lampland's photographs. Much more impressive, as Hale acknowledged, was the testimony of Nobel lauriate Albert Michelson; he had visited Flagstaff, and he could not detect the canals on photographs shown to him.[27]

By 1910 Hale, with Barnard's support, had decided to take the issue to the next level. "At the National Academy meeting I shall pay my respects to Lowell by describing my recent observations." As it turned out, the response to his talk was more gratifying than he had initially expected. At the meeting he encountered two of Lowell's defenders, Edward Morse and Elihu Thomson. In the discussion of his paper, Thomson admitted to Hale that "under the best conditions these [the canals] appeared to break up." Morse, however, was not converted to Hale's position.[28]

Despite these triumphs, Hale still worried that he had not yet matched the quality of Lampland's photographs of Mars. To provide further evidence refuting the existence of the canals, he invited Barnard during the next opposition to "attack the subject seriously with the 60-inch reflector." In point of fact, E. E. Barnard was already a supporter of Hale's 1909 images of Mars and, in any case, would have been reluctant to challenge someone of Hale's stature. Nonetheless, Barnard hoped that the new opposition would provide an opportunity to "make out some reason for the existence of the canal system as reported from other quarters." By late November he had completed the observations of Mars that confirmed his 1894 version of the planet. "No one could accurately delineate the remarkable complexity of detail of the features which were visible in moments of steadiness," he announced. Of greatest importance: the canals were nowhere in evidence.[29]

Even as the establishment was attacking Lowell's views of Mars, Lowell himself had turned his attention to Saturn and reported the appearance of knots, or tores (doughnut-shaped surfaces) on the planet's rings. This claim brought out the fire power of the establishment once again and led back to reflections on Lowell's Mars observations. Barnard wrote Hale as follows: "I see that Lowell is still on a 'tore' or should I say a 'tare.' His 'invisaging' is 'forthrightly' doing up things in the shape of Mars and Saturn—poor Saturn! One has no sympathy for Mars who has always been guilty of lending himself passively to all the Brenners, etc. that have held out their hand to him."[30]

The fact that the establishment and Lowell let pass a 1913 invitation from W. H. Pickering to observe Mars simultaneously from different observatories may indicate that the issue was now regarded as closed, or that there was little trust in Pickering as a reliable administrator of such a project. Both Campbell and Hale joined Lowell in rejecting Pickering's invitation.[31]

Water Vapor on Mars

While some astronomers were challenging Lowell's observational and photographic enterprise, others were working to undermine Slipher's findings on water vapor. As soon as he received the Mars-moon spectra, Campbell showed them to Charles G. Abbot and Hale, neither of whom could make out any difference in the critical small *a* band. These skeptical responses no doubt encouraged Campbell to attempt to settle the issue once and for all and thus finally to discredit Lowell's Martian project. He determined to make a high-altitude test where the moisture of Earth's atmosphere would be less likely to affect the outcome of the spectrographic work. To ensure the success of the expedition, Campbell made an exploratory trip to Mount Whitney in 1908 and another visit to photograph Mars during its opposition in 1909. For this project he had the support of both Frost and Hale. Indeed, Frost wrote to volunteer the assistance of James Wallace, who had helped Lowell with Martian photographs. The expedition was also endorsed by University of California president Wheeler, whom Campbell warned, however, to avoid newspaper publicity, which "would be unfortunate and make my life miserable."[32]

The results of the expedition, as Donald Osterbrock has shown, indicated clearly that there was no difference in band *a* in the moon and Mars spectra. These results in turn suggested a very low upper limit for water vapor on Mars. As the news of Campbell's findings reached establishment astronomers, they responded with uniformly lavish praise. Hale wrote: "It is immensely satisfactory to have the question of water vapor settled once for all. I am curious to know what attitude Brother Lowell will take." From Chicago, F. R. Moulton wrote facetiously: "I am quite surprised at them [the results] as from the earlier accounts I had seen I could only draw the conclusion that you had gone up the mountain for the express purpose of verifying Lowell's conclusions. I also was inclined to get the impression that you regarded Lowell 'as the foremost astronomer of the world to day' as he is described in the September number of Popular Mechanics." Frost requested that Campbell publish his results in the *Astrophysical Journal*, as Slipher had done, but Campbell preferred to use the *Lick Observatory Bulletin* as the vehicle for his findings, to avoid direct controversy with Slipher.[33]

A second test of the presence of water vapor on Mars using the Doppler method confirmed Campbell's original finding. He chose this method because it offered a way to separate the lines on the spectrogram resulting from water vapor in Earth's atmosphere from those representing water vapor on Mars. Such a separation occurs when Mars has a high radial velocity relative to Earth. Campbell timed his spectrogram accordingly. The confirmation of the Mount Whitney findings that he achieved was very satisfying to Hale, who wrote Campbell on no less than three occasions to praise him for settling the issue definitively. Significantly, while Hale was delivering a paper on his 1909 photographs of Mars to the National Academy of Sciences, he brought along the two Campbell papers dealing with the water vapor question to present to the same audience. However, as he reported to Campbell, time ran out and he was unable to deliver "your very important papers."[34]

Not surprisingly, Lowell was far from pacified by Campbell's intervention in the Martian question. The Mount Whitney expedition, according to Lowell, had been inconclusive because of poor atmospheric conditions on the mountain during the spectrographic work. Abbot, in response, explained that the spectrographic tests had been made only after the bad weather had abated, and Campbell wrote scornfully: "It should be seen by a blind man, even though he carry a brief for intelligent life on Mars, that the more moisture there was in our own atmosphere, the less there could have been in the Martian atmosphere."[35]

With regard to the Doppler test, Lowell, through his friend George Agassiz, was able to embarrass Campbell by revealing that his observatory had beaten Lick to the punch; as reported in a *Lowell Observatory Bulletin*, Slipher and Lowell had been the first astronomers to try the Doppler method for purposes of detecting water vapor on Mars, though their results had been inconclusive. Campbell explained in somewhat red-faced fashion that he had been in Europe when that issue of the *Bulletin* had arrived at Mount Hamilton in 1905. As further justification for his failing to take note of Lowell's initiative, he pointed out that the *Lowell Observatory Bulletin* had no table of contents.[36]

The water vapor issue became, in effect, another occasion for collaboration among the establishment scientists, particularly the California branch. Campbell undertook his expedition with the full support of Hale, who eventually trucked Campbell's papers reporting the results to Washington. And Campbell's collaborator Charles Abbot, although he was the Smithsonian's solar astronomer, spent much of his time working at Hale's Mount Wilson facility.

The Harvard Telegraphic Service

Lowell had found himself up against an alliance of E. C. Pickering and Simon Newcomb in the matter of psychology of perception. He again faced Pickering in the struggle to determine which observatory would control the telegraphic service providing rapid distribution of information about new discoveries to observatories around the world.

The telegraphic service, originally a Smithsonian prerogative, was given over to Harvard Observatory in 1883. Pickering then selected John Ritchie, Jr., a science journalist who wrote for various Boston newspapers, to administer the service. However, in 1905, when it became apparent that Ritchie was assuming too much independence and distributing telegrams without crediting Harvard, Pickering fired him and managed the distribution directly from the observatory. After Campbell was informed of Ritchie's dismissal, he expressed his approval of the decision on the grounds that announcements sent by Ritchie duplicated those from Harvard.[37]

The very next year, Pickering moved to transform the service because of the increasing numbers of astronomers "interesting themselves in observation of comets such as [C. A.] Lovett, [H. R.] Morgan and Lowell." Pickering proposed to limit the number of reports on comet positions, ostensibly to avoid the expense of sending out too many telegrams. However, Pickering wished to clarify that these limitations should not do "anything which would deprive subscribers of any of the Lick observations, since they are much more valuable than most of the others which are received." Some of Lowell's observations, by contrast, were apparently considered less valuable. In 1907 Pickering notified Lowell that the Harvard service would not distribute notices from Lowell Observatory regarding the appearance of surface features on Mars. This decision was interpreted by Lowell as a deliberate effort to limit recognition for his Martian work. Pickering justified his decision as follows: "The object is to communicate to correspondents announcements concerning objects which require immediate observation by other Observatories. It is not intended to cover announcements of general interest which do not call for immediate action on the part of observers elsewhere." Such notices, he explained, would be sent as bulletins by Harvard.[38]

By return mail, Lowell cancelled his subscription to the telegraphic service and that same day invited John Ritchie to take up, on his behalf, the transmission of telegrams to other observatories in the United States and also to Kiel, Germany, where H. Kobold, the editor of *Astronomische Nachrichten,* had been sending out news of discoveries in the United States to European observatories. Lowell knew that he would find in Ritchie a sympathetic response, since Ritchie was a staunch opponent of Pickering's astrophysical research program. And sure enough, in his reply to Lowell Ritchie complained: "Pickering has grabbed the distribution business—and shoved me out into the cold." All the same, Ritchie did not wish to confront Pickering directly but proposed that, if Lowell should offer a prize to discoverers of comets, he could wrest the distribution service from Pickering. While Lowell applauded the idea, Ritchie's refusal to take up the distribution service himself left Lowell with no one to implement the awards for comet discoveries.[39]

Instead, Lowell contacted Kobold to arrange for the distribution of planetary news from America, supplied and paid for by Lowell, to European observatories; Lowell himself would take up the distribution of planetary news in America. At this point, Lowell did not envision infringing on Pickering's distribution of cometary news. In late 1909, after Lowell refused another offer by Pickering to distribute planetary news by mail bulletin, Pickering—with collaboration from Campbell—sought to take over the distribution of planetary news. Lowell, in response, brought the matter to the attention of the newly inaugurated Harvard president—his brother—and then to the *Boston Post*.[40]

The letter to the president asked him to intervene to prevent Pickering from "poaching" on his (Lowell's) new planetary distribution system. In his reply to Percival, Lawrence Lowell mailed a copy of an undated article from the *Post* and noted in pointed fashion that Harvard had not been the source of the information on which the article was based. Lawrence Lowell's letter carried the unmistakable implication that the *Post* had received information directly or indirectly from Percival Lowell, and the tenor of the article bore out the implication. The writer explained that Harvard had been the "graveyard" of astronomical announcements from the Lowell Observatory because of "friction" between Lowell and Pickering "as to the value of the work in planetary discoveries by the former." The article went on to discuss Harvard's deliberate plan to discredit the work of the Lowell Observatory—a plan that had been frustrated by the observatory's growing prestige. Harvard, in turn, had been discredited by the success of Lowell's distribution arrangement with Kobold in Germany. Having revealed the long-standing rivalry between the Lowell and Harvard observatories, the article then clarified the implication of this rivalry for Harvard's new president. "Whether the astronomer will have the moral support of his brother or whether the president of Harvard will decide for duty as against brotherly love is one of the questions that will confront the new president." Although the article was unsigned, it perfectly reflected the anti-Pickering bias often expressed by John Ritchie, Jr.[41]

The significance of the Pickering-Lowell struggle is clear enough. Pickering—with the support of Campbell—hoped to use his telegraphic service in such a way as to deny recognition to Lowell's Martian work, while Lowell hoped to circumvent Pickering by launching his own distribution center, thus calling attention not only to discoveries at Lowell but also to the indispensable service he was providing to the astronomical world. In this particular episode, Lowell succeeded in the short run, but his efforts cemented the alliance between Campbell and Pickering that would soon thwart Lowell in his campaign to control elections at the American Academy of Arts and Sciences.

The American Academy of Arts and Sciences

Lowell and Pickering became involved in yet another struggle for prestige and recognition, this time in connection with elections to the American Academy of Arts and Sciences. Both men had been members for years of this organization, which had strong regional and local roots. Their contest concerned the election of a nonresident astronomer to replace Simon Newcomb, who died in 1910. Of course, the American Academy by 1900 had taken a backseat to the National Academy of Sciences, in which membership conferred much greater prestige. But since Lowell and his associates had been virtually locked out of that academy by the opposition of the establishment, the American Academy provided a next best avenue of recognition for the observatory, and one over which Lowell could exercise some influence. He had, after all, been a member since 1891 and was related to, or a close friend of, at least half a dozen other members. Indeed, by 1910 he had already succeeded in having V. M. Slipher elected to the academy.[42]

To replace Newcomb, Lowell proposed that the membership select C. O. Lampland. However, his proposal was not to go uncontested. E. C. Pickering chose to advance the nomination of his California colleague W. W. Campbell. Lowell responded to Pickering's efforts by circulating a petition to establish as academy policy the election of individuals solely on the basis of their achievements as opposed to the official positions they occupied. This petition gained some support, but W. E. Story, who circulated the petition in Worcester, Massachusetts, and found all seven of the city's academy members in support of the principle but opposed to signing the petition, suggested that Lowell abandon the campaign. (It is interesting to note that among the eight signers of the petition were Lowell; his cousins Eliot Clarke and A. L. Rotch; his best friend, Barrett Wendell; his journalist crony, Ritchie; and the curator of the Lowell Institute, W. T. Sedgwick.) Story advised Lowell instead to raise the issue of the proper criteria for candidates for the academy in a meeting of the membership and in that session to clarify the nature of his reservations about Campbell. Unfortunately, academy records and the Lowell Observatory correspondence do not reveal whether Lowell pursued this option.[43]

Lowell's campaign against Campbell also featured a charge that Campbell had deliberately misrepresented the Lowell Observatory in various conflicts—especially the water vapor issue—over the past few years. To provide chapter and verse on this point, Lowell urged Slipher not only to write in favor of Lampland, which Lowell himself did as well, but also to dig up as much dirt as possible on Campbell. This request, however, yielded no effective ammunition for Lowell's campaign. Apparently tiring of the endless combat between Lowell and other astronomers, Slipher wrote a rather perfunctory letter noting such minor points as the confusion over whether astronomers could easily shift telescope sizes and still make pertinent observations. The council of the academy was not persuaded that Campbell had deliberately misrepresented Lowell's argument and did not realize, thanks to a misunderstanding of Elihu Thomson, that Lowell wished Lampland to fill Newcomb's seat.[44]

In his eagerness to strengthen Lampland's candidacy, Lowell took a number of other steps. He pestered his brother to find someone to propose Lampland. Lowell solicited letters from George Davidson, the retired California astronomer and geodesist, as well as from S. W. Burnham and E. E. Barnard of Yerkes. Alerted by Davidson that E. C. Pickering had left Lampland's name off some of the ballots, Lowell protested, only to discover to his embarrassment that Davidson's ballot included Lampland's name. The election, conducted in late 1910, resulted in a decisive victory for Campbell, with Lampland finishing last among the six candidates. In the aftermath of the election, Lowell wrote Lampland a letter of consolation, touted his "brilliant inventions about the 24-inch," and assured him that his election to the academy would materialize in short order. Lowell nominated Lampland again, without success, in 1913 and again in 1915, when he was finally elected.[45]

Meanwhile, Lowell worked on an entirely different scheme within the academy to maximize his influence in the organization and the astronomical profession. He proposed to donate a gold medal to the academy that would be awarded annually to a young astronomer. Lowell specified, however, that astrophysicists would not be eligible for the award. "They have usurped . . . the lime-light to the exclusion of the deeper and more profound parts of astronomy proper . . . Astrophysics I thoroughly welcome but as it is now getting the lion's share of endowment and help I wish to bring back the recognition due to its elder sister whom I may add is still young." The choice of the award recipient Lowell would have preferred to vest in the hands of a single person, consistent with family practice and his own governance of the observatory. That being impossible, he tried to stack the committee with two men, Burnham and Story, on whom he could rely. When Professor John Trowbridge, speaking for the academy, insisted on putting E. C. Pickering on the selection committee, Lowell agreed, but only if he himself were made chair of the committee. Otherwise, he wrote, the medal would be withdrawn, as it was. Once again Pickering thwarted Lowell. However, Lowell had the consolation in 1915 of seeing his assistant Lampland at last elected to membership in the academy, and he took the occasion to launch a proposal for transforming the whole governance system of the organization, which, he said, was characterized by "stagnation and oldfogyism." Officers, he insisted, should be elected every two years, and the council should be reduced in size.[46]

The remarkable thing about Lowell's efforts was that—as influential in Boston as he was—he failed over such a long time to achieve the election of his assistant to the academy. As John Lankford has suggested, this failure indicates, among other things, how successful the establishment had become in imposing its will not only in the National Academy of Sciences but also in a more regional organization like the American Academy in Boston, where the local elite—Lowell, Agassiz, et al.—had control. And as the gold medal episode indicates, the triumph of the establishment signified the victory of astrophysics within the astronomical community.[47]

Lowell's campaign for Mars clearly had unintended consequences. By undertaking a research project at the limits of seeing and making what appeared to be excessive claims about his evidence, Lowell generated a united front among opponents who were otherwise much at odds with each other. Together, they orchestrated a campaign of their own that challenged virtually every dimension of Lowell's work, from claims about the atmosphere in Flagstaff to the reality of the canals and the presence of water vapor on Mars. They curbed his efforts to disseminate his findings regarding the surface of Mars and denied professional recognition to members of his staff.

The timing of Lowell's challenge helps to explain the force of the reaction. His work coincided with the rise of the community of astrophysicists. In their fledgling condition, they were understandably alarmed by Lowell's ability to dominate discussion of matters astronomical in the press. The attention given to his work and his insistence that stellar astronomy was less significant than planetary research raised the possibility that sources of funding might dry up. They also worried that he, rather than they, would set the standards for truth in the profession.

The declining public faith in Lowell's claims after 1909 cannot be laid entirely to the energy and scope of the establishment campaign, however. As it happened, the work coincided with challenges from Lowell's opponents in Britain and the observations of the French astronomer E. M. Antoniadi, who demonstrated convincingly that with a large reflecting telescope the canals of Mars dissolved into a myriad of details. Together, the American establishment and European astronomers helped to lay Lowell's claims to rest.[48]

12

The Search for Recognition

Lowell and his assistants experienced frustration and a strong sense of isolation in the face of the astronomy establishment's united front. In fashioning a response to their rivals, the Lowell observers pursued a complicated course of action, and, inevitably, the perspective of the director diverged from that of his assistants. Lowell's assistants were young, professionally trained observers of the new order; as such, they were troubled by his confrontational posture and lamented the controversy surrounding the work at the observatory. Though loyal to him, they sought to move on from the Mars campaign to other projects. By contrast, Lowell was strongly invested in his controversial project and continued to defend the existence of the canals. Nonetheless, he too realized the importance of redeeming the Lowell Observatory in the eyes of its rivals by undertaking more conventional research. Director and assistants thus struggled to find common ground in order to lift the standing of the observatory. In this endeavor, they were partially successful.

In the aftermath of the 1909 Martian opposition, Lowell continued to use the media to assert the reality of the canals, even though this course risked greater isolation for the observatory and the possibility of undermining public confidence in science. At the same time, Lowell took on a new research project that was intended to alter dramatically the image of the observatory: the search for Planet X. He launched this search in a cautious, even discreet, fashion and uttered no word of his activity to the public or other astronomers until the publication of his memoir in 1915. In contrast to the uncertainties of the Mars project, conducted at the limits of seeing, the search for Planet X was based on rigorous mathematical calculations to determine the probable location of the planet, which was then subject to definite confirmation through a photographic search.[1]

Though this new project sat well with Lowell's assistants, they were understandably gun-shy about his confrontational tendencies. Indeed, V. M. Slipher and C. O. Lampland experienced a twofold sense of frustration in the face of the establishment campaign against the observatory: they were frustrated with their own director, whose reputation had brought into question the quality of all research done at the observatory, and with establishment astronomers, who refused to distinguish between their work and Lowell's. They responded to this frustration in different ways at different times. They occasionally confronted the establishment, in public or in private, but they usually preferred other options. One option was to ignore the attacks and continue the research program on Mars, the results of which were published in observatory bulletins or in the popular media. On other occasions, they undertook work more easily appreciated by mainstream astronomers. There is no evidence, however, that either Slipher or Lampland considered resigning from the observatory to take another position. Instead, they sought council from their Indiana University mentor, astronomy professor John Miller, who advised the two men to keep their respective noses to the grindstone and let their work speak for itself.

The result of this strategy was a growing appreciation of V. M. Slipher—and, to a lesser extent, Lampland—among establishment astronomers that gradually and imperceptibly pulled the observatory toward the mainstream. By the time of his death in 1916, and with the failure of the Planet X search, Lowell resigned himself to satisfying vicariously his craving for recognition from established astronomers. He continued to confront the establishment but encouraged his assistants to present their work to leading journals and to accept the honors that came their way. He came increasingly to identify the observatory to outsiders in terms of the work of his assistants rather than his own research on Mars. In this backward fashion, the observatory achieved a measure of recognition in the astronomical world at the end of Lowell's lifetime.[2]

Slipher and Lampland found themselves in an awkward position as the controversy mounted over the Martian canals. Caught between their loyalty to Lowell and their eagerness to achieve recognition, they looked for helpful guidance to Miller, who was at this time professor of astronomy at Swarthmore College. Miller had gained the respect of the establishment, which considered him for membership a number of times in the National Academy of Sciences. He saw the advantages of working at a small observatory, like Lowell, that was endowed with good instruments and good atmosphere. Accordingly, Miller helped his former students navigate their troubled waters.[3]

V. M. Slipher reached a crisis point in 1908 as he faced the campaign of the power elite to isolate the observatory and turn it into a pariah institution. Suffering from Campbell's skepticism regarding his Martian spectrograms and from his attack in the *Outlook* on Lowell's claims for seeing, Slipher was determined to publicly defend the observatory. He wrote John Miller to explain why he had reluctantly decided to take the path of confrontation: "We have stood enough of such [criticism] in recent years . . . but the time has come when it seems to me we might now take a stand and have it out. Both in print and direct with Professor Campbell." Only in that way, he believed, would it be possible to undermine the public faith in Lick Observatory's criticisms. He added significantly: "I know you will deplore this—the same as I do, but this Observatory has been getting it so generally that I have come to the conclusion that where we can defend ourselves we shall have to do it or otherwise everything we publish will be discredited." Slipher promised Miller he would write "in a dignified manner" and "give the facts fairly." And he added that the Lowell Observatory was also handicapped by its much smaller size vis-à-vis Lick and would therefore have to exploit good seeing—and public recognition of good seeing—to achieve proper recognition for its work.[4]

On this occasion, Miller reluctantly agreed that Slipher should proceed with a response to Campbell. "I do deplore controversy as you know but you are more than justified in answering Campbell in regards to the star-fields. Good seeing anywhere but at Lick is a sore point for Lick observers . . . Let your answer be as you have said made in a dignified way."[5]

As we have seen, however, Slipher soon came to regret his public encounter with Campbell. It raised tempers without solving the issues at hand. In the aftermath of the confrontation, Slipher apparently decided to abandon this strategy for dealing with the establishment. His decision was precipitated by Miller's report of the reception accorded to Slipher's paper on the Mars spectrograms at the 1909 meetings of the American Association for the Advancement of Science. According to Miller, who read the paper, it had not received the attention it deserved from his peers. "I want to say this *confidentially* and I am sure you will treat it so, the objection to your paper had its origin in the critical temper of men toward Lowell's work. Do your work as carefully you have before [*sic*] and you can allay all criticisms."[6]

Just a month earlier, Miller had offered encouragement to Slipher that clearly reflected his own approval of Lampland's and Slipher's work as well as the advantages of the observatory site and instruments. "Lowell Observatory has been criticised but no one observatory has been gaining scientific favor more than Lowell Observatory. At this I rejoice. The results obtained there recently sufficiently proves the contention that you have good seeing and good apparatus." Clearly Miller expected that his students' work would speak for itself so long as they were not caught up in Lowell's claims for life on Mars. To avoid contamination in the eyes of the establishment, both men would have to walk a fine line. They would have to sidestep the public controversy surrounding Mars, while at the same time maintaining good relations with their director.[7]

By late 1909, aided by the end of favorable Martian oppositions for more than a decade, Slipher turned the corner. He would now be able to devote more time to his spectrographic work, which, as Miller pointed out, had already received recognition. Furthermore, Slipher had come to realize that the more distance he could put between Lowell and himself, the more likely he would be to succeed. Indeed, in apologizing to Campbell for going public with the comparison of seeing at Lick and Lowell, he explained that his main object had been "to receive an unprejudiced hearing"—precisely the point that Miller had made earlier in the year. The experiment in confrontation gave way rapidly to another: letting his work speak for itself.[8]

To be sure, Slipher did not immediately banish his feelings of anger and frustration toward Campbell and company, as is evident from his reaction to criticism of his discovery that star dust reflects light. He wrote Lowell in 1913: "What I should like to do would be to just bait the enemy so as to get them to take the field publically [*sic*] and after they had paraded their power and wisdom, and then come out with such direct and definite observational evidence as to make an example of them and their attitude. There are those as you know who are always ready to jump upon something from here and I would like to see them 'get burned.' So why not now let them have a little time to find fault, meanwhile we shall be loading our guns. This is not the right kind of spirit I admit but it does seem excusable." It is significant, however, that this time, as opposed to 1908, Slipher vented his hostile feelings in private to avoid antagonizing his rivals. He may also have been using this letter to assure Lowell that, despite his recent recognition, he had not forgotten the observatory's treatment at the hands of the establishment. Certainly Lowell did not pass up opportunities to remind his staff of past humiliations. In the summer of 1914 he wrote Slipher that John Duncan, another Indiana astronomy graduate, would not be visiting Lowell Observatory. Moreover, averred Lowell, his absence would not be missed, because he was the "dupe of those who wish us ill" for having reported that the Lick research on the speed of nebulae "was equal to yours! Let this be confidential."[9]

Unhappily for Slipher and Lampland, Lowell never fully dropped his public claims for the canals of Mars, though they were delivered less frequently after 1909. He reported to his assistants in 1910 that the British public "is simply mad over Mars," that "hundreds of people were turned away from the lecture at the Royal institution," and that the *Daily Mail* had been receiving hundreds of letters on the subject of Mars, while he himself had been beseiged by reporters and photographers. From Slipher's faint praise of another such Lowell performance, we can infer his fear that all research at Lowell Observatory would be tainted by association with the claims for the canals, especially those appearing in the popular media. Slipher wrote confidentially to John Miller that Lowell Observatory would have to "keep away from telegraphic announcements . . . especially as we have been accused of doing startling things or things in a startling manner."[10]

To more quickly gain recognition from mainstream astronomers, Miller encouraged his former students to break their isolation by appearing at national meetings. He urged both Slipher and Lampland to participate in the American Association for the Advancement of Science (AAAS) and thus "to make the personal acquaintance of the scientists of the country," and he also pleaded with Slipher to speak before the American Philosophical Society as a way of enhancing his prestige in the astronomical community. These recommendations may have had some effect; Slipher did attend the AAAS meetings in 1914, but his preoccupation with directorial responsibilities after 1916 left him with little time to devote to such activities. Moreover, it seems clear that neither Slipher nor Lampland found much satisfaction in attending professional meetings.[11]

Though Miller's advice clarified the most advantageous approach for Slipher and Lampland, Lowell and his staff still struggled to determine the best course of action in dealing with the establishment. In numerous specific instances they had to make difficult decisions as to how they could most effectively achieve recognition for their work from rival observatories, as well as from the Astronomical and Astrophysical Society of America (AASA) and the *Astrophysical Journal.* No single strategy was adopted for all occasions, nor did the observatory always speak with one voice. Lowell never fully abandoned his strategy of confrontation, while his assistants generally preferred a more accommodating stance toward the power elite. And, as we shall see, when the latter approach succeeded in achieving recognition for the observatory, they managed to bring Lowell along with them.

Three cases illustrate the ways in which Lowell Observatory attempted to deal with its more powerful rivals: the matter of publication in the *Astrophysical Journal,* the question of whether to cooperate with the comet committee of the AASA, and the response to the meeting of the International Solar Union. In point of fact, the observatory had a long and tangled history with the *Astrophysical Journal* dating back to 1895, when George Ellery Hale and James Keeler, disturbed by the delay and expense associated with revisions of Lowell's "On Martian Longitudes," asked him not to submit additional articles to their publication. From that time on, A. E. Douglass, out of loyalty to Lowell, had refrained from sending his own articles to the *Journal.* In so doing, however, he had been "barred out of the highest standard magazine in the world on that subject." It is significant, then, that Slipher had begun to publish his results in the *Astrophysical Journal* in late 1905. Clearly this step signaled a change, however reluctant, in the observatory's posture toward the mainstream. Slipher, no doubt with Lowell's approval, was seeking recognition from establishment astronomers even as Lowell Observatory persisted in the research on the canals of Mars that so offended rival astronomers. In this way, Lowell and his assistants adopted an ambivalent course, affirming the standards of the establishment and resisting them at the same time. Indeed, as late as 1911 Lowell was still feuding with the journal. At that time, he instructed V. M. Slipher to return the latest issue to Chicago inscribed as follows: "Do not wish to continue subscription." Louise Leonard added: "Dr. L has no use for it here."[12]

In the summer of 1910, a group of distinguished astronomers from the United States and Europe, all members of the International Union for Cooperation in Solar Research, met in the United States. The union had been organized seven years earlier by Hale and included among its members W. W. Campbell, E. C. Pickering, and other leading astrophysicists. The conferees were to gather at Mount Wilson, and they had scheduled a stop at Flagstaff to visit the Lowell Observatory while en route. At the time,

Percival Lowell was in the midst of conflicts with the power elite over the telegraphic distribution of astronomical news as well as the Martian water vapor issue. Nonetheless he belatedly issued an invitation, and the conferees stopped at Flagstaff. They were then joined on their westward trek by V. M. Slipher and C. O. Lampland, who must have decided to break their isolation and take advantage of this opportunity to gather with the likes of Pickering, Campbell, and Hale.[13]

The appearance of Halley's comet in 1910 offered all observatories an opportunity to establish their prowess at observing and photographing an object of interest both to astronomers and the larger public. It also created problems for the observatories, for a variety of reasons. Among other concerns, establishment observatories were worried that Lowell would sensationalize the event. E. E. Barnard, a member of the Astronomical and Astrophysical Society's comet committee, wrote Hale: "I think it will be well to get some observations of the comet as a check on some that will be made elsewhere." For Lowell, the arrival of Halley's comet posed a variation on the usual dilemma: whether the observatory should collaborate with its rivals or attempt to upstage them by operating independently.[14]

The occasion for Lowell's quandary was the comet committee's plan to publish photographs collected from various observatories. Initially Lowell wrote to Slipher to advise that the observatory's photographs should not be sent to the AASA to "help that lame duck make a showing." Nonetheless Lowell reluctantly considered the possibility of offering up a few photographs to the collaborative project if they were clearly labeled "taken by permission from the Lowell Observatory Bulletin which see for a complete memoir on the subject." Apparently V. M. Slipher took Lowell's advice to heart, because he wrote Lampland to say that he had originally intended to comply with the request from the AASA, but was now uncertain. However, Lowell's letter had also requested of Slipher "any suggestions in the matter," indicating the director's growing respect for his assistant's views. In response, Slipher decided that, despite the impersonal and therefore improper nature of the communication from the AASA, it would be advantageous to have the Lowell Observatory represented in its publication.[15]

Lowell then resolved the issue consistent with his own and Slipher's views. In a letter to the comet committee, he announced that the observatory was publishing two large memoirs on the comet but would allow the committee to reproduce some of the plates from those memoirs. In return, the comet committee would have to agree to publicly acknowledge their solicitation of the plates from the observatory. Sounding a bit like the observatory director he would become, Slipher complained to Louise Leonard that, while Yerkes and Lick were getting newspaper notices for their work on Halley's comet, Lowell had been neglected. Clearly, Lowell was showing increasing respect for Slipher's views, while Slipher was beginning to think more carefully about the position of the observatory in the larger astronomical community.[16]

Lowell moderated—but did not abandon—his confrontational strategy as he witnessed the growing recognition from the power elite for his own assistants. He gave an address on Venus at the Clark University twentieth-anniversary celebration in 1909 that was a deliberate effort to persuade listeners that the observatory was engaged in research projects other than the study of Mars and extraterrestrial life, and to downplay the most controversial aspect of Lowell Observatory's operation. This message was apparently reaching at least Lowell's friends, if not the general public. In a letter to Henry Cabot Lodge, William Sturgis Bigelow lavished praise on Lowell while implying strong reservations about the work on Mars. "It is one of the crack observatories in the world. Nobody questions this, whether or not they endorse his private views about the canals of Mars, the study of which is only a small part of the work of the observatory."[17]

Lowell was all the more inclined to put Mars on the back burner because Slipher was receiving accolades for his spectrographic work. In 1911, for example, Lowell's archenemy, W. W. Campbell, against whom Slipher had also been pitted in the controversy over water vapor on Mars, invited Slipher to join a committee of astronomers from different observatories, formed to increase knowledge of stellar radial velocities. Three years later Campbell congratulated Slipher on his pathbreaking discovery of the radial velocities of spiral nebulae. And in 1916 Campbell wrote Slipher and Lampland to request up-to-date information on radial velocities of spiral nebulae that he expected to use in his presidential address to the American Association for the Advancement of Science. Nearly a year earlier, Hale and Frost, on the recommendation of Campbell, had considered inviting Slipher to contribute to proceedings they were organizing. Their only hesitation was that the invitation would have to go through Percival Lowell, whom they clearly did not wish to include. This episode suggests that, although Slipher was able to achieve recognition for his work, Lowell remained an obstacle he and other astronomers had to work around.[18]

The work of the Lowell Observatory, through the efforts of Slipher and Lampland, was thus gaining the respect of the astronomical establishment.

Percival Lowell hoped to make his own contribution to raising the standing of the observatory through his search for Planet X. He expected a successful search, conducted in a well-established discipline like celestial mechanics, to win over professional astronomers who had been alienated by his sensational claims for the canals of Mars; in contrast to his work on that planet, the search for Planet X was based largely on Newtonian celestial mechanics, a universally accepted methodology. Moreover, Lowell could be certain that the discovery of Planet X would bring recognition not only from astronomical circles but also from the larger public.

The trans-Neptunian project also appealed to Lowell because it enabled him to display his versatility—here, within astronomy—as he had already done in other fields. The search for Planet X would require an entirely different set of skills from those he had used in his Martian work. To discover the new planet, Lowell would have to solve mathematical equations to determine a probable planetary orbit as a prelude to conducting a photographic search, rather than drawing surface detail of a known planet from telescopic observations.

In most respects, Lowell's new project was a perfect fit for a small observatory working on planetary research in an age of factory observatories. It certainly belonged on the agenda for solar-system research, which Lowell had designated as the proper domain for his observatory. And, at least in its initial phases, the search could be carried out by the small staff employed at Lowell, without acquiring complicated, esoteric, and expensive equipment.

Accordingly, it is not surprising to discover that Lowell's interest in locating a trans-Neptunian planet was shared by other mavericks stamped from the same mold as he, and well known to him. Former assistants T. J. J. See and W. H. Pickering both offered predictions: See weighed in with several trans-Neptunian candidates in 1904, while Pickering identified the first of seven such planets in 1908. Much earlier, Lowell's French colleague Camille Flammarion and his old friend David Todd had made their own predictions. Flammarion had noted in 1879 the possible existence of a planet at a point around which streams of comets and meteors orbited. Two years earlier, David Todd had predicted a somewhat more distant orbit for a trans-Neptunian planet. In all probability, Lowell had spoken about the prospects of a search for this planet with Todd in late 1899 and early 1900 during his sojourn in Amherst and Algeria, two years before his first public discussion of Planet X.[19]

By contrast, the project had no appeal to astrophysicists immersed in stellar research. In this regard, it is interesting to note that when Walter Adams, the assistant director of Lick Observatory, and Harlow Shapley, his young assistant, received W. H. Pickering's 1919 prediction of the position of a trans-Neptunian planet, neither wanted anything to do with the search for it. Instead, they passed the project on to the janitor of the observatory, Milton Humason, who took on the search as a way of demonstrating his competence for astronomical work.[20]

Lowell launched his search for Planet X in 1905, but he had clearly been thinking about a trans-Neptunian planet for several years. His MIT lectures in 1902 twice pointed to the likelihood of the existence of such a body, based on cometary orbits like those identified by Flammarion. Lowell remarked: "It may seem to you strange to speak thus confidently of what no mortal eye has seen, but the finger of the signboard of phenomena points so clearly as to justify the definite article. The eye of analysis has already suspected the invisible." Further along in the lectures, Lowell identified a different location for the unknown planet, thus suggesting that the "signboard" was not so precise.[21]

In any event, Percival Lowell was clearly captivated by celestial mechanics, which he had first studied under Benjamin Peirce. He was, as Peirce insisted, an excellent mathematician. Furthermore, like many of his predecessors, Lowell was exhilarated by the idea of using the laws of celestial mechanics to discover unseen bodies in the universe. Accordingly, he celebrated one of the dramatic scientific events of the nineteenth century, the discovery of Neptune. What particularly impressed him about this event was the confident prediction of Neptune's location by the French astronomer Urbain J. J. Le Verrier, based on unexplained perturbations in Uranus's orbit. The fact that Le Verrier's invitation had galvanized Johann Gottfried Galle to search successfully for the planet very near the predicted location suggested to Lowell the power of celestial mechanics as a research tool.[22]

The discovery of Neptune might have been less dramatic to Lowell had he not been a disciple of Benjamin Peirce. In the wake of the discovery, Peirce, along with the American astronomer Sears Cook Walker, took up the question of whether Neptune was actually the planet Le Verrier had predicted. Based on calculations of Neptune's orbit, they disputed Le Verrier's claim. In this way Lowell had a direct link to one of the great scientific discoveries of the century. His fascination with celestial mechanics was, no doubt, also reinforced by the unsuccessful search for an intra-mercurial planet that preoccupied astronomers in the wake of Le Verrier's prediction of such a body in 1859 and the report from the amateur astronomer Dr. Edmond Modeste Lescarbaut that he had actually seen the planet.[23]

Lowell's search for Planet X was complicated by his multiple responsibilities. In the first four years of the project, Lowell gave primary attention to observations of Mars during the favorable oppositions of 1905, 1907, and 1909. Accordingly, he was unable to use the telescope to search for Planet X at those times. Furthermore, he had to squeeze his own calculations designed to identify the orbit of the missing planet into periods between oppositions. Even after 1909, Lowell continued to use every opposition to accumulate data for his Martian work.

During the first phase of his new project, Lowell separated the photographic search for Planet X from the mathematical search. He obtained photographs and astrographic charts from Flammarion and E. C. Pickering for star fields along the ecliptic. Since this was clearly an appropriate place to begin the search, he had graduate assistants from Indiana University search that area of the sky and take photographs. Meanwhile, Lowell himself was engaged in the entirely separate mathematical search for the planet.

By 1910 Lowell had decided that the photographic search would be more likely to succeed if he used his calculations to determine the regions of the sky where Planet X would most probably be found. Drawing from the long history of celestial mechanics, Lowell actually tried three different methods of calculation, two of which were shortcuts. He first tested Flammarion's scheme of using cometary and asteroidal orbits to indicate the location of the missing planet; he soon supplemented this approach with a method of graphical analysis pioneered by David Todd, who argued that one planet acted on another decisively when they reached the same celestial longitude in their orbits. When neither of these methods proved satisfactory, Lowell returned to Le Verrier's tried and true method of calculating the unexplained perturbations in Uranus's orbit as a basis for computing the orbit of X. In the end, completing these calculations proved to be a Herculean task.[24]

While he struggled with the orbital calculations, Lowell worked with his assistants to develop an effective strategy for photographing the missing planet. He suggested appropriate locations for the photographic search, and he provided his assistants with good equipment. On two separate occasions new cameras were obtained. When the 5-inch Brashear proved to be inadequate, Lowell borrowed a 9-inch Brashear photographic doublet from John Miller at Swarthmore in the spring of 1913. Two years later, photographs taken with this instrument actually recorded Pluto, but the images were dim and the planet went unrecognized until after its discovery in 1930, when the plates were reexamined. Apparently, by 1915 Lowell was considering replacing the 9-inch doublet with a 16-inch Bruce photographic objective to record a dimmer planet than the one he had originally predicted. Lowell also belatedly bought a Zeiss blink comparator, at Lampland's recommendation. Using this instrument, it was possible to compare two plates of the same section of the sky to determine whether any celestial objects had shifted position or changed brightness in the interval between photographs.[25]

Aside from Lowell's fascination with the mathematical approach to discovering previously unknown celestial objects, and the likelihood that the discovery of X would enhance his reputation, a number of other factors contributed to the appeal of the new project. It had a solidity and respectability that contrasted sharply with the sensational quality of the Martian research. In the search for Planet X, Lowell was working in the established discipline of celestial mechanics, which was based on gravitational principles pioneered by Isaac Newton. In keeping with this solidity, Lowell conducted his research with a great deal more discretion than he had shown in the Mars work. He had published his claims for the likelihood of intelligent life on Mars even before he had left Boston to make his first observations of the planet in Flagstaff. By contrast, the first public notice of the trans-Neptunian endeavor was published in 1915, after he had been at work on the project for nearly ten years. Even though Lowell had written two earlier papers, one in January 1909 and the other in June 1909, assigning different values to the distance, period, size, and location of Planet X, he had published neither of them.

Even more surprising was the tentative statement of his claims for Planet X, again contrasting with the certainty of his predictions for life on Mars. "Owing to the inexactitude of our data, then, we cannot regard our results with the complacency of completeness we should like. Just as Lagrange and Laplace believed that they had proved the eternal stability of our system, and just as further study has shown this confidence to have been misplaced; so the fine definiteness of positioning of an unknown by the bold analysis of Le Verrier or [John Couch] Adams appears in the light of subsequent research to be only possible under certain circumstances. Analytics thought to promise the precision of a rifle and finds it must rely upon the promiscuity of a shot gun after all."[26]

Lowell acknowledged in the search for Planet X what he refused to face in the search for life on Mars: that he had no conclusive proof for his thesis. The sober and cautious tone of his memoir suggests not only his deep disappointment but also a willingness to mute his claims in the light of the accepted standards for scientific proof. No doubt the absence of conclusive proof for the location of Planet X accounts for the refusal of the American Academy of Arts and Sciences to publish Lowell's "memoir" on the subject. He was thus left to underwrite the publication himself. Despite this further disappointment, he tried to put a positive spin on the search. "To learn the general solution and the limitations of a problem is really instructive and important as if it permitted specifically of exact prediction. For that too means advance."[27]

In one important respect, the new project resembled Lowell's Martian work. He was once again, though quite unexpectedly, operating at the limits of seeing. Initially Lowell had assumed that Planet X would be a large and therefore rather bright object in the sky. As the search proceeded, however, it became clear that the undetected planet was much smaller and dimmer than he had initially supposed. Pluto, at that moment an object of the sixteenth magnitude, strained the limits of his photographic equipment.

Another interesting twist was that, despite taking up a maverick project, Lowell veered toward big science in the later phases of his search. As the research reached a critical point in 1913, Lowell sought frantically to solve the equations that would clarify the location of Planet X. This required hiring additional calculators beyond the one person he had employed up to 1909. By 1912 the number of calculators in his Boston office had reached four. Meanwhile he had added a graduate student to his staff, beginning in 1905, to handle the photographic search for Planet X. Thus what had started as an old-fashioned project suited to a small observatory was rapidly transformed into a substantial data-analysis operation. As the project wound down, however, the staff was rapidly reduced to a single calculator.

Despite prodigious labor by Lowell and calculators, the search failed and with it Lowell's hopes of redeeming himself within the astronomical community. It was, as his brother noted, the "sharpest disappointment of his life." Nonetheless the eventual discovery of Pluto in 1930, fourteen years after Lowell's death, must be partially credited to Lowell, even though his Planet X bore little resemblance to the newly discovered planet. Without question, Lowell's search laid the groundwork for the resumption of the Planet X project in 1930 that, in turn, resulted in Clyde Tombaugh's discovery of Pluto.[28]

In many respects, the search for a trans-Neptunian planet brought Lowell full circle, to the beginning of his astronomical career. He had launched his observatory with the assistance of fellow maverick W. H. Pickering. The two men had separated and pursued their careers independently, but both took up the search for a trans-Neptunian planet. Although Lowell apparently launched his search first, Pickering's paper delivered to the American Academy of Arts and Sciences in November of 1908 spurred Lowell on. And Pickering, on that occasion, tried unsuccessfully to persuade Lowell to search for Planet O. Moreover, Pickering remained in the hunt for a trans-Neptunian planet after Lowell's death and lobbied at the time of Pluto's discovery to receive partial credit for it.[29]

Meanwhile, as it became apparent to Lowell in 1914 that his search for Planet X would fail, he sought consolation in the favorable reception of his assistants' work. It reflected well on the observatory even if it brought him, personally, little respect. Increasingly Lowell spoke of observatory achievements more in terms of the work of his assistants than with regard to the canals of Mars.

The observatory correspondence demonstrates Lowell's own pride in the accomplishments of his men, almost always expressed as a function of observatory rivalry. He either praised their achievements directly to his assistants or used the media to spread the word. In 1907, for example, Lowell wrote to Slipher to commend his spectrograms, which he compared to work at Lick, "to the offering [*sic*] you my heartiest congratulations in the way you have beaten them." In 1913 he applauded Slipher's discovery that the Pleiades nebulosity was the result of star dust reflecting light. Lowell took this to be evidence in support of the nebular hypothesis and treated with scorn F. R. Moulton and E. C. Pickering, who qualified their acceptance of Slipher's achievement. In making his case before the trustees of the Boston Public Library for a permanent exhibit of materials from the observatory, Lowell used as a primary illustration of its value not the detection of Martian canals, but "Slipher's spectrographic discovery of the rotation of a nebula. It has excited world-wide interest and has spurred the next best equipped place, the Lick Observatory, to follow in its footsteps." He also subtly shifted the grounds of his support for Lampland's election to the American Academy of Arts and Sciences from Lampland's success in photographing the canals of Mars to his assistance to Slipher in "working out the parallax of a nebula; which has important bearing upon Dr. V. M. Slipher's detection of nebular rotation." Lowell clearly identified those achievements as far more compelling to the establishment than his own claims for the canals.[30]

Lowell's satisfaction with the work of his assistants was well expressed just one year before his death, in a letter to E. C. Slipher praising his article in the *American Museum Journal.* "It is so good both in matter and style that it beckons you for the future. I look forward to reading your articles, a solace to my old age in prospect." These remarks suggest both the satisfaction he took in the quality of his assistants' work, and the frustration at the lack of recognition for his own projects. He had founded the observatory with the hope of proving the existence of life on Mars and then of finding Planet X. He had failed at both of these endeavors and, on occasion, had made himself and his observatory a laughingstock. Now, as his career was about to end, he could experience only vicarious pleasure in the work of his assistants. Ironically, the Sliphers and Lampland were drawn from the same class of university-trained, professional astronomers who had rejected his work in the first place. In this sense, the observatory succeeded by redefining itself to better reflect the methods and approaches of its establishment rivals. Over the course of twenty-two years, and despite the failure of his own projects, Lowell succeeded in building and staffing an astronomical observatory that would be a significant player in the world of twentieth-century science.[31]

Lowell's astronomical career was shaped by both his personality and his interactions with the American astronomical community. He brought to astronomy the same individualism, adventurous spirit, imaginative thinking, combativeness, and desire for recognition that had marked his travels in Japan. The observatory he founded and the projects he undertook bore the imprint of these personal traits. In the face of the increasingly organized character of American astronomy, Lowell was able to satisfy his passion for personal independence by founding and managing his own institution. In the process, he set the research agenda and hired a staff willing to conduct the investigations on his chosen projects.

Lowell's penchant for adventure was reflected in almost every aspect of his astronomical work. By entering the field without any previous experience or long-range plan for an astronomical career, he risked embarrassment at the hands of his professional rivals. In the seven years following the first expedition to Flagstaff in 1894, Lowell continued to search for a better location and an appropriate research agenda. Only after this lengthy period of instability did Lowell make a commitment to a permanent observatory on Mars Hill and hire assistants for the long haul.

In attempting to prove the existence of the Martian canals, Lowell necessarily adopted an imaginative research strategy to supplement the uncertain empirical evidence he and his assistants gathered from visual experiments, drawings, photographs, and spectrograms. His use of a thought experiment comparing the evolution of Earth and Mars indicates the highly theoretical underpinning of his case for life on Mars. Moreover, his project, which excited the public imagination, gave him ample opportunity to present the case for the canals of Mars not only in scientific journals but also in the popular media. And in the face of skepticism from rival astronomers about the results of his research, he felt compelled to exercise his combative instincts even to the point of publicly proclaiming the superiority of seeing at his observatory. Belatedly, and unknown to many astronomers, Lowell also launched a search, employing more conventional methods, to discover a trans-Neptunian planet. In this endeavor, he sought to redeem himself in the eyes of astronomers and the larger public.

Despite the fact that Lowell founded a small institution in the age of factory observatories, he created a stir in the astronomical community. The power elite of astrophysicists, who directed the factory observatories, managed the Astronomical and Astrophysical Society, and published the *Astrophysical Journal,* found Lowell a threat to their newly won hegemony. His success in diverting public attention from their own projects to the canals of Mars endangered the private sources of support on which they depended. Accordingly, mainstream astronomers joined forces to attack Lowell's credibility and assure the continued recognition of their position as authorities in the field.

That Lowell's rivals regarded his work as a serious threat is evident from the zeal with which they opposed him. To challenge his findings, establishment astronomers had to divert time and resources from their own projects and set aside issues that had divided them. They then devised a campaign to cast doubt on Lowell's claims. In the process of investigating facets of the Mars project that Lowell and his assistants were researching, they raised questions about his approach to issues of perception, the visibility of the canals through the telescope and in photographs, and the sufficiency of water vapor on Mars to support intelligent life. The results of these investigations, which differed from Lowell's, eventually persuaded other scientists and segments of the public that Lowell's findings were flawed. The establishment also took advantage of its control over the distribution of news of recent astronomical discoveries to limit the communication of Lowell's results. In addition, they used the *Astrophysical Journal* and the AASA, as well as the popular media, to publicize their critiques of his position. And they succeeded in exploiting the reward system to recognize establishment astronomers in preference to Lowell and members of his staff.

By joining forces, mainstream astronomers isolated Lowell and his colleagues. Given their resources and numbers, it is not surprising that over time they gained the ear of the public and managed to diminish public confidence in Lowell's findings. However, they could not contain the widespread interest in the nature and extent of extraterrestrial life that Lowell had generated. And they were clearly surprised that C. O. Lampland's impressive photographs of spiral nebulae and V. M. Slipher's brilliant research on the radial velocity of spiral nebulae were produced at a maverick observatory founded to establish the existence of life on Mars.

◆ ◆ ◆

Conclusion: A Proper Bostonian on Mars Hill

Percival Lowell not only pursued a career in Flagstaff, Arizona, but created a life there as well. He built a home, became part of the community, and chose to be buried on the grounds of his observatory. In the process, he became a new man who resembled his Brahmin father more than his own younger self. In the Arizona setting, he came to appreciate the role of an intelligent elite in establishing a community. His rebellious inclinations were channeled into dealings with mainstream astronomers, while in Flagstaff itself he advocated "the gospel of the rights of community." In this way, Lowell's geographical separation from Boston was accompanied by an acceptance of the Brahmin ideal of elite action to promote a culturally enlightened community.[1]

To be fully at home in Arizona, Lowell fashioned a small empire of his own on Mars Hill. He boasted to Frederic Stimson in 1915 that the observatory "is a veritable colony, three observing domes, four astronomers' houses, the new building, etc. You as a diplomat will understand the style one must keep up as envoy of Mars to Earth." He had transformed his four-room house, built in 1894, into the fully heated, twenty-five-room "Baronial Mansion," with a tennis court and a view to the San Francisco Mountains. It was maintained while Lowell was in residence by servants whom he brought from Boston. And despite the distance, he recreated the Boston social scene by hosting friends from the East as well as members of the local elite. Dinners were enhanced by fine wines and cigars.[2]

From this splendid domain, he ventured into the surrounding terrain. His best friend in Arizona, Judge Edward Doe, introduced Lowell to the White Mountains and the San Francisco Peaks as well as Oak Creek Canyon. They traveled to the Moqui (Hopi) Reservation to observe the snake dance and to the Grand Canyon with Edward Morse. On many of their trips, they hunted and fished. On one occasion Lowell sent Doe a lure "to catch the elusive trout" and wrote: "I go in spirit with you to the paradise of Oak Creek." More often, however, they went together in the flesh, as a telegram from Lowell proclaimed: "Expect the worst blizzard of the season: Me—by the limited March 8th at 9:20 . . . I expect to be with you about a month; so kill all the calves you can lay hands on—if fat—and make arrangements to go with me early in April to the Petrified Forest. No *no* accepted."[3]

On their camping expeditions Lowell and Doe often pursued different game. While Doe hunted animals in Sycamore Canyon, Lowell bagged trees, including "an absolutely unknown and unsuspected kind of willow, habitats extended, etc. to say nothing of a record rattler who undertook to address me without an introduction." In fact, Lowell was following instructions from the director of the Arnold Arboretum, Charles Sprague Sargent. Under his tutelage, Lowell had already found new species of willow, ash, and a mesa oak that was later named *quercus lowelli.* He had also extended the range of species of sycamore, poplar, and oak, and correlated the distribution of various sorts of junipers by altitude. Just the day before his death, Lowell and C. O. Lampland were examining a species of willow discovered in Sycamore Canyon.[4]

Even in the farthest corners of Sycamore Canyon, however, Boston was never far from his mind. Collecting his tree specimens, Lowell imagined how they would enlarge Professor Sargent's collection at the Arnold Arboretum. And despite his growing attachment to the Arizona landscape, he remained a vigorous partisan of the old urban landscape of Boston. Thus, Lowell joined his sister Amy to argue against the relocation of the Boston Athenaeum, which had achieved the status of "an heirloom from the past."[5]

Lowell's love of the Arizona terrain was in tension with his eagerness to join the Flagstaff elite to promote, as he put it, "a new community on our globe." Institutional imperatives, after all, required a strong community to create a supportive environment for the observatory. And, conversely, the observatory could elevate the tone of cultural life in a frontier community.[6]

Accordingly, Lowell offered material and intellectual support to the community while seeking to shape its growth. He arranged for distinguished visitors at the observatory to offer public lectures and delivered some of his own. He supported local charities and invited children to the observatory at Christmas time, when he donned a Santa Claus outfit. He shaped the editorial views of the *Coconino Sun* by threatening to take it over and then by lending money to the editor.[7]

Lowell and other community leaders supported the construction of a new opera house in Flagstaff and approached the Santa Fe Railroad to build a large tourist hotel and a tram to the top of the San Francisco Mountains, where a second hotel would be located. Recognizing the possible detrimental effects of the project on the environment, he justified the proposal all the same. "I dislike of course to spoil the virgin simplicity of our beloved Arid Zone! But I think it will be a good thing for Flagstaff." While the project failed, Flagstaff did become the takeoff point for visits to the Grand Canyon.[8]

Increasingly, Lowell used Arizona as a platform for promoting a Brahmin ideology of elitism and strident nationalism. He lashed out against women's suffrage, labor unions, and such Progressive proposals as initiative and recall, because they would give decision-making power to "the lowest classes of community—on the pleasing principle that he who possesses least knowledge of, and least interest in, the country is the most eminently equipped to govern it." Prohibition he regarded as an invasion of individual liberty. To weather the crisis, Lowell ordered twelve cases of wine along with scotch, beer, and sherry—and then could not accommodate them in the wine cellar of the Baronial Mansion.[9]

Elite power, along with national unity, he believed, was being undermined by immigration of the "ruck and rabble" of Europe. To address this problem, following Henry Cabot Lodge and other Brahmin friends, Lowell proposed patriotism as an antidote. He warned the miners in Kingman, Arizona, not to challenge their bosses. "If oneness of purpose is vital to Mars, it is no less imperative for us. In disunion lies inefficiency." He would turn the country into "a solid homogeneous whole" by imposing immigration restriction; and, since only the "very ablest intelligences" could solve the problems of the country, he briefly considered taking the nomination of the Republican party to become senator from Arizona.[10]

Like other Brahmins, and despite his advocacy of Martian pacifism, Lowell urged Americans to join the Allied cause during the First World War. As the war progressed, he denounced the Germans as beastly, uncivilized, and a threat to American security. In their treatment of the Belgians, he proclaimed, the Germans "have shown themselves much lower [on the scale of civilized peoples] than had been supposed."[11]

Lowell's death by stroke in 1916 gave further evidence of his feelings about Flagstaff. His decisions to be buried on Mars Hill and to leave his fortune to preserve the observatory indicate the strong allegiance he had developed to Flagstaff. Equally significant, however, was the arrangement Lowell had made for governing the observatory through a single trustee, following the model of the Lowell Institute. In effect Flagstaff would have its observatory, but the Lowell family would decide the future direction of his institution.[12]

In important ways, the arrangements for Lowell's funeral made a similar statement. The service was conducted by the local Episcopalian and Catholic priests, and the pallbearers included Lowell's observatory colleagues and various local dignitaries; the faculty of Arizona Normal School as well as the Putnams and Lowell's wife were in attendance. While Lowell was laid to rest in Flagstaff, his presence in the family plot at Mount Auburn Cemetery near Boston was assured by a chunk of petrified wood announcing the Flagstaff burial. The marriage of town and observatory, of East and West, was thus accomplished in Lowell's unique fashion.[13]

Following Lowell's death, V. M. Slipher, his assistant of sixteen years and hand-picked successor, wrote of his former boss: "He had unbounded energy, enthusiasm and perseverance, somewhat the pioneer spirit and courage. He entered science by an unusual course but he made and filled a large place in astronomy and in the world as well." Slipher's remarks direct our attention to Lowell's passion for innovation, but fail to acknowledge his sometimes conservative, even backward-looking approach to astronomy. As a polymath in an age of increasing specialization, Lowell had not always been attuned to the most advanced techniques and instrumentation. Moreover, in eulogizing Lowell Slipher necessarily ignored his director's dubious methods of investigation, which had often marred his results.[14]

Nonetheless, Slipher was correct in identifying Lowell's pioneering spirit as the key to understanding the man. This spirit came to the fore as he departed for East Asia at age twenty-eight. On that occasion and others, he challenged traditional Brahmin ways, which prescribed a Boston and investment-centered life. As a cosmopolitan and a polymath, Lowell became a Brahmin with a difference. In the process, he and his friends developed an effective response to the challenge of the newly rich New York and Chicago industrialists, whose wealth and power threatened to render Boston's Brahmins irrelevant.

On the home front, Lowell aspired to liberate his fellow Brahmins from the tyranny of traditional practices and expectations and thus to revitalize the upper-class lifestyle in Boston. He and his friends brightened Brahmin life by downplaying the work ethic and spending more time at leisure. Their parties, heavily flavored with alcohol, provided carefree and expressive diversions. At the same time, they strove for a healthier and more adventurous lifestyle. Polo playing and mountain climbing, which required strength, skill, and endurance, contributed in a literal sense to the revitalizing of the upper class. Meanwhile, intellectual and artistic activities to stretch the mind provided the appropriate counterpart to more demanding physical exercise. Clubs for scientists, authors, and artists offered intellectual challenges and prepared their membership for careers in those fields. The new lifestyle demanded time and money but served as a prelude and an accompaniment to careers beyond the business world. By adopting both a new lifestyle and new careers, Lowell's Brahmin cohort positioned itself to play a significant leadership role in twentieth-century America.

Lowell's work was innovative in a number of ways. As an investor, he was drawn to the new scheme of selling off cotton stocks and investing in distant railroads and utilities. This adventurous strategy anticipated the cosmopolitan yearnings that caused Lowell to drop his promising business career in favor of travel and writing. He set out for East Asia along with other globe-trotters, but vowed to eschew their superficial wanderings. Accordingly, he sought to discover the core truths of a distant culture, both to understand it better and to shed light on his own country and culture. In the process, Lowell engaged in an almost obsessive search for the few remaining remote and unexplored corners of East Asia, including Korea, and Japan's Noto Peninsula and Mount Ontake. He alerted Americans to the importance of East Asia and to the beauties of East Asian art. But he was most truly a pioneer in attempting to systematize the study of East Asia. This effort was part of an ambitious scheme to compare and contrast civilizations around the globe and more particularly to clarify the significant differences between feudal and industrializing societies. Thus Lowell's books on East Asia contributed to the continuing debate about the strengths and weaknesses of industrial society.

Lowell's innovative research on trance-practicing Shinto sects verified the reality of the trances, and he also used his findings to support his claim that the Japanese people were impersonal in character. Unfortunately, like so many of his contemporaries, Lowell confused folkways with constitutional defects of the brain. Thus, many of his claims about East Asia have suffered the fate of Herbert Spencer's racial ranking system, on which they were based. His condescending assumptions about the intellectual limitations of East Asians were rejected, even in his own time, by experts on that part of the world.

But of course Lowell's most daring project was the attempt to offer scientific proof of the existence of intelligent life on Mars. He was the first to put the search for extraterrestrial life on a scientific basis, by proposing a telescopic campaign that would yield solid evidence for the existence of such life. His search generated one of the great scientific debates of the century, and the controversy was eventually resolved in favor of his rivals. Seven years before his death, Lowell's sensational claims for the evidence of life on Mars were virtually laid to rest. Despite his errors in this investigation, however, Lowell captured the imagination of the larger public with his lucid and compelling writings on the subject. In the process, he influenced the pioneers of the space program, who either read his books or the science fiction works of Edgar Rice Burroughs and H. G. Wells, both of whom were inspired by Lowell.

As the first astronomer to build an observatory in the high desert country of the American Southwest, Lowell was quite literally a pioneer. Not only did he take advantage of the site to test the prospects for good seeing, but he also came to enjoy exploring the region's exotic natural life. In both respects, Lowell played an important role in the westward movement of American science in the late nineteenth century.

The timing of Lowell's decision to enter astronomy was, however, more problematic than the location of the observatory. After all, he launched his Arizona expedition just as factory observatories were emerging and astrophysicists were taking control of the discipline. His decision to establish a small observatory devoted to planetary research ran counter to the dominant trend among leaders in the profession, who were giving priority to stellar research. In conducting his Martian campaign, Lowell deliberately reduced the size of the telescopic lens he used, and he initially preferred to sketch rather than photograph the surface features of the planets he viewed. In these respects, we could say that Lowell's conception of astronomy was backward-looking rather than innovative. Moreover, despite his cosmopolitan aspirations on Earth, Lowell devoted the bulk of his observing time to searching for surface details on nearby planets while his colleagues were using large reflecting telescopes to expand the boundaries of the universe. His observations, he believed, revealed a civilization that was identical to the technocracy that his MIT colleagues were attempting to promote, so in this respect, as well, the search for extraterrestrial life became an exercise in parochialism.

Lowell's work, on both Japan and Mars, was distorted by scientific practices that violated the norms of his professional colleagues, provoked challenges to his findings, and threatened to undermine public confidence in scientific research. Lowell chose exotic research topics for which conclusive evidence was difficult to collect, and, in its absence, accepted unreliable data or offered strained interpretations of his findings. His unscientific handling of evidence was due in part to the deductive approach that he borrowed from Herbert Spencer. Assuming the truth of Spencer's evolutionary views, Lowell was certain that the great diversity of organic and inorganic elements, as well as intelligent beings and societies in the solar system, had a common ancestry in a gaseous nebula. His research, designed to affirm rather than to test this proposition, invariably yielded the desired results.

Lowell's colleagues were disturbed by his persistent use of flawed evidence and strained interpretations. For example, they questioned the existence of lines on the Martian surface that Lowell saw at a distance of tens of millions of miles through a relatively small telescopic lens, and the claim that East Asians entered into trances more easily than Westerners and were thus lacking in individualism. Colleagues also protested Lowell's use of methods that were either out of date, inconclusive, or inappropriate. His psychical research, for example, was based on observations of many groups of occult practitioners rather than a few individuals tracked over a period of time, and he encouraged friends and amateurs to confirm the reality of the Martian canals, despite the fact that they had no experience looking through the telescope. Finally, Lowell often published his results in popular books and periodicals as opposed to the new professional journals, as if public opinion, rather than that of the experts, was the best way to decide the truth of his claims.

Despite his own failings as an astronomer, Lowell succeeded in founding an observatory that advanced the astronomical research agenda of the period. In this regard, he must be credited with selecting as his assistants, V. M. Slipher and C. O. Lampland, whose work won the respect of mainstream astronomers at other observatories. Lowell himself spent most of his last years on the search for Planet X, a far more conventional research project than his Martian work. And though he never located Pluto, his own determination to find the planet inspired a new campaign that culminated in its discovery fourteen years after his death.

Successes and failures aside, what particularly draws our attention to Lowell, as Slipher reminds us, is the way in which he entered science. It is still stunning to consider that, in the winter and spring of 1894, Lowell was simultaneously finishing the proofs for a book on occult practices in Japan and making plans to conduct an astronomical expedition to the Arizona territory. Even at the time, Lowell's contemporaries were struck by the rapidity with which he appeared to put aside his calling as a man of letters to take up a new one as a man of science. After all, the gap between the two cultures was already growing in the 1890s as the age of specialization and professionalization dawned.

We can gain an understanding of Lowell's eagerness to cross disciplinary boundaries by paying attention to significant factors in his own development. Not incidental to his polymath inclinations was the fact that he never had to worry about making a living. Even so, many of Lowell's contemporaries, who were equally wealthy, preferred the staid career of the Brahmin investor. It took courage in this era to put one's credibility on the line in taking up a new career every ten years or so. It required versatility as well. Lowell was not only gifted in science and letters, but he was also eager to exercise his skills in these areas. He enjoyed trying out and contributing to new fields. And, as we have seen, he managed to create successive careers, culminating in his idiosyncratic astronomical work, which drew on the skills he had developed in earlier endeavors.

Lowell was also deeply influenced by his Harvard undergraduate education, which emphasized the unity of all knowledge. The curriculum suggested the ease and necessity of crossing disciplinary boundaries to gain a comprehensive view of knowledge. In the course of this education, Lowell became equally well versed in science and the humanities. Shortly after graduation, he looked to his own friends and classmates as models to demonstrate the viability of a succession of careers or the simultaneous practice of two or more. All the same, none of his friends, not even Henry Adams, attempted, as Lowell did, to cross the line between science and culture.

Lowell was able to accomplish this feat in large part because he disregarded the divide between the two realms. Science, in his view, was a methodology perfectly designed to reveal the core truths of civilizations even when they were geographically and culturally distant. To realize his cosmopolitan aspirations, Lowell investigated increasingly distant cultures, whether in Japan or on Mars, by exploiting scientific methods. His research on trances was intended to reveal the secrets of traditional Japanese culture, and he hoped the telescopic investigation of the surface features of Mars would provide a penetrating glimpse of the Martian way of life.

Lowell's reading of Herbert Spencer encouraged his refusal to delineate science from culture. After all, Spencer's theory required the orderly investigation of natural phenomena—astronomical, geological, and biological—to account for the rise of intelligent life on Earth or other planets. According to Spencer and Lowell, mind and morals were shaped by the matter and forces from which they derived. Thus sociology and ethics were merely subfields of the natural sciences. Cultures, as well, were products of the evolution of brain and body and of the natural environment in which they were located.

While Lowell's bold effort to join science to culture at first blush seems an innovative project, it was in reality an outmoded approach to knowledge of the universe. The methodology he borrowed from Spencer was flawed by the assumption that current conditions were the predictable outcome of a single principle working gradually and inevitably from past to present. In this system there was no room for the cataclysms, chance events, or choice, which a new generation of intellectuals would regard as essential components of the universe.

Furthermore, Spencer's hierarchy of disciplines, in which the experts in the master discipline, cosmology, were accorded special powers, came increasingly under scrutiny. In the new age of disciplinary specialization, the privileging of cosmology contradicted the belief that each discipline was autonomous and responsible for settling its own methodological issues. The crossing of disciplinary boundaries in this setting required either the assistance of a colleague from the other discipline or knowledge of the latest findings in the field. Accordingly, Lowell's and Spencer's approach to knowledge became increasingly outmoded.

The resulting attacks on Lowell's work by his colleagues in astronomy and geology initially aroused Lowell's competitive instincts. Adopting the rhetoric of the day, he insisted on his own expertise in planetary astronomy to buttress the claims he made. Over time, however, his failure to achieve recognition from authorities in his own field was deeply disappointing. As Frederic Stimson noted after Lowell's death, "Mars went back on him and was a disappointment." Lowell himself was convinced that his own brilliance made it difficult for less intelligent men to recognize his work. This rationalization served to blind him to the fact that much of the opposition to his work was generated by his own failure to employ recognized scientific methods or to acknowledge changing norms in the disciplines.[15]

But Lowell's unhappiness almost certainly arose from other sources as well. His efforts to respond to the generational crisis in Boston brought on by challenges from the newly rich industrialists of Chicago and New York exacted a psychological toll. On the one hand, as we have seen, Lowell sought to challenge the traditional norms of Brahmin culture. He carried forward this rebellious posture in his struggle with the astronomical establishment, which became the psychological equivalent of his father. On the other hand, as he settled into his life in Flagstaff, Lowell reinvented himself as a Brahmin in the mold of his father and effectively reverted from the prodigal to the true firstborn son. This vacillation between challenges to authority and efforts to promote and defend it no doubt caused a great deal of psychic tension. For an individualist like Lowell with an intensely self-conscious and introspective nature, such fluctuation must have been difficult to bear. So was the personal isolation brought on by his obsessive efforts to protect his own independence. Lowell passed up opportunities to develop a richer and more intimate personal life with women who were his social and intellectual equals. His staunch individualism thus kept Lowell from savoring the adventurous life he led and enjoying his accomplishments, which would have satisfied lesser men.[16]

NOTES

INDEX

Notes

◆ ◆ ◆

Abbreviations

BPL	Boston Public Library, Boston
DHS	Dedham Historical Society, Dedham, Mass.
GM	Isabella Stewart Gardner Museum, Boston
HL	Houghton Library, Cambridge, Mass.
HUA	Harvard University Archives, Cambridge, Mass.
LOA	Lowell Observatory Archives, Flagstaff, Ariz.
MDLC	Manuscript Division of the Library of Congress, Washington, D.C.
MHS	Massachusetts Historical Society, Boston
MLSA	Mary Lea Shane Archives of the Lick Observatory, Santa Cruz, Calif.
MWPOL	Mount Wilson and Palomar Observatory Library, Pasadena, Calif.
PMS	Peabody Museum of Salem, Salem, Mass.
YUA	Yale University Archives, New Haven, Conn.

In addition, I have used William Graves Hoyt, ed., *Early Correspondence of the Lowell Observatory, 1894–1916: Microfilm Edition,* 10 rolls (Flagstaff, Ariz.: Lowell Observatory, 1973).

Introduction

1. Louise Leonard characterized Lowell as "the athletic type" and noted his passion for doing things "at once." *Percival Lowell: An Afterglow* (Boston: Richard G. Badger, 1921), pp. 30, 41.
2. Barrett Wendell to Frederic Stimson, Nov. 14, 1916, Wendell papers, HL. On Lowell and Roosevelt, see J. D. Fernie, "The Roosevelt of Astronomy," *Journal of the Royal Astronomical Society of Canada* 71 (Feb. 1977): L9–L10. Edwin Hall to A. Lawrence Lowell, April 10, 1908, A. L. Lowell papers, HUA.

3. On Lowell's arrogance, see Stimson to Wendell, undated (context indicates Jan. 1, 1917), Wendell papers, HL.
4. On Lowell's moods, see Leonard, *Percival Lowell,* p. 19. Stimson wrote Wendell: "If P. L. had been without the faculty of introspection, he would have been happy and a complete success," undated (context indicates Jan. 1, 1917). On Lowell's admiration for Wilde, see Lowell to Wendell, Nov. 20, 1883; see also Wendell to Stimson, Nov. 14, 1916; Ralph Curtis to Wendell, Dec. 11, 1914, Wendell papers, HL.
5. On Lowell and Mars, see William Graves Hoyt, *Lowell and Mars* (Tucson: University of Arizona Press, 1976); Michael Crowe, *The Extraterrestrial Life Debate, 1750–1900: The Idea of a Plurality of Worlds from Kant to Lowell* (Cambridge: Cambridge University Press, 1986); Steven Dick, *The Biological Universe: The Twentieth-Century Extraterrestrial Life Debate and the Limits of Science* (Cambridge: Cambridge University Press, 1996); William Sheehan, *Planets and Perception: Telescopic Views and Interpretations, 1609–1909* (Tucson: University of Arizona Press, 1988). On Lowell and the Brahmins, see Ferris Greenslet, *The Lowells and Their Seven Worlds* (Boston: Houghton Mifflin Co., 1945); Ernest Samuels, *Bernard Berenson: The Making of a Legend* (Cambridge, Mass.: Harvard University Press, Belknap Press, 1987); R. W. B. Lewis, *Edith Wharton: A Biography* (New York: Harper and Row, 1975). On Lowell and East Asia, see T. J. Jackson Lears, *No Place of Grace: Antimodernism and the Transformation of American Culture, 1880–1920* (New York: Pantheon Books, 1982); Lawrence Chisolm, *Fenellosa: The Far East and American Culture* (New Haven, Conn.: Yale University Press, 1963).
6. On the Brahmins, see the previous note and note 1 in the introduction to Part I.
7. On the canals debate, see especially Hoyt, *Lowell and Mars.*

I. The Private Lowell

1. Existing biographies of Lowell give short shrift to his private life and the Brahmin milieu. See William Graves Hoyt, *Lowell and Mars* (Tuscon: University of Arizona Press, 1976); A. L. Lowell, *The Biography of Percival Lowell* (New York: Macmillan Co., 1935); Ferris Greenslet, *The Lowells and Their Seven Worlds* (Boston: Houghton Mifflin Co., 1945), pp. 345–367; Louise Leonard, *Percival Lowell: An Afterglow* (Boston: Richard G. Badger, 1921); T. J. Jackson Lears, *No Place of Grace: Antimodernism and the Transformation of American Culture, 1880–1920* (New York: Pantheon Books, 1981), pp. 234–237; on the Brahmins, see also Frederic Cople Jaher, "The Proper Bostonians in the Age of Industrial Capitalism," in Jaher, ed., *The Age of Industrialism in America* (New York: Free Press, 1968), pp. 188–262; Martin Green, *The Problem of Boston: Some Readings in Cultural History* (New York: W. W. Norton and Co., 1966); E. Digby

Baltzell, *Puritan Boston and Quaker Philadelphia: The Protestant Ethic and the Spirit of Class Authority and Leadership* (Boston: Beacon Press, 1979); Cleveland Amory, *The Proper Bostonians* (New York: E. P. Dutton and Co., 1947); Ronald Story, *The Forging of an Aristocracy: Harvard and the Boston Upper Class, 1800–1870* (Middletown, Conn.: Wesleyan University Press, 1980); and Betty G. Farrell, *Elite Families: Class and Power in Nineteenth-Century Boston* (Albany: State University of New York Press, 1993).

2. On expressive individualism, see Robert Bellah, et al., *Habits of the Heart: Individualism and Commitment in American Life* (Berkeley: University of California Press, 1993), and Colin Campbell, *The Romantic Ethic and the Spirit of Modern Consumerism* (Oxford and New York: Oxford University Press, 1987), pp. 183–184, 193–195.
3. Lowell to William Lowell Putnam, May 10, 1901, Lowell family papers, HL. Douglass Shand-Tucci, *The Art of Scandal: The Life and Times of Isabella Stewart Gardner* (New York: Harper, Collins, 1997); John H. Conover, *The Blue Hill Meteorological Observatory: The First 100 Years—1885–1985* (Boston: American Meteorological Society, 1990). Interestingly, Lowell was a cousin of Abbott Lawrence Rotch, Blue Hill's founder, and of Mrs. Gardner's husband, John Lowell Gardner.
4. On Boston culture and finance, see Story, *The Forging of an Aristocracy;* Robert F. Dalzell, Jr., *Enterprising Elite: The Boston Associates and the World They Made* (Cambridge, Mass.: Harvard University Press, 1987); Arthur M. Johnson and Barrie E. Supple, *Boston Capitalists and Western Railroads: A Study in the Nineteenth-Century Railroad Investment Process* (Cambridge, Mass.: Harvard University Press, 1967); Laurence F. Gross, *The Course of Industrial Decline: The Boott Cotton Mills of Lowell, Massachusetts, 1835–1955* (Baltimore: Johns Hopkins University Press, 1993). Lowell to Putnam, April 12, 1916, Lowell family papers, HL. On the Zuni link, see the introduction to Curtis M. Hinsley and David Wilcox, ed., *The Southwest in the American Imagination: The Writings of Sylvester Baxter, 1881–1889* (Tucson: University of Arizona Press, 1996).
5. On his versatility, see Lowell, *The Biography of Percival Lowell,* pp. 49–50. In the spring of 1889, for example, Lowell wrote the Phi Beta Kappa poem delivered annually at Harvard's graduation, and *Noto,* served as treasurer of Lowell Bleachery, managed his finances, and played polo.

1. The Making of an Improper Bostonian

1. See Jean Gould, *Amy: The World of Amy Lowell and the Imagist Movement* (New York: Dodd, Mead and Co., 1975), p. 275.
2. On Jackson, see *The Autobiography of Nathaniel Southgate Shaler with a Supplement by His Wife* (Boston: Houghton Mifflin Co., 1909), p. 109.
3. Lowell to Barrett Wendell, Nov. 20, 1883, Wendell papers, HL. On Wilde, see

Richard Ellmann, *Oscar Wilde* (New York: Alfred A. Knopf, 1988), pp. 182–183; J. E. Chamberlin, *Ripe Was the Drowsy Hour: The Age of Oscar Wilde* (New York: Seabury Press, 1977), pp. 12, 14; and Mary Warner Blanchard, *Oscar Wilde's America: Counterculture in the Gilded Age* (New Haven, Conn.: Yale University Press, 1998).

4. On birth order, see Frank Sullowell, *Born to Rebel: Birth Order, Family Dynamics and Creative Lives* (New York: Pantheon Books, 1996). Lowell to Elizabeth Lowell, Oct. 2, 1882, and undated letter (context indicates spring 1883), Lowell family papers, HL.
5. See A. L. Lowell, *The Biography of Percival Lowell* (New York: Macmillan Co., 1935), p. 92; Katharine Bowlker, "Percival Lowell," p. 7, Lowell family papers, HL.
6. Katharine B. Lowell to Lowell, undated (written in Aug. 1894), Lowell papers, LOA. See also Charles K. Hofling, "Percival Lowell and the Canals of Mars," *British Journal of Medical Psychology* 37 (1964): pp. 38–39, and Lowell's letters to his mother from Silligs School in Vevez, Switzerland, spring 1866, Lowell papers, Boston Atheneum.
7. On Lowell's State Street qualities, see Ralph Curtis to Wendell, Dec. 7, 1916, Wendell papers, HL. On Augustus Lowell, see Wendell to Robert White-Thompson, June 23, 1900, in M. A. DeWolfe Howe, ed., *Barrett Wendell and His Letters* (Boston: Atlantic Monthly Press, 1924), p. 137. On self-assertion, see Peter Gay, *The Bourgeois Experience: Victoria to Freud,* vol. 3: *The Cultivation of Hatred* (New York: Oxford University Press, 1986), p. 425.
8. Ferris Greenslet, *The Lowells and Their Seven Worlds* (Boston: Houghton Mifflin Co., 1945), p. 366. For Lowell's view of personality, see his *The Soul of the Far East* (Boston: Houghton Mifflin Co., 1888), pp. 15–28, 201, and *Occult Japan or the Way of the Gods: An Esoteric Study of Japanese Personality and Possession* (Boston: Houghton Mifflin Co., 1894), pp. 278–289.
9. See Lowell, "Augustus Lowell," *Proceedings of the American Academy of Arts and Sciences* 37 (Aug. 1902): 635–654. Lowell identified his father's Brahmin traits as "judiciousness," "will, ability, integrity"; "his abilities were solid, not showy," ibid., pp. 650–652. On Lawrence and Augustus Lowell, see Cleveland Amory, *The Proper Bostonians* (New York: E. P. Dutton and Co., 1947), pp. 44, 62–63, 70, 91, 324. Henry Yeomans's *Abbott Lawrence Lowell* (Cambridge, Mass.: Harvard University Press, 1948) is not helpful on Lowell family life.
10. "Augustus Lowell," *Proceedings,* pp. 644, 650, 652. On Amy Lowell's objections, see Lowell to Amy Lowell, Aug. 30, 1901, Amy Lowell papers, HL.
11. See G. Peabody to Augustus Lowell, April 12, May 9, 1897; A. G. Cumnock to Augustus Lowell, June 16, Oct. 2, 1882; Lowell to Elizabeth, Sept. 9, Oct. 2, 1882, Oct. 9, 1884, July 11, 1887; Lowell to Putnam, March 7, 1893, Lowell family papers, HL.

12. See "The Technology Merger," *Boston Evening Transcript,* April 25, 1905, p. 10; Samuel C. Prescott, *When MIT Was "Boston Tech," 1861–1916* (Cambridge, Mass.: The Technology Press, 1954), p. 196. A. L. Lowell to Charles Lowell, May 29, 1896. Lawrence wrote Frederic Stimson that the observations of Mars "had been carried nearly as far as it was possible to carry them," Feb. 6, 1917, A. L. Lowell papers, HUA. On the flap over planetary news, see chap. 11.
13. On differences between the brothers, see Wendell to Stimson, Aug. 26, 1880, Wendell papers, HL; on the inauguration, see Lowell to V. M. Slipher, June 7, 1909, Lowell papers, LOA; on the funeral, see Wendell to Stimson, Nov. 14, 1916, Wendell Papers, HL.
14. Lowell to Wendell, April 24, 1877, Wendell papers, HL. Louise Leonard, *Percival Lowell: An Afterglow* (Boston: Richard G. Badger, 1921), p. 20.
15. Lowell to Putnam, Sept. 24, 1891, Lowell family papers, HL. On Alfred Roosevelt's death, see Edmund Morris, *The Rise of Theodore Roosevelt* (New York: Coward, McCann and Geoghegan, 1979), p. 441.
16. Lowell, "Oliver Wendell Holmes," *Proceedings of the American Academy of Arts and Sciences* 30 (1895): 557.
17. See Robert Grant, *Fourscore: An Autobiography* (Boston: Houghton Mifflin Co., 1934), p. 131.
18. Lowell to Elizabeth, May 10, 1882, Lowell family papers, HL; Lowell, "A Visit to Shirane San," *Appalachia* 5 (June 1888): 102.
19. Lowell, "At Giverny," handwritten manuscript in LOA. Lowell to Elizabeth, Aug. 24, Sept. 13, 1882, Lowell family papers, HL.
20. Lowell to Elizabeth, Nov. 28, 1883, Jan. 13, 1896, Lowell family papers, HL; Lowell to Amy, Jan. 15, 1893, Amy Lowell papers, HL; Lowell, *Chosön: The Land of the Morning Calm: A Sketch of Korea* (Boston: Ticknor and Co., 1886), p. 359.
21. Charles Eliot to E. C. Pickering, Nov. 22, 1894, E. C. Pickering papers, HUA. "HWD" blamed the conflict on the fact that "Boston society . . . was very narrow. He (Lowell) offended the sensibilities of this small group, and he was dealt with most severely," *Boston Transcript,* Nov. 19, 1916, in Percival Lowell file, HUA. See also Curtis to Stimson, Nov. 16, 1916, Wendell papers, HL.
22. On self-consciousness, see *The Soul of the Far East,* pp. 21–22. Lowell's prototype for excessive self-consciousness was the hero of Edward Bulwer-Lytton's novel *Kenelm Chillingly: His Adventures and Opinions* (New York: Harper and Bros., 1893). See "Charging Records," Harvard College Library, p. 376, HUA.
23. Lowell to Elizabeth, June 19, 1882, Lowell family papers, HL.
24. On Lowell's breakdown, see Holfing, "Percival Lowell and the Canals of Mars," p. 37. On Lowell's work ethic, see Louise Leonard to V. M. Slipher, Feb. 3, 1908, April 3, 1911, V. M. Slipher papers, LOA. On neurasthenia, see Janet Oppenheim, *"Shattered Nerves": Doctors, Patients, and Depression in Victorian Eng-*

land (Oxford: Oxford University Press, 1991), pp. 99, 16–17, 127–129, 152, 168–169, and Peter Gay, *The Bourgeois Experience: Victoria to Freud,* vol. 2: *The Tender Passion* (New York: Oxford University Press, 1986), pp. 329–350.

25. Putnam to Leonard, Sept. 29, 1897, Lowell papers, LOA. Leonard to Slipher, Aug. 1, 1913, V. M. Slipher papers, LOA. A. L. Lowell to W. R. Thayer, Oct. 31, Nov. 1, 1899; Lowell to A. L. Lowell, Nov. 1, 1899, A. L. Lowell papers, HUA.
26. Lowell to A. E. Douglass, Sept. 17, 1897; T. J. J. See to Douglass, Sept. 13, 1897; See to Douglass, Oct. 31, 1897, Douglass papers, LOA. On Lowell's criticism of his father, see Lowell, *The Biography of Percival Lowell,* p. 98.
27. Putnam to Douglass, Jan. 6, March 4, 1898, March 29, 1899; Leonard to Douglass, Dec. 4, 1899, Douglass papers, LOA. Putnam to See, Jan. 19, 1898, May 11, 1898, Putnam papers, LOA. Putnam to Lowell, April 29, 1898, Lowell papers, LOA. See also Leonard to David Todd, undated letter; Mabel Loomis Todd Journal (May 1900), pp. 112–113, Todd papers, YUA.

2. Lowell and His Peers

1. On men's and women's separate spheres, see Elaine Showalter, *Sexual Anarchy: Gender and Culture at the Fin de Siecle* (London: Penguin, 1990), pp. 8, 9, 24–26, and Brian Harrison, *Separate Spheres: The Opposition to Women's Suffrage in Britain* (New York: Holmes and Meier Publishers, 1978), pp. 97–99. On separate spheres in Boston, see Frederic J. Stimson, *My United States* (New York: Charles Scribner's Sons, 1931), pp. 113, 111. On bachelors, see Howard Chudacoff, *The Age of the Bachelor: Creating an American Subculture* (Princeton, N. J.: Princeton University Press, 1999).
2. Howe later commented: "No Tavern Club could be made from a set of Rotarians"; he added, however, that "we never sank into Bohemianism," Mark DeWolfe Howe *A Partial (and Not Impartial) Semi-Centennial History of the Tavern Club, 1884–1934* (Boston: The Tavern Club, 1934), pp. 3, 24. On Boston gays, see Roger Austen, *Genteel Pagan: The Double Life of Charles Warren Stoddard* (Amherst: University of Massachusetts Press, 1991), pp. 140–141; Douglass Shand-Tucci, *Boston Bohemia, 1881–1900,* vol. 1 (Amherst: University of Massachusetts Press, 1995), pp. 206, 210, 216–217; idem, *The Art of Scandal: The Life and Times of Isabella Stewart Gardner* (New York: Harper Collins, 1998), pp. 87–88. Elaine Showalter argues that Anglo-American men's clubs, despite their facade of aggressive heterosexuality, "existed on the fragile borderline that separated male bonding from homosexuality," *Sexual Anarchy,* pp. 8, 9.
3. On the *Lampoon,* see Robert Grant, *Fourscore: An Autobiography* (Boston: Houghton Mifflin Co., 1934), p. 113.
4. For biographical details on Lowell's three friends, see class reports for the

classes of 1876 and 1877, HUA. On Curtis, see David McKibbon, *Sargent's Boston with an Essay and a Biographical Summary and a Complete Check List of Sargent's Portraits* (Boston: Museum of Fine Arts, 1956), pp. 34–36. Wharton's story was entitled "The Verdict." See R. W. B. Lewis, *Edith Wharton: A Biography* (New York: Harper and Row, 1975), p. 193. See also Barrett Wendell, *A Literary History of America* (New York: Charles Scribner's Sons, 1900).

5. Lowell, Curtis, and Wendell joined the nonexclusive Hasty Pudding Club and the more exclusive Delta Kappa Epsilon fraternity. They, along with Stimson, were members of the Institute of 1770. See *Catalogue of the Alpha Chapter, ΔKE, Institute of 1770,* and *11th Catalogue of the Officers and Members of the Hasty Pudding Club in Harvard College* (Cambridge, 1891), HUA. On law practices, see Stimson, *My United States,* p. 95; and Grant, *Fourscore,* p. 119. Grant and A. Lawrence Lowell served on the committee that upheld Sacco and Vanzetti's conviction in 1927.
6. Lowell's parents entertained his friends regularly. See Barrett Wendell to Frederic Stimson, July 31, 1876; Stimson to Wendell, July 5, 1880, Aug. 1, 1882, Wendell papers, HL; Mrs. Lowell recruited Stimson and Wendell for the Place aux Dames Drama Group in 1887. Katharine Lowell to Stimson, Nov. 2 (year not indicated), Stimson papers, MHS.
7. On bachelor apartments, see Stimson to Wendell, Aug. 26, Sept. 1, 1879, Wendell papers, HL; and Grant, *Fourscore,* p. 189. On Lowell's bachelor life, see Lowell to Wendell, Aug. 11, 1877, Wendell papers, HL.
8. Lowell to Wendell, July 3, 1879, Nov. 29, 1876; Wendell Papers, HL. H. C. Mercer to Lowell, July 14, 1890, Lowell family papers, HL. Lowell, *Occult Japan: The Way of the Gods: An Esoteric Study of Japanese Personality and Possession* (Boston: Houghton Mifflin Co., 1894), p. 342.
9. Bigelow invited Wendell to join Lowell, Frank Chadwick, and Morton Prince at Tuckernuck, undated, Wendell papers, HL. Lowell to A. E. Douglass, Sept. 3, 1895, Douglass papers, LOA; William Sturgis Bigelow to Lowell, Aug. 14, 1894, Lowell Papers, LOA. On Tuckernuck, see John W. Crowley, "Eden off Nantucket: W. S. Bigelow and 'Tuckernuck,'" *Essex Institute Historical Studies* 109 (1973): 3–8.
10. Wendell is quoted in Stow Persons, *The Decline of American Gentility* (New York: Columbia University Press, 1973), p. 274; Stimson to Wendell, July 16, Sept. 24, Nov. 11, 1878, Jan. 17, 1879, June 10, 1880. Wendell to Stimson, Feb. 27, 1879, undated, HL; On upper-class "ennui," see T. J. J. Lears, *No Place of Grace: Antimodernism and the Transformation of American Culture, 1820–1920* (New York: Pantheon, 1982), pp. 4–58.
11. Lowell to Wendell, Jan. 16, Aug. 11, 1877; Wendell to Stimson, undated (context indicates 1878); Stimson to Wendell, Sept. 24, July 16, Oct. 14, 1878, Jan. 17, 1879, Wendell papers, HL.

12. Lowell to Wendell, Oct. 21, 1878, Wendell papers, HL.
13. On Boston clubs, see Grant, *Fourscore,* p. 23; Cleveland Amory, *The Proper Bostonians* (New York: E. P. Dutton and Co., 1947), p. 300; Howe, *A Partial,* p. 5; Alexander W. Williams, *A Social History of the Greater Boston Clubs* (Boston: Borne Publishing Co., 1970). Stimson to Wendell, Nov. 14, 1883, Wendell papers, HL. Wendell and Bigelow were members of both Somerset and St. Botolph. See "St. Botolph Club," and "History of the Somerset Club, 1852–1913," BPL. Grant, *Fourscore,* p. 199. On the Sargent painting, see Shand-Tucci, *The Art of Scandal,* pp. 70–72. Since Lowell joined St. Botolph in 1884, he probably did not meet Wilde there in 1882.
14. On the Mathematical and Physical Club, see "James Mill Peirce," *Science* 24 (July 13, 1906): 47. Lowell also participated in the American Academy of Arts and Sciences on a regular basis.
15. Lowell to T. W. Higginson, April 28, 1887, Authors Club papers, BPL.
16. See Allan Forbes, *Sport in Norfolk County* (Boston: Houghton Mifflin Co., 1938), pp. 11, 14, 21, 28. On the origins of the Dedham Polo Club, see Lowell to Stimson, Nov. 11, 1886, Stimson papers, MHS.
17. *The Vaccination Book,* Dec. 25, 1887, DHS. Stimson to Wendell, April 21, 1879, Wendell papers, HL. On Lowell and Stimson in the Smokies, see Stimson, *My United States,* pp. 61, 178–180; Lowell to Stimson, April 5, 1884, Stimson papers, MHS.
18. *The Vaccination Book,* Aug. 5, 1888, Feb. 22, 1890, DHS; Forbes, *Sport in Norfolk County,* pp. 29. 45. On Brandeis's relations with Brahmins, see Allon Gal, "Brandeis, Progressivism, and Zionism: A Study in the Interaction of Ideas and Social Background," Ph.D. diss., Brandeis University, 1975, pp. 244–249. On Samuel Warren, see Martin Green, *The Mount Vernon Street Warrens: A Boston Story, 1860–1910* (New York: Charles Scribner's Sons, 1989), pp. 105–107.
19. Lowell to Amy Lowell, March 21, 1893, Amy Lowell papers, HL; Jack and Isabella Gardner both signed the *Vaccination Book* on May 5, 1894; on the polo club party, see *The Vaccination Book,* Dec. 31, 1889, DHS, and Forbes, *Sport in Norfolk County,* p. 26. On the Lowell-Gardner friendship, see their correspondence and the *Guest Book for Green Hill, Brookline,* Isabella Stewart Gardner papers, GM. On Gardner, see Shand-Tucci, *The Art of Scandal.* Sports and the classics were two masculine interests that kept women at a distance; see Harrison, *Separate Spheres,* p. 99.
20. The Appalachian Mountain Club, which Lowell joined in 1887, also promoted manliness, health, and autonomy. Lowell delivered papers at club meetings and served as vice president in 1890 and president in 1916. See *Appalachia* 6 (1890): 103; 14 (1916): 104, 108.
21. Lowell to Basil Hall Chamberlain, June 18, 1905, January 26, 1916; J. Orange to Lowell, June 26, 1907, letter in possession of the author; Chamberlain to

Sugiura Toshiro, Nov. 1916, Sugiura papers, Aichi Normal University Archives, Japan. T. J. J. See to Douglass, Sept. 18, 1897, Douglass papers, LOA.

22. Brian Harrison identifies imperialism as a "masculine business which frequently entailed the celibate life or at the very least the neglected wife," *Separate Spheres,* p. 104.
23. On Lowell's ideas regarding marriage and family, see Lowell, *The Soul of the Far East* (Boston: Houghton Mifflin Co., 1888), p. 35, "The Revelation of Evolution: A Thought and Its Thinkers," *Atlantic Monthly* 104 (Aug. 1909): 182, and "Oliver Wendell Holmes," *Proceedings of the American Academy of Arts and Sciences* 30 (1895): 558; Lowell to Elizabeth Lowell Putnam, Jan. 7, 1896, Lowell family papers, HL; and the *Coconino Sun,* May 30, 1907, p. 1.
24. See *Occult Japan,* pp. 280–282.
25. On Victorian sexual practices, see Michael Mason, *The Making of Victorian Sexuality* (New York: Oxford University Press, 1994), p. 6; Karen Lystra, *Searching the Heart: Women, Men, and Romantic Love in Nineteenth-Century America* (New York: Oxford University Press, 1989), p. 116. Lowell wavered as to whether he believed women had sexual needs that had to be fulfilled; see Aug. 26 entry in "Journal—1904," HL. See also postcards from Lowell to Todd, March 13, April 10, and May 5, 1900, Todd papers, YUA.
26. Lowell to Elizabeth, July 9, Aug. 2, Sept. 9, 1882, Lowell family papers, HL.
27. Ferris Greenslet, *The Lowells and Their Seven Worlds* (Boston: Houghton Mifflin Co., 1945), pp. 352–355.
28. See Ellery Segwick, ed., *The Atlantic Harvest* (Boston: Little, Brown, 1947), pp. 603–607. The *Massachusetts General Hospital Book* records on Jan. 27, 1887, Chapman's "amputation for self-inflicted burn up to the wrist" of the left hand, but offers no explanation of how the wound was inflicted. The letters of J. J. Chapman to O. W. Holmes, Jr., March 16 and 19, 1887, are similarly silent, Holmes papers, Harvard Law School. Apparently Lowell and his assistants had a brief fling with some Flagstaff women in 1894. Lowell asked Douglass to extend greetings to the "fairer, the picnic, half of our world" and spoke of "wild oats planted in Flagstaff." See Douglass to Lowell, July 6, 1894, Lowell to Douglass, July 17, July 28, 1894, Lowell and Douglass papers, LOA.
29. Lowell found Berenson "acute and agreeable," see Aug. 23, Aug. 27, Sept. 28, Oct. 4 entries in his "Journal—1904," HL. Berenson wrote Wharton (spring 1912) that he was "enjoying solid Bostonian talks with the astronomer Percival Lowell," as quoted in Ernest Samuels, *Bernard Berenson: The Making of a Legend* (Cambridge, Mass.: Harvard University Press, Belknap Press, 1987), p. 139. On Adams and Cameron, see Patricia O'Toole, *The Five of Hearts: An Intimate Portrait of Henry Adams and His Friends, 1880–1918* (New York: C. Potter, 1990), p. 200.
30. Ralph Curtis to Wendell, March 15, 1915, Wendell papers, HL. Louise Leonard,

Percival Lowell: An Afterglow (Boston: Richard G. Badger, 1921), p. 20. Lowell, "At Giverny," manuscript, LOA. Lowell asked for 400 Temptation Key West cigars to be sent to him in southern France in 1900. Douglass to Charles Deuel, April 9, 1900, Douglass papers, LOA.

31. While Wendell appreciated Lowell's style ("He looked very beautiful and happy"), Curtis, who was "horrified by his appearing in yellow shoes and a silk hat . . . in the evening to go with morning overcoat," lamented that "savants are hors la loi." Curtis to Wendell, May 1, 1912, July 1914, Aug. 18, 1914; Wendell to Stimson, Sept. 11, Oct. 29, 1915, Wendell papers, HL. Leonard, *Percival Lowell*, pp. 107, 121.
32. Aug. 29, Sept. 13 entries in "Journal—1904," HL.
33. Aug. 25, Sept. 28 entries in ibid.
34. Lowell to Irva Struthers, letters number 13, 19, 21, and 23, Lowell papers, LOA. Note that the Strutherses are listed in the *Social Register, Philadelphia, 1903* 17 (Nov. 1902): 142. On the exhibitions of her painting, see the Historical Society of Pennsylvania to author, June 26, 1998. Though Lowell claimed to "stick to the church (doubtless from my early religious training)," he was an agnostic and hostile to Christianity. See Lowell to George Colby Chase, president of Bates College, Feb. 8, 1902, Lowell papers, LOA.
35. See Edith Pettit, "Frederick MacMonnies: Portrait Painter," *International Studio* 29 (Oct. 1906): 319–324. Miss Pettit was listed in the *Social Register, Philadelphia, 1906,* 20 (Nov. 1905): 131. Her father was a member of the Union League, the Merion Cricket Club, and the Penn Club. Historical Society of Pennsylvania to author, June 26, 1998.
36. July 25, Aug. 15 entries in "Journal—1904," HL.
37. Aug. 12, 13, 14, 15, 19, 22, 28 entries in ibid.
38. Sept. 6, 19, 27, 28, Oct. 2 entries in ibid. On American artists at Giverny, see Mary Smart, *A Flight with Fame: The Life and Art of Frederick MacMonnies (1863–1937)* (Madison, Conn.: Sound View Press, 1996), and William H. Gerdts, *Monet's Giverny: An Impressionist Colony* (New York: Abbeville Press Publishers, 1993).
39. Aug. 17, 22, 24, 28, Sept. 1, 11, 26, 27, 30 entries in "Journal—1904," HL. Lowell flirted with other women in the summer of 1904. See entries for July 31, Aug. 1, Aug. 17, Sept. 11, Sept. 16, Oct. 6.
40. Sept. 5, 7 entries in "Journal—1904," HL; Lowell to Struthers, Dec. 7, 1903 (letter #25), LOA. Ralph Curtis agreed that Lowell was unhappy and overworked; Curtis to Wendell, Feb. 2, 1910, Wendell papers, HL.
41. There is no record of Lowell's feelings about his mother's death. Lowell purchased 11 West Cedar Street from Keith on Nov. 30, 1896. See purchase agreement, LOA.
42. On Leonard and Lowell, see Jan Hollis, "Wrexie," *Griffith Observer* 56 (Jan.

1992): 10–17, as well as Leonard's letters, LOA. See also her "Drawings of Mars," *Popular Astronomy* 15 (Aug.–Sept. 1907): 387–388. Observatory viewing records indicate that Leonard was a frequent observer, fall, 1896, LOA.

43. Lowell to Eliot Clarke, Jan. 23, 1896 (two letters), Feb. 10, 1896, Eliot Clarke papers, HL. For Lowell's letters to Leonard, see Leonard, *Percival Lowell.* Leonard to Lowell, June 23, 1907, Lowell papers, LOA. Leonard to V. M. Slipher, Dec. 9, 1907, V. M. Slipher papers, LOA.
44. On Mrs. Lowell's loyalty, see Henry Giclas's memoir, "Lowell Observatory" (author's possession). Curtis to Wendell, Nov. 29, 1916, Wendell papers, HL.
45. J. K. Macdonald to Lowell, May 17, 1905, Lowell to G. E. Kissel, Oct. 27, 1906, Lowell papers, LOA. See also Oct. 10 entry in "Journal—1904," HL. Macdonald to Leonard, Sept. 1909. On the Lowells' lawsuits, see *Boston Post,* Dec. 9, 1915, *Boston Transcript,* Aug. 28, 1915, *Boston Herald,* Oct. 7, 1915, *Boston Transcript,* n.d., Lowell file, HUA. The *Boston Post* claimed that Mrs. Lowell "secured control of several estates on Beacon Hill," June 10, 1908, HUA.
46. Leonard to E. C. Slipher, June 1, 1908, LOA. Constance Lowell to V. M. Slipher, Nov. 29, 1916, V. M. Slipher papers, LOA. Though Lawrence Lowell failed to mention Louise Leonard in the biography of his brother, he made monthly payments to support her after the 1929 crash destroyed her savings. A. L. Lowell to Helen Goodell, May 27 and June 26, 1937, LOA.

3. Preparation of a Polymath

1. Men of modest means, including Simon Newcomb, Joseph Henry, and William Whewell, were genuine polymaths.
2. For the text of his poem, see HUA. On Holmes, see Eleanor M. Tilton, *Amiable Autocrat: A Biography of Oliver Wendell Holmes* (New York: Henry Schuman, 1947). On Peirce, see Carolyn Eisele, "Benjamin Peirce," *Dictionary of Scientific Biography* (New York: Charles Scribner's Sons, 1979–80), vol. 10, pp. 478–481.
3. On his courses, see "Yearly Returns, 1872–76," HUA; on his essays, see HUA. Lowell to John Trowbridge, Nov. 7, 1910, Lowell papers, LOA. On the polymath phenomenon, little work has been done. See, for example, S. S. Schweber, "Scientists as Intellectuals: The Early Victorians," in James Paradis and Thomas Postlewait, ed., *Victorian Science and Victorian Values: Literary Perspectives* (New Brunswick, N.J.: Rutgers University Press, 1981), pp. 1–37; Martin J. S. Rudwick, *The Great Devonian Controversy: The Shaping of Scientific Knowledge among Gentlemanly Specialists of Knowledge* (Chicago: University of Chicago Press, 1985), pp. 17–18.
4. On curricular change, see Laurence R. Veysey, *The Emergence of the American University* (Chicago: University of Chicago Press, 1965); Hugh Hawkins, *Between Harvard and America: The Educational Leadership of Charles W. Eliot*

(Oxford: Oxford University Press, 1972); Paul Buck, ed., *Social Sciences at Harvard, 1860–1920* (Cambridge, Mass.: Harvard University Press, 1965); Frederick Rudolph, *Curriculum: A History of the American Undergraduate Course of Study since 1636* (San Francisco: Jossey-Bass, 1977); Robert A. McCaughey, "The Transformation of American Academic Life: Harvard University, 1821–1892," *Perspectives in American History* 8 (1974): 239–332. On the decline of prescription, see Hawkins, *Between Harvard and America,* p. 95.

5. Francis Bowen, *The Principles of Political Economy Applied to the Conditions, the Resources and the Institutions of the American People,* 5th ed. (Boston: Little, Brown and Co., 1856), p. 68.
6. Joseph Henry papers, Smithsonian Institution, as cited in Albert E. Moyer, *Joseph Henry: The Rise of an American Scientist* (Washington, D.C.: Smithsonian Institution Press, 1997), p. 156.
7. See *The Education of Henry Adams* (Boston: Houghton Mifflin Co., pp. 54–69. On another Harvard polymath, see James Turner, *The Liberal Education of Charles Eliot Norton* (Baltimore: Johns Hopkins University Press, 1999).
8. Concentrations (Majors) were introduced at Harvard after 1900. See Samuel Eliot Morison, *Three Centuries of Harvard* (Cambridge, Mass.: Harvard University Press, 1936), p. 446. On Eliot's conception of electives see Hawkins, *Between Harvard and America,* pp. 88–90.
9. See A. L. Lowell, *The Biography of Percival Lowell* (New York: Macmillan Co., 1935), p. 192. Frederic Stimson to Barrett Wendell, May 6, 1916, Wendell papers, HL; Paul Bourget, *Outre-mer* (Paris: Lemerre, 1893), as quoted in M. D. Howe, *A Partial (And Not Impartial) Semi-Centennial History of the Tavern Club 1884–1934* (Boston: Tavern Club, 1934), p. 70.
10. On these five graduates, see their Harvard College class biographies and Henry Aaron Yeomans, *Abbott Lawrence Lowell, 1856–1943* (Cambridge, Mass.: Harvard University Press, 1948); Frederic J. Stimson, *My United States* (New York: Charles Scribner's Sons, 1931); M. A. DeWolfe Howe, *Barrett Wendell and His Letters* (Boston: Atlantic Monthly Press, 1924); Robert Grant, *Fourscore: An Autobiography* (Boston: Houghton Mifflin Co., 1934).
11. See especially chaps. 5 and 6 in Betty G. Farrell, *Elite Families: Class and Power in Nineteenth-Century Boston* (Albany: State University of New York Press, 1993).
12. Lowell, *The Biography of Percival Lowell,* p. 8; Stimson to Wendell, June 20, 1880; Wendell to Stimson, June 5, 1915, Wendell papers, HL.
13. Lowell, *The Biography of Percival Lowell,* p. 49. On Lowell's wealth, see his Aug. 1900 assets, Lowell papers, LOA. On Augustus's gift to his sons, see Augustus Lowell will, Feb. 7, 1900, Norfolk County Probate Court, Dedham, Mass.
14. Lowell to Wendell, May 2, 1884, Wendell papers, HL. Lowell, *The Biography of Percival Lowell,* p. 49; *Harvard College Class of 1876: 5th Report* (1894), pp. 34–35.

15. Lowell to Elizabeth Lowell, Sept. 2, 1887, Lowell family papers, HL.
16. On the Lowells, Putnam, and leading Boston investors, see Lowell to T. J. Coolidge, Nov. 10, 1902; Lowell to William Lowell Putnam, Jan. 25, 1903, Feb. 17, 1903; and Lawrence Lowell to Lowell, May 24, 1907, Lowell papers, LOA. On Putnam, see "Biography Prepared for the Permanent Exhibit of MIT, March 21, 1927," Schlesinger Library, Radcliffe College; and Farrell, *Elite Families,* p. 130.
17. Lowell to Emlin Roosevelt, Dec. 11, 1901, Feb. 25, 1902, April 7, 1908; Lowell to James Roosevelt, April 9, April 13, 1908, Nov. 23, Dec. 9, 1914, Feb. 4, 1915, Lowell papers, LOA.
18. On Putnam's fee, see Sept. 23, 1903, account of Lowell income; Lowell to Putnam, May 22, 1903; J. K. Macdonald to Lowell, March 21, 1907; Lowell to William B. Dana Co., undated. Lowell to Lawrence, Aug. 4, 1903; Lowell to Putnam, July 17, 1903, Lowell papers, LOA.
19. On Lowell and J. P. Morgan, see Lowell to Putnam, March 18, 1901, Oct. 9, 1911, March 16, 1903 Lowell family papers, HL. Lowell and Morgan met at annual meetings of the Half Moon Club; Lowell to Morgan, July 9, 1915. On Lowell's interest in holding companies, see Lowell to Mr. Lewis (official with United Gas Improvement Co.) March 23, 1911; Lowell to James Roosevelt, June 28, Aug. 20, 1915. See also Lowell to Putnam, Sept. 28, 1900, May 10, 1901, May 13, Aug. 18, 1901, April 8, April 20, May 11, 1903, March 30, 1905, April 12, 1916. On Lowell's belief in bold management, see Lowell to Putnam, undated, re Plymouth Cordage: "directors are too old fogeyish," and Sept. 2, 1911, Lowell papers, LOA.
20. Lowell to Putnam, Aug. 12, 1907, Lowell papers, LOA.
21. Lowell to Putnam, Feb. 22, April 20, May 11, 1903, Lowell papers, LOA. Lawrence to Lowell, May 6, 1907, LOA. *New York Times,* Jan. 3, 1909, p. 7; see also Arthur M. Johnson and Barrie E. Supple, *Boston Capitalists and Western Railroads: A Study in the Nineteenth-Century Railroad Investment Process* (Cambridge, Mass.: Harvard University Press, 1967), pp. 292–325, and Farrell, *Elite Families,* p. 159. On Lowell's view of the Santa Fe, see Lowell to Putnam, April 12, 1916, Lowell Family papers, HL.
22. *New York Times,* Jan. 3, 1909, p. 7. See also Wade H. Wright, *History of the Georgia Power Company, 1855–1956* (Atlanta: Georgia Power Co., 1957), pp. 25, 75, 94–96.
23. See *New York Times,* Jan. 3, 1909, p. 7; see also "In the Matter of the Estate of Percival Lowell: Inventory and Appraisement," Coconino County, Arizona. Lowell to P. S. Arkwright, October 29, 1908, July 28, 1910; Lowell to H. M. Atkinson, March 23, 1911, Jan. 18, 1906, Sept. 8, 1907; Lowell to president and directors of Atlanta and Birmingham Construction Company, Jan. 13, 1909, Lowell papers, LOA. On the family involvement in the southern investment strategy, see Lowell to Farmer's Loan and Trust, July 18, 1910; Lowell to Katie

(Katharine) Lowell, Oct. 2, 1909; and Lowell to F. W. Notman, April 4, 1907, Lowell papers, LOA. Lowell, Putnam, and Lawrence Lowell oversaw Katharine, Amy, and Elizabeth's trusts and bought stocks for themselves and the trusts at the same time, after consultation with each other and sometimes with the women. Lowell to Jackson and Curtis; Nov. 4, 1909; Lowell to Amy Lowell, Nov. 4, 1909; Lowell to Emlin Roosevelt, Dec. 11, 1901, Feb. 25, 1902; Lowell to T. J. Coolidge, Nov. 18, 1902, May 9, 1911; Lowell to Lee, Higginson and Company, Nov. 3, 1904, Lowell papers, LOA. Lowell to Amy, July 8, 1911, Amy Lowell papers, HL.

24. On future governance of the observatory, see the will drawn up by Putnam and signed on Dec. 15, 1915, Lowell papers, LOA. Lowell's father shared management of the observatory during his son's illness. On Putnam as director, see Putnam papers, LOA.
25. On observatory history after 1916, see William Lowell Putnam, et al., *The Explorers of Mars Hill: A Centennial History of Lowell Observatory, 1894–1994* (West Kennebunk, Me.: Phoenix Publishing, 1994); on the search for Pluto, see William Graves Hoyt, *Planets X and Pluto* (Tucson: University of Arizona Press, 1980).
26. The $25,000 figure is the estimated income from an investment of $500,000 at 5 percent interest. Observatory financial statements, LOA. See Putnam to A. E. Douglass, Aug. 1, 1898, Douglass papers, LOA.
27. Lowell's income for six and one-half months was listed on Sept. 23, 1903, as $55,722. Prorating that sum yields an annual income of about $100,000. No full list of investments is available for this period, but Lowell's estate was valued in 1916 at just over $2,000,000. See "In the Matter of the Estate of Percival Lowell: Inventory and Appraisement." See also Hoyt, *Lowell and Mars*, pp. 103, 187, 189.
28. Lowell to U.S. Treasury Dept., Sept. 25, 1915; Lowell to William Gibbs McAdoo, Jan. 26, 1916; Lowell to Senator Murray Crane, Jan. 27, 1916, LOA. William Sturgis Bigelow to Henry Cabot Lodge, Feb. 7, 1916, in Akira Murakata, "Selected Letters of Dr. William Sturgis Bigelow," Ph.D. diss., George Washington University, 1971, p. 406.
29. Wendell to Stimson, Oct. 25, 1881, Wendell papers, HL; A. G. Cumnock to Augustus Lowell, June 16, 1882, Oct. 2, 1882, Lowell family papers, HL. See Lawrence F. Gross, *The Course of Industrial Decline: The Boott Cotton Mills of Lowell, Massachusetts, 1835–1955* (Baltimore: Johns Hopkins University Press, 1993), pp. 45–48.
30. For details of staffing, see William Graves Hoyt, *Lowell and Mars* (Tucson: University of Arizona Press, 1976), pp. 123–125, 129–130, and George Webb, *Tree Rings and Telescopes* (Tucson: University of Arizona Press, 1983), chap. 3. For years after he was fired in 1901, Douglass sought unsuccessfully to restore rela-

tions with Lowell. Douglass to Lowell, Nov. 11, 1906, Douglass papers, University of Arizona Archives.

31. Lowell to V. M. Slipher, June 6, 1904, Aug. 4, 1904; Slipher to Lowell, Aug. 22, 1905; Lowell to Carl Lampland, Sept. 26, 1908; Lowell to Harry Hussey, June 14, 1915, Lowell papers; Lampland to Slipher, March 28, 1909, V. M. Slipher papers, LOA. See also Robert Smith, "Red Shifts and Gold Medals," and Henry L. Giclas, "The Instrumental Man, 1894–1956," in Putnam et al., *The Explorers of Mars Hill,* pp. 43–66, 107–124.
32. Lawrence to Lowell, June 13, 1901. Lowell to Henry S. Pritchett, June 30, 1902, July 2, 1902, undated, Lowell papers, LOA.
33. Pritchett to Lowell, Oct. 7, Oct. 25, 1902, Jan. 17, April 5, 1903; Lowell to Pritchett, Oct. 7, Oct. 25, 1902, Lowell papers, LOA. See "Biography," E. L. Putnam papers, Schlesinger Library, Radcliffe College.
34. Lowell to Elizabeth, Sept. 9, 1882, Lowell family papers; Lowell, *Chosön: The Land of the Morning Calm: A Sketch of Korea* (Boston: Ticknor and Co., 1886), p. 59.
35. Lowell to Elizabeth, Aug. 18, 1882, Lowell family papers.
36. Lowell, *The Biography of Percival Lowell,* p. 3.
37. Lowell to Hussey, Dec. 16, 1901, Lowell papers, LOA. Hussey was a former member of the British Merchant Marine.
38. *Chosön,* 59. On "mere tourists," see Lowell to Wendell, April 15, 1890, Wendell papers, HL. On "globe-trotters," see Lowell, *Occult Japan: or the Way of the Gods; An Esoteric Study of Japanese Personality and Possession* (Boston: Houghton Mifflin Co., 1894), p. 101.
39. Lowell to Wendell, Nov. 29, 1876, Wendell papers, HL; Lowell to Elizabeth, May 17, 1886, Lowell family papers, HL.
40. *Chosön,* p. 299; Lowell, *Mars and Its Canals* (New York: Macmillan Co., 1906), pp. 5–8, 11.
41. Lowell, *Mars* (Boston: Houghton Mifflin Co., 1895), pp. 85, 88, 92–93; Lowell to Amy, Nov. 11, 1916, Amy Lowell papers, HL.
42. Lowell to Wendell, Aug. 24, Nov. 29, 1876, Oct. 21, 1878, Jan. 12, 1884, Oct. 1, 1879, Wendell papers, HL. Lowell to Stimson, Jan. 12, 1884, Stimson papers, MHS. *Chosön,* p. 107.
43. *Mars and Its Canals,* pp. 382–3; Lowell, *Mars as the Abode of Life* (New York: Macmillan Co., 1908), p. 216; for Lowell's view of Martians, see chap. 6 of *Mars.*
44. Lowell, "Oliver Wendell Holmes," *Proceedings of the American Academy of Arts and Sciences* 30 (1895): 558.
45. *Chosön,* p. 84; Lowell to Elizabeth, June 20, 1883, Lowell family papers, HL.
46. See Lowell to Wendell, May 2, 1884, Wendell papers, HL; Lowell to Elizabeth, Feb. 7, 1889, Lowell family papers, HL.

47. See John Gardner diary (June 19, 21, 22, July 4, 14, 1883), Isabella Stewart Gardner papers, Gardner Museum. Reports on Lowell's knowledge of Japanese vary. Bigelow marveled at his proficiency; See Ferris Greenslet, *The Lowells and Their Seven Worlds* (Boston: Houghton Mifflin Co., 1945), p. 349. By contrast, W. B. Mason denied that Lowell spoke Japanese "with comparative ease" or that he could write it at all; see "*Percival Lowell: An Afterglow* (1921)," *Japan Weekly Chronicle,* Dec. 8, 1921, p. 815. Lowell used a translator in his interviews with the head of the Shinto sect in Tokyo. Lowell to Putnam, Aug. 29, 1891, Lowell family papers, HL. See also *A Handbook for Travelers in Japan,* 3rd ed., revised and for the most part rewritten by Basil Hall Chamberlain and W. B. Mason (London: J. Murray, 1891).
48. On Takamine's hospitality, see Edward Sylvester Morse, *Japan Day-by-Day: 1877, 1878–79, 1882–83* (Boston: Houghton Mifflin Co., 1917), vol. 2, p. 212. On Miyaoka as European, see Lowell to Elizabeth, Dec. 20, 1883, Lowell family papers, HL.
49. On Miyaoka's English, see Lowell to Elizabeth, May 13, 1884, Lowell family papers, HL. On Takamine, Miyaoka, and Takenaka, see Morse, *Japan Day-by-Day,* 1, pp. xii–xiii, 388; 2, pp. 212, 324.
50. Lowell to Wendell, May 2, 1884, Wendell papers, HL; Lowell to Amy, April 10, 1891, Amy Lowell papers, HL.
51. Lowell to Wendell, May 2, 1884, Wendell papers, HL; Lowell to Putnam, April 22, 1891; Lowell to Elizabeth, Jan. 14, 1889, Lowell family papers, HL.
52. Lowell to Anna Lowell, April 25, 1884, A. L. Lowell papers, HUA; Lowell to Elizabeth, April 22, 1891, May 28, 1884, Lowell family papers, HL. Lowell to Amy, April 10, 1891, Amy Lowell papers, HL.
53. Lowell to Katharine, July 16, 1883; Lowell to Alfred Roosevelt, April 29, 1889; Lowell to Elizabeth, June 20, 1883, Lowell family papers, HL; Lowell to Wendell, July 14, 1883, April 13, 1884, Wendell papers, HL; Lowell to Stimson, April 5, 1884, Stimson papers, MHS. Lowell to Elizabeth, May 1, 1889, Lowell family papers. Lowell to Wendell, June 23, 1884, Wendell papers, HL.
54. Lowell, *Noto: An Unexplored Corner of Japan* (Boston: Houghton Mifflin Co., 1891), p. 9; Lowell to Lawrence, Aug. 14, 1883, A. L. Lowell papers, HUA; Lowell to Wendell, July 14, 1883, Wendell papers, HL; Lowell to Katharine, July 29, 1883; Lowell to Putnam, May 25, 1891, Lowell family papers, HL. *The Biography of Percival Lowell,* p. 14; *Chosön,* p. 69. Lowell to Elizabeth, May 16, 1889, Dec. 5, 1891, July 22, 1883, Lowell family papers, HL. On the Korean mission to the United States to ratify the treaty establishing relations between the two countries, see Gary Walter, "The Korean Special Mission to the United States of America," *Journal of Korean Studies* 1 (1969): 89–142.
55. Lowell to the Roosevelts, July 29, 1883, Lowell family papers; Lowell, *The Biography of Percival Lowell,* p. 14; *Noto,* p. 9; *Chosön,* p. 80.

56. Lowell to Elizabeth, July 3, 1884, May 16, 1889, July 22, 1883, Lowell family papers, HL; *Chosön,* p. 71; Lowell to Amy, Aug. 4, 1883, Amy Lowell papers, HL; Lowell to Wendell, Aug. 31, 1883, Wendell papers, HL.
57. Lowell to Elizabeth, July 22, 1883, July 29, 1883, Lowell family papers, HL.
58. On his mountain climbing, see Lowell to Putnam, Aug. 16, 1891, Aug. 19, 1891, Lowell family papers, HL; *Noto,* chaps. 14–17; *Occult Japan,* chap. 1; Lowell "A Visit to Shirane San," *Appalachia* 5 (June 1888): 87–108; and Rosewell B. Lawrence, "Bandai-san, Miomote, and Matsushima," *Appalachia* 6 (Dec. 8, 1891): 30–39. See also Lowell, "An Ascent of Fuji," *Appalachia* 7 (March 9, 1892): 91, and "The Mountains of Mars," *Appalachia* 7 (Feb. 13, 1895): 367. On Lowell's interest in volcanoes, see "A Visit to Shirane San," p. 102; on his river floats, see *Noto,* pp. 234–261, and Lowell to Putnam, April 23, 1889, Lowell family papers, HL.

4. New Careers

1. On Lowell's versatility, see A. L. Lowell, *The Biography of Percival Lowell* (New York: Macmillan Co, 1935), pp. 48–49.
2. On his rejection of realism, see Lowell to Frederic Stimson, Jan. 12, 1884, Stimson papers, MHS.
3. C. David Heymann claims that the Lowells' "dinner table conversation, even when the family dined alone, was intense and often brilliant," *American Aristocracy: The Lives and Times of James Russell, Amy and Robert Lowell* (New York: Dodd, Mead and Co., 1980), p. 161. See Katharine Bowlker, "Percival Lowell" (pp. 5–6), Lowell family papers, HL; on Lowell's poetry, see Ferris Greenslet, *The Lowells and Their Seven Worlds* (Boston: Houghton Mifflin Co., 1945), p. 347; Lowell to Barrett Wendell, Nov. 19, 1876, Wendell papers, HL.
4. Lowell to Elizabeth Lowell, Feb. 17, 1884, Lowell family papers, HL. On the letter books, see Lowell to Stimson, May 30, 1883, Stimson papers, MHS; Lowell, *Biography of Percival Lowell,* p. 29.
5. Lowell to Stimson, April 5, Jan. 12, 1884, Stimson Papers, MHS. Lowell to Wendell, July 31, 1876, Sept. 3, May 21, 1884; Wendell to Stimson, July 31, 1876, Wendell papers, HL. Lowell to Stimson, undated (the context indicates early 1884), as quoted in Greenslet, *The Lowells and Their Seven Worlds,* p. 354. See also Lowell to Elizabeth, Sept. 9, 1887, Lowell family papers, HL.
6. Lowell to his mother, Oct. 6, 1884, as quoted in Lowell, *Biography of Percival Lowell,* p. 31; he read "a little of mine" to the Alfred Roosevelts. Lowell to Elizabeth, May 3, 1886. Lowell family papers, HL.
7. On the New York Authors Club, see "The Authors' Club," *Literary World* 20

(March 30, 1889): 112–113; "The Authors' Decennial Dinner," *Critic* 22 (March 4, 1883): 131–133. See also Edward Avery to Stimson, Jan. 8, 1886, Stimson papers, MHS; *Boston Daily Advertizer,* April 13, 1887, p. 1. Lowell to T. W. Higginson, April 28, 1887, Authors Club papers, Boston Public Library. Lowell delivered a "Smoke Talk" on "Japanese Trances" on March 21, 1894, and another on "The Planet Mars" at St. Botolph on Jan. 2, 1895, St. Botolph Club papers, BPL.

8. See *Boston Evening Transcript,* April 13, 1887, p. 4. See also "Authors Club Circular" and response, Authors Club papers, BPL.
9. Lowell to Arthur Wentworth Eaton, undated, Eaton papers, HL. *Boston Daily Advertizer,* April 13, 1887, p. 1 and editorial. *Boston Evening Transcript,* April 11, 1887, p. 2; April 12, 1887, p. 1; April 13, 1887, p. 4. Authors Club papers, BPL.
10. J. T. Wheelwright to Higginson, May 6, 1887; Lowell to Higginson, April 26, 1887, Authors Club papers, BPL. Mildred Buchanan Flagg, *Notable Boston Authors: Members of the Boston Authors Club, 1900–1966* (Cambridge, Mass.: Dresser, Chapman and Grimes, 1965), p. 3.
11. Lowell to Stimson, early Sept. 1884, as quoted in Lowell, *Biography of Percival Lowell,* pp. 30–31.
12. Lowell to William Lowell Putnam, Jan. 24, 1903, Lowell family papers, HL. Lowell, *Mars and Its Canals* (New York: Macmillan Co., 1906), p. viii.
13. Lafcadio Hearn to Yrjo Hirn, Aug. 1902, in Elizabeth Bisland, ed., *The Life and Letters of Lafcadio Hearn* (Boston: Houghton Mifflin Co., 1906), vol. 2, p. 479. *Atlantic Monthly* 57 (May 1886): 690–697.
14. Lowell read Edward Bulwer-Lytton's *The Parisians* (New York: Harper and Bros., 1875) and *Kenelm Chillingly: His Adventures and Opinions* (New York: Harper and Bros., 1873) as an undergraduate. See "Charging Records," Harvard College Library (1875–76), p. 369, HUA; on his passion for Stevenson, see Basil Hall Chamberlain to Hearn, Aug. 5, 1893, in Kazuo Koizumi, ed., *Letters from Basil Hall Chamberlain to Lafcadio Hearn* (Tokyo: Hokuseido Press, 1936), p. 33. On novelists and male romances, see Elaine Showalter, *Sexual Anarchy: Gender and Culture at the Fin de Siècle* (London: Penguin, 1990), pp. 80, 108.
15. See *Mars and Its Canals,* p. viii; Lowell, "The Revelation of Evolution," *Atlantic Monthly* 104 (Aug. 1909): 177–178; Lowell, *The Evolution of Worlds* (New York: Macmillan Co., 1909), pp. 136–137.
16. "The Revelation of Evolution," p. 182.
17. Lowell, *Mars,* (Boston: Houghton Mifflin Co., 1895), p. 86; *Mars and Its Canals,* p. 204. Lowell insisted that all great scientists faced public hostility to their discoveries. "The Revelation of Evolution," pp. 180–181.
18. Lowell, *Noto: An Unexplored Corner of Japan* (Boston: Houghton Mifflin Co.,

1891), chaps. 14–17; Lowell, *Chosön: The Land of the Morning Calm: A Sketch of Korea* (Boston: Ticknor and Co., 1886), p. 76.

19. Lowell, "A Korean Coup d'État," *Atlantic Monthly* 58 (Nov. 1886): 599–618 (p. 616); Lowell, "The Tale of a Japanese Reformer," *Atlantic Monthly* 66 (Nov. 1890): 680–693.
20. On Schiaparelli, see Piero Tempesti, "Schiaparelli poeta," *L'astronomia* 12 (March 1990): 25–31; on Lowell's taste in poetry, see Koizumi, ed., *Letters from Basil Hall Chamberlain,* p. 33.
21. Lowell's letter to Stimson is quoted in Greenslet, *The Lowells and Their Seven Worlds,* pp. 352–353.
22. See William Sheehan, *Planets and Perception: Telescopic Views and Interpretations, 1609–1909* (Tucson: University of Arizona Press, 1988), pp. 181–182.
23. "Sakura no Saka," Phi Beta Kappa Poem, HUA; see Greenslet, *The Lowells and Their Seven Worlds,* p. 352; "Ontake," The First Book of the Authors Club, *Libre Scriptorum* (New York, 1893), pp. 375–385; "Mars" is dated from internal evidence, LOA; *Can Grande's Castle* (New York: Macmillan Co., 1919), pp. 57–58.
24. Kazuo Koizumi, ed., *More Letters from Basil Hall Chamberlain to Lafcadio Hearn* (Tokyo: Hokuseido Press, 1937), pp. 96–97, 101.
25. For Lowell's letter and reactions to it, see *Japan Weekly Mail,* June 3, 1893, pp. 648, 653–654; June 10, 1893, pp. 664–665, 676–677; June 17, 1893, pp. 692, 705–706; June 24, 1893, p. 731.
26. *Boston Commonwealth,* March 2, 1895, p. 137. *Boston Globe,* Nov. 9, 1906.
27. Lowell to Putnam, May 11, 1903, Lowell family papers, HL.
28. Michael L. Carlebach, *The Origins of Photojournalism in America* (Washington, D. C.: Smithsonian Institution Press, 1992), chap. 4.
29. Lowell to Wendell, Sept. 30, 1883, Wendell papers, HL; Lowell to Elizabeth, April 19, June 11, 1884, Lowell family papers, HL; Lowell to Lawrence Lowell, Aug. 10, 1883, A. L. Lowell papers, HUA. On the popularity of the dry plate, see Beaumont Newhall, *The History of Photography* (New York: Museum of Modern Art, 1982), p. 124. *Chosön,* pp. 153–161.
30. Lowell to Amy Lowell, May 11, 1893, Amy Lowell papers, HL.
31. Greenslet, *The Lowells and Their Seven Worlds,* pp. 352–353; Lowell to Wendell, Feb. 18, 1884, Wendell papers, HL; Lowell to Lawrence, April 24, 1884, A. L. Lowell papers, HUA. *New York Times,* Jan. 4, 1886, p. 3.
32. *Nation* 42 (Jan. 28, 1886): 84; W. H. Pickering, Lowell's future collaborator, was president of the Boston Society of Amateur Photographers. See the Exhibit at the Lowell Observatory Museum. On Lowell's photographs, see *Book Buyer* 3, series 3 (1886): 16–18; and *Spectator* 59 (April 24, 1886): 551–552.
33. On cultural centers, see Frederic J. Stimson, *My United States* (New York: Charles Scribner's Sons, 1931), pp. 76–84; on Lowell's views of publishers, see Lowell to Elizabeth, Oct. 5, 1907, Lowell family papers, HL.

34. *Records of the Board of Overseers of Harvard College,* vol. 11, 1871–1882, p. 462; vol. 12, 1882–1891, pp. 17, 295; vol. 13, 1891–1900, pp. 77B, 135B, 191B, 303E, 347B, 405D, 447D. Lowell to the Putnams, Feb. 28, 1891, Lowell family papers, HL.
35. John Lankford, *American Astronomy: Community, Careers, and Power, 1859–1940* (Chicago: University of Chicago Press, 1997), pp. 200–203, 370.
36. Mary Jo Nye, "Aristocratic Culture and the Pursuit of Science: The De Broglies in Modern France," *Isis* 88 (Sept. 1997): 397–421; see especially, p. 403. On amateurs and professionals, see John Lankford, "Amateurs: Astronomy's Enduring Resource," *Sky and Telescope* 75 (Nov. 1988): 482–483.
37. On Lowell's contempt for conservatives, see Lowell, *Mars as the Abode of Life* (New York: Macmillan Co., 1908), p. 184, and *The Evolution of Worlds,* pp. 180, 202. Lowell to A. E. Douglass, Aug. 9, 1897; Lowell to David Todd, June 9, 1907; Lowell to Douglass, Nov. 21, 1900, Lowell papers, LOA.
38. Lowell to Lester Frank Ward, March 26, 1907; Ward to Lowell, March 31, 1907; T. J. J. See to Lowell, Oct. 5, 1907, Lowell papers, LOA.
39. Sept. 24 entry, "Journal—1904," Lowell papers, HL.
40. On the linguistic question, see Sheehan, *Planets and Perception,* pp. 3–4.
41. *Mars as the Abode of Life,* p. 135; Lowell, *Occult Japan: The Way of the Gods: An Esoteric Study of Japanese Personality and Possession* (Boston: Houghton Mifflin Co., 1894), p. 14.
42. "Ontake," p. 371; *Occult Japan,* p. 60; *The Evolution of Worlds,* p. 221.
43. Lowell, *The Soul of the Far East* (Boston: Houghton Mifflin Co., 1888), p. 112; *Chosön,* p. 210; *Occult Japan,* p. 17.
44. *The Evolution of Worlds,* p. 238.
45. On Lowell's approach, see Lowell, *Mars,* pp. 8, 31, 75, 149; see also Henry Norris Russell, "Percival Lowell and His Work," *Outlook* Dec.6, 1916, p. 783, and Victor R. Baker, *The Channels of Mars* (Austin: University of Texas Press, 1982), p. 13.
46. On Lowell's search for good seeing, see William Graves Hoyt, *Lowell and Mars* (Tucson: University of Arizona Press, 1976), pp. 51–52; on viewing from Japan, see Lowell to Putnam, undated (approximate date is Jan. 15, 1893), Lowell family papers, HL.
47. See Hoyt, *Lowell and Mars,* pp. 109–121, 134–137, 145, 237–239.
48. See William Graves Hoyt, *Planets X and Pluto* (Tucson: University of Arizona Press, 1980).
49. *Mars as the Abode of Life,* p. 184; *Mars,* p. 174; *Mars and Its Canals,* p. 167.
50. Chamberlain to Hearn in Koizumi, ed., *Letters from Basil Hall Chamberlain to Lafcadio Hearn,* pp. 32–35; Douglass to Putnam, March 12, 1891, Douglass papers, University of Arizona Archives.
51. Lowell, *The Solar System* (Boston: Houghton Mifflin Co., 1903), p. 71; *The Evolution of Worlds,* p. 128.

52. *Mars as the Abode of Life*, pp. 138–139; *The Solar System*, p. 17; *Mars and Its Canals*, p. 376; *The Evolution of Worlds*, p. 28.
53. *Mars*, p. 6.
54. *Mars and Its Canals*, p. 127.
55. *Mars and Its Canals*, p. 289.
56. *Mars*, pp. 111, 158, 160.
57. Ralph Curtis to Mrs. Gardner, Dec. 10, 1916, Gardner papers, GM.

II. Lowell as Spencerian

1. See Don DeVere Walker, "The *Popular Science Monthly*, 1872–1878: A Study in the Dissemination of Scientific Ideas in America," Ph.D. diss., University of Minnesota, 1956.

5. Cosmic Philosopher

1. Alexander von Humboldt, *Cosmos: A Sketch of the Physical Description of the Universe*, 5 vols., trans. E. C. Otté (New York: Harper Bros., 1859). Spencer's *Synthetic Philosophy* appeared in ten volumes between 1862 and 1893.
2. Spencer disliked the term *cosmic philosophy* in reference to his work. See John Fiske, *Outlines of Cosmic Philosophy* (1874; Boston: Houghton Mifflin Co., 1902), vol. 1, p. x. On extraterrestrial life, Spencer was noncommittal in "The Nebular Hypothesis"; Fiske, however, argued strongly for life on Mars. *Outlines of Cosmic Philosophy*, 1, p. 384.
3. On Darwin and evolution, see Peter J. Bowler, *Evolution: The History of an Idea* (Berkeley: University of California Press, 1984), and Robert J. Richards, *Darwin and the Emergence of Evolutionary Theories of Mind and Behavior* (Chicago: University of Chicago Press, 1987). Whether Darwin viewed natural selection as progressive is now a disputed point; see Matthew H. Nitecki, ed., *Evolutionary Progress* (Chicago: University of Chicago Press, 1988).
4. Lowell drew on Fiske, *Outlines of Cosmic Philosophy*, 4, p. 127, in *Mars and Its Canals* (New York: Macmillan Co., 1906), p. 381; he also ordered Fiske's two-volume *Essays Historical and Literary* (New York: Macmillan Co., 1902) for his library. Houghton Mifflin to Lowell, Feb. 2, 1903, Lowell papers, LOA. Lowell's analogy between a scientific hypothesis and a criminal investigation, in his *Mars as the Abode of Life* (New York: Macmillan Co., 1908), chap. 6, was previously used by T. H. Huxley in "On Our Knowledge of the Causes of the Phenomena of Organic Nature," *Darwiniana: Essays* (New York: D. Appleton and Co., 1896), pp. 368–375, a passage read and marked by Lowell. He sent Ernst Haeckel *Mars and Its Canals*, praised his "masterly . . . explanation of life," and thanked him for a recent publication that had made "so profound and wide-

spread an impression," Lowell to Haeckel, March 26, July 5, 1907, Lowell papers, LOA.

5. Relations between Huxley, Spencer, and Haeckel have been considered most recently in Adrian Desmond, *Huxley: The Devil's Disciple* (London: Michael Joseph, 1995); Huxley, Spencer, and other scientists were members of the elite "X" club, see pp. 327–330.
6. Fiske, *Outlines of Cosmic Philosophy,* 1, p. xi; 2, p. 242. Huxley, *Darwiniana: Essays,* p. 213; T. H. Huxley, *Discourses Biological and Geological: Essays,* vol. 8 (New York: D. Appleton and Co., 1896), p. 325. Herbert Spencer, *First Principles* (1860; New York: D. Appleton and Co., 1907), pp. 332, 396.
7. Spencer, *First Principles,* pp. 380–381; T. H. Huxley, *Science and the Christian Tradition* (New York: D. Appleton and Co.), p. 72.
8. T. H. Huxley, *Science and the Christian Tradition: Essays,* p. 49; Ernst Haeckel, *The Wonders of Life,* trans. Joseph McCabe (New York: Harper and Bros., 1904), pp. 97–98; Fiske, *Outlines of Cosmic Philosophy,* 2, p. 303.
9. Herbert Spencer, *Principles of Psychology,* vol. 1 (New York: D. Appleton and Co., 1906), pp. 137–138. Huxley and Haeckel affirmed this hierarchical approach. Huxley, *Science and the Christian Tradition,* p. 43, and *Discourses Biological and Geological,* pp. 318–319. Ernst Haeckel, *Last Words of Evolution,* trans. from the 2nd ed. by Joseph McCabe (New York: Peter Eckler Publishing Co., 1920) p. 24, and *The Wonders of Life,* p. 360.
10. Lowell to William Lowell Putnam, July 22, 1891, Lowell family papers, HL. James Croll, *The Philosophical Basis of Evolution* (London: Edward Stanford, 1890).
11. Lowell read "The Nebular Hypothesis" in Herbert Spencer, *Essays: Scientific, Political and Speculative* (London: Williams and Norgate, 1868), vol. 1, pp. 237–299. See "Charging Records," Harvard College Library (1875–76), p. 369, HUA. Lowell's views may also have been shaped by Benjamin Peirce's version of cosmic evolution, presented in Peirce's Lowell Institute lectures (1879) and his *Ideality in the Physical Sciences* (Boston: Little, Brown and Co., 1883).
12. Lowell to Putnam, Aug. 19, 1891, Lowell family papers, HL; notes in margins of Lowell's copy of Croll, *The Philosophical Basis of Evolution,* p. 26, Lowell Observatory Library; Spencer, *First Principles,* p. 110.
13. Lowell to Putnam, Aug. 19, 1891, Lowell family papers, HL; Spencer, *First Principles,* p. 101.
14. Lowell to Putnam, July 22, 1891, Lowell family papers, HL; notes on blank page at the back of Croll, *The Philosophical Basis of Evolution.* For Spencer's somewhat different version of Kant's argument, see *First Principles,* pp. 40–42.
15. Lowell's notes on blank page at the back of Croll, *The Philosophical Basis of Evolution.*
16. See Lowell to Putnam, Aug. 19, 1891, Lowell family papers, HL; Spencer, *First Principles,* p. 159 (footnote).

17. Notes in margins of Croll, *The Philosophical Basis of Evolution,* pp. 58–59. See also Richardson and Kane, "Orthogenesis and Evolution in the 19th Century," in Nitecki, ed., *Evolutionary Progress,* pp. 150, 166.
18. Lowell, "The Revelation of Evolution," *Atlantic Monthly* 104 (Aug. 1909): 174.
19. Lowell's only reference to Spencer occurs in *Occult Japan: The Way of the Gods: An Esoteric Study of Japanese Personality and Possession* (Boston: Houghton Mifflin Co., 1894), p. 18; ibid., p. 378; Lowell, *The Soul of the Far East* (Boston: Houghton Mifflin Co., 1888), p. 195; Spencer, *First Principles,* pp. 179, 308; "The Revelation of Evolution," p. 178.
20. "The Revelation of Evolution," p. 179. Lowell, *The Evolution of Worlds* (New York: Macmillan Co., 1909), pp. 129, 149–51, 167, 180, 208.
21. *The Evolution of Worlds,* p. 156; *Mars as the Abode of Life,* pp. 11, 71.
22. Huxley, *Science and the Christian Tradition: Essays,* p. 72; Spencer, *First Principles,* p. 381; *The Evolution of Worlds,* p. 153; *Mars as the Abode of Life,* p. 43.
23. *The Evolution of Worlds,* p. 187. See Huxley, *Discourses Biological and Geological,* p. 326; Haeckel, *Last Words on Evolution,* p. 95. On radioactivity and the debate over the age of the Earth, see Ernest Rutherford, "Radium: The Cause of the Earth's Heat," *Harper's Monthly Magazine* 110 (Feb. 1905): 390–396.
24. C. Hart Merriam, *Results of a Biological Survey of the San Francisco Mountain Region and Desert of the Little Colorado,* North American Fauna no. 3 (Washington, D.C.: Government Printing Office, 1890), U.S. Department of Agriculture, Division of Ornithology and Mammology; Huxley, *Discourses Biological and Geological,* pp. 69–109, 256; *Mars as the Abode of Life,* pp. 56–59, 37–38, 76; *Mars and Its Canals,* pp. 18–19, 378–379, 352, 356; Haeckel, *The Wonders of Life,* p. 38.
25. *Mars as the Abode of Life,* pp. 96–98, 87–88; *The Evolution of Worlds,* p. 160. Lowell, *Mars* (Boston: Houghton Mifflin Co., 1895), p. 4.
26. *Mars and Its Canals,* p. 364; *Mars as the Abode of Life,* p. 205. See also T. H. Huxley, *Methods and Results: Essays* (New York: D. Appleton and Co., 1904), p. 164; Haeckel, *Last Words on Evolution,* pp. 134, 104, 315.
27. *The Evolution of Worlds,* p. 158; *Mars as the Abode of Life,* pp. 64, 108.
28. *Mars as the Abode of Life,* pp. 64, 108; *Mars,* p. 212.
29. *Mars and Its Canals,* p. 213, 16, 157–158; *The Evolution of Worlds,* p. 236.
30. On the superior authority of astronomy, see *The Evolution of Worlds,* pp. 184, 172–173, and *Mars as the Abode of Life,* p. 13.

6. Image-Maker

1. On Lowell and East Asia, see Robert S. Ellwood, "Percival Lowell's Journey to the East," *Sewanee Review* 78 (Spring 1970): 285–309, and Allen E. Tuttle, "Lafcadio Hearn and the Soul of the Far East," *Contemporary Japan* 23 (July–Sept. 1955): 529–552. Although Lowell spent very little time outside of Japan,

he generalized from that experience to all of East Asia; consider, for example, his title: *The Soul of the Far East* (Boston: Houghton Mifflin Co., 1888). The impact of Lowell's work may be judged in various ways. Two of his books, *The Soul of the Far East* and *Noto: An Unexplored Corner of Japan* (Boston: Houghton Mifflin Co., 1891), had been published previously in serial form, in *Atlantic Monthly* 60 (Sept.–Dec. 1887): 405–413, 515–524, 614–624, 836–845, 845–849, and 67 (Jan.–June 1891): 1–15, 175–191, 364–379, 482–500. New editions of *The Soul of the Far East* were published in 1888, 1893, 1896, 1908, 1910, and 1911, and a German translation appeared in 1911.

2. William Elliot Griffis, *The Mikado's Empire,* 2 vols. (1876; New York: Harper Brothers, 1903. On the popularity of Griffis's book, see Basil Hall Chamberlain's *Things Japanese* (Rutland, Vt.: Charles E. Tuttle Co., 1971; reprint of 5th ed. published in 1905), which ranked it first among general works on Japan in English in 1904. Hearn's *Glimpses of Unfamiliar Japan,* 2 vols. (1894; Rutland, Vt.: Charles E. Tuttle, 1976) was ranked second, and Lowell's *The Soul of the Far East* sixth. On Griffis's experience in Japan, see Robert Rosenstone, *Mirror in the Shrine: American Encounters with Meiji Japan* (Cambridge, Mass.: Harvard University Press, 1988). Foreign travelers' accounts of Japan are discussed in John Ashmead, Jr., *The Idea of Japan, 1853–1895: Japan as Described by American and Other Travellers from the West* (New York and London: Garland Publishing, 1987); Jean-Pierre Lehmann, *The Image of Japan: From Feudal Isolation to World Power, 1850–1905* (London: George Allen and Unwin, 1978); Yokoyama Toshio, *Japan in the Victorian Mind: A Study of Stereotyped Images of a Nation, 1850–80* (Basingstoke, Hampshire: Macmillan, 1987); Ian Littlewood, *The Idea of Japan: Western Images, Western Myths* (Chicago: Ivan R. Dee, 1996); and Carl T. Jackson, *The Oriental Religions and American Thought* (Westport, Conn.: Greenwood Press, 1981).
3. Among scholarly accounts of ranking are John S. Haller, Jr., *Outcasts from Evolution: Scientific Attitudes of Racial Inferiority, 1859–1900* (Urbana: University of Illinois Press, 1971), and George W. Stocking, Jr., *Victorian Anthropology* (New York: Free Press, 1987).
4. Herbert Spencer, *The Principles of Biology* (New York: D. Appleton, 1898–1900), vol. 2, pp. 529–531; idem, *The Principles of Psychology* (New York: D. Appleton, 1906), vol. 1, p. 137, and 2, pp. 524–525; idem, *The Principles of Sociology* (New York: D. Appleton, 1900–1901), vol. 1, p. 86. Lowell may also have encountered Spencer's ranking system in John Fiske's *Outlines of Cosmic Philosophy* (1874; Boston: Houghton Mifflin Co., 1902), vol. 4, pp. 3, 8–9, 23–24, 27, 29, 33–34, 40–49, 53, 91.
5. Herbert Spencer, "The Comparative Psychology of Man," *Popular Science Monthly* 8 (Jan. 1876): 260–261; in his *First Principles,* Spencer identified Japan as a civilization whose development had been arrested (1860; New York: D. Appleton and Co., 1907), pp. 520–521. On specialization as a form of prog-

ress, see Herbert Spencer, "Progress: Its Law and Cause," pp. 15, 52–53, and "The Social Organism," pp. 393, 404, in *Essays: Scientific, Political and Speculative,* vol. 1 (London: William and Norgate, 1868).

6. See Spencer, *Principles of Psychology,* 2, pp. 534–535.
7. Fiske, *Outlines,* 4, pp. 23–24.
8. On industrialization, see Thomas R. Hietala, *Manifest Design: Anxious Aggrandizement in Late Jacksonian America* (Ithaca, N.Y.: Cornell University Press, 1985), pp. 96–110, and Drew L. McCoy, *The Elusive Republic: The Political Economy in Jeffersonian America* (Chapel Hill: University of North Carolina Press, 1980), pp. 140–158.
9. On westward expansion and antebellum racial views, see Hietala, *Manifest Design,* pp. 133–172, and William Stanton, *The Leopard's Spots: Scientific Attitudes toward Race in America, 1815–59* (Chicago: University of Chicago Press, 1960), pp. 33, 131, 158.
10. Griffis, *The Mikado's Empire,* 2, p. 430.
11. Ibid., 1, pp. 7, 320, 324; 2, p. 370.
12. Ibid., 2, pp. 419, 440.
13. Ibid., 1, p. 161; 2, pp. 384, 555, 561.
14. See Joseph Henning, "The American Search for Japan's Racial Origins and Religious Intentions, 1894–1905," paper presented at the American Historical Association meeting, Seattle, Jan. 11, 1998.
15. On world's fairs, see Neil Harris, "All the World a Melting Pot? Japan at American Fairs, 1876–1904," in Akira Iriye, ed., *Mutual Images: Essays in American-Japanese Relations* (Cambridge, Mass.: Harvard University Press, 1975), pp. 22–54.
16. On Japanese-American rivalry, see Akira Iriye, *Across the Pacific: An Inner History of American–East Asian Relations* (New York: Harcourt, Brace and World, 1967), pp. 59–61.
17. On elite anti-industrialism, see Lawrence Chisolm, *Fenollosa, the Far East and American Culture* (New Haven, Conn.: Yale University Press, 1963); Van Wyck Brooks, *Fenollosa and His Circle, with Other Essays in Biography* (New York: E. P. Dutton and Co., 1962), pp. 4–7, 16–18; T. J. J. Lears, *No Place of Grace: Antimodernism and the Transformation of American Culture, 1880–1920* (New York: Pantheon Books, 1982), pp. 225–234. On the contrast theme, see Lehmann, *The Image of Japan,* p. 53; and Littlewood, *The Idea of Japan,* p. 10. On Americans and East Asian art, see Warren I. Cohen, *East Asian Art and American Culture: A Study in International Relations* (New York: Columbia University Press, 1992).
18. See Basil Hall Chamberlain to Lafcadio Hearn, Aug. 27, 1893, in Kazuo Koizumi, ed., *More Letters from Basil Hall Chamberlain to Lafcadio Hearn* (Tokyo: Hokuseido Press, 1937), pp. 96–97.
19. On Morse's lectures, see *Boston Evening Transcript,* Jan. 12, 17, Feb. 6, 7, 14, 18,

1882, and Dorothy Wayman, *Edward Sylvester Morse* (Cambridge, Mass.: Harvard University Press, 1942); on the Lowell Institute, see Edward Weeks, *The Lowells and Their Institute* (Boston: Little, Brown and Co., 1966).

20. On Lowell's collecting of Japanese art, see Barrett Wendell's letter to his wife, July 16, 1886, in M. A. DeWolfe Howe, ed., *Barrett Wendell and His Letters* (Boston: Atlantic Monthly Press, 1924), pp. 64–65. See also Lowell to Edward Morse, April 24, 1884, Morse papers, PMS; Lowell to Elizabeth Lowell, July 7, Nov. 28, 1883, Lowell family papers; Lowell to Barrett Wendell, Jan. 24, 1884, Wendell papers, HL. *Chosön: The Land of the Morning Calm: A Sketch of Korea* (Boston: Ticknor and Co., 1886), p. 299. "A Visit to Shirane San," *Appalachia* 5, (June 1888): 98. Lowell, *Noto,* pp. 1–2, 110. Basil Hall Chamberlain and W. B. Mason, *A Handbook for Travelers in Central and Northern Japan* (London: John Murray Publisher, 1903).
21. Lowell, *Occult Japan: The Way of the Gods: An Esoteric Study of Japanese Personality and Possession* (Boston: Houghton Mifflin Co., 1894), pp. 1–2, 14, 275.
22. *Dial* 9 (April 1889): 329; *Japan Weekly Mail,* Feb. 9, 1889, p. 135.
23. On hostility to modernization, see Lears, *No Place of Grace.*
24. *The Soul of the Far East,* pp. 3–5, 80, 16–17, 24.
25. Ibid., pp. 20–26, 9, 195. Lowell first discussed "impersonality" in *Chosön,* chaps. 12–14.
26. *The Soul of the Far East,* pp. 35–36, 76.
27. Ibid., pp. 88, 90, 106. Lowell's use of language as an index of social advancement may have been borrowed from Spencer's essay "Progress: Its Law and Cause," p. 17. Lowell presented his evolutionary approach to language more fully in an essay comparing Japanese and Burmese; his views were attacked by E. H. Parker, a member of the Asiatic Society of Japan and a student of Asian languages, who considered Lowell "merely a pleasant and thoughtful writer, with no profound knowledge of either Burmese or Japanese." See Lowell, "A Comparison of the Japanese and Burmese Languages," *Transactions of the Asiatic Society of Japan* 19 (June 13, 1891): 584–585, and E. H. Parker, "Touching Burmese, Japanese, Chinese, and Korean," *Transactions of the Asiatic Society of Japan* 21 (1893): 144–145.
28. See Ferris Greenslet, *The Lowells and Their Seven Worlds* (Boston: Houghton Mifflin Co., 1945), pp. 352–354. *Chosön,* pp. 370, 346; *Noto,* p. 56.
29. *The Soul of the Far East,* p. 111, 114, 129, 141, 143.
30. Ibid., pp. 111–113, 148, 157, 125, 122.
31. Ibid., pp. 169, 176. On the conversions of Fenellosa and Bigelow to Buddhism, see Lears, *No Place of Grace,* pp. 229–232.
32. *The Soul of the Far East,* p. 218. *Chosön,* p. 250.
33. *The Soul of the Far East,* pp. 216, 213, 225–226. See also Spencer, *Principles of Psychology,* 2, pp. 534–535.

34. On stereotypes of "Mongolians," see Stanton, *The Leopard's Spots,* p. 33. On the value of cooperation, see Spencer, *The Principles of Ethics,* 2, p. 10; 1, p. 241; Fiske, *Outlines,* 4, pp. 8–9.
35. On Hearn and Spencer, see Lafcadio Hearn, "About Ancestor Worship," in *Kokoro: Hints and Echoes of Japanese Inner Life* (Boston: Houghton Mifflin Co., 1896), pp. 266–306. See also Tuttle, "Lafcadio Hearn and the Soul of the Far East," pp. 531, 541–543.
36. See Hearn, *Kokoro,* p. 11.

7. Psychical Researcher

1. See William James, "The Confidences of a 'Psychical Researcher,'" in James, *Essays in Psychical Research,* ed. Frederick H. Burckhardt (Cambridge, Mass.: Harvard University Press, 1986), pp. 366, 374. On Lowell's contributions, see Carmen Blacker, *The Catalpa Bow: A Study of Shamanistic Practices in Japan* (London: George Allen and Unwyn, 1975), pp. 279–280, and Winston Davis, *Dojo: Magic and Exorcism in Modern Japan* (Stanford, Calif.: Stanford University Press, 1980), p. 88.
2. William James to Henry James, April 11, 1900, William James papers, HL. William James, *The Principles of Psychology,* 2 vols. (New York: Henry Holt and Co., 1890), vol. 1, p. 396. (Unless otherwise noted, page numbers are from the original edition, though I have used the notes from *The Works of William James,* 3 vols. (Cambridge, Mass.: Harvard University Press, 1981). In Lowell, *Occult Japan: The Way of the Gods: An Esoteric Study of Japanese Personality and Possession* (Boston: Houghton Mifflin Co., 1894), there are direct references to James (pp. 314, 316, 339) and the Society for Psychical Research (p. 247).
3. Lowell to William Lowell Putnam, Aug. 16, 29, Sept. 9, 24, 30, 1891, Lowell family papers, HL. On Lowell and the high priest, see *Occult Japan,* pp. 139, 170.
4. Ibid., Feb. 16, 1893, Lowell family papers, HL.
5. Lowell also delivered two papers on Shinto trances to the Yokohama Literary Society; see *Japan Weekly Mail,* April 1, 1893, p. 377, and Oct. 16, 1893, p. 512. On James's lectures, see Eugene Taylor, *William James on Exceptional Mental States: The 1896 Lowell Lectures* (Amherst, Mass.: University of Massachusetts Press, 1984), pp. 189–193. James cited *Occult Japan,* p. 153, on demoniacal possession. Henri F. Ellenberger, *The Discovery of the Unconscious: The History and Evolution of Dynamic Psychiatry* (New York: Basic Books, 1970), pp. 4–15, cites other early studies of possession. "Esoteric Shinto," *Transactions of the Asiatic Society of Japan* 21 (Feb. 15, 1893): 106–135; 21 (March 8, 1893): 152–197; 21 (May 17, 1893): 241–270; 22 (Oct. 18, 1893): 1–26. See also *Journal of*

the Society for Psychical Research 15 (Oct. 1900): 434–437, and Henry K. Beauchamp, "Fire-Walking Ceremonies in India," ibid. 15 (Nov. 1900): 321.

6. On Lowell's scientific agenda, see *Occult Japan,* p. 14. Lowell wrote Putnam that his study of Shinto would be "strictly scientific," Aug. 29, 1891, Lowell family papers, HL.
7. *Occult Japan,* p. 14.
8. Ibid., "Shinto" and "Miracles" chaps.
9. See *Occult Japan,* pp. 102, 280, 288. I have used quotation marks to set off the first usage of judgmental terms that express Lowell's own view. In a similar fashion, I place in quotes terms acceptable in Lowell's time that now appear to be derogatory.
10. On late-nineteenth-century doubts about science, see Janet Oppenheim, *The Other World: Spiritualism and Psychical Research in England, 1850–1914* (Cambridge: Cambridge University Press, 1985); Frank Miller Turner, *Between Science and Religion: The Reaction to Scientific Naturalism in Late Victorian England* (New Haven, Conn.: Yale University Press, 1974); R. Laurence Moore, *In Search of White Crows: Spiritualism, Parapsychology, and American Culture* (New York: Oxford University Press, 1977).
11. See Seymour H. Mauskopf, "The History of the American Society for Psychical Research: An Interpretation," *Journal of the American Society for Psychical Research* 83 (Jan. 1989): 7–29; Eugene Taylor, "The American Society for Psychical Research, 1884–1889," *Proceedings of Presented Papers: The Parapsychological Association,* 28th convention (1985), pp. 187–196; On the Asiatic Society of Japan, I have used minutes and membership lists from the society's journal, *Transactions.*
12. James was influenced by the founders of the Society for Psychical Research, especially Frederic Myers; see his work on subliminal consciousness in *Proceedings of the Society for Psychical Research* 7–9 (1891–94). These articles have been collected in Myers, *The Subliminal Consciousness* (1889; New York: Arno Press, 1976). On the movement in France, see Ellenberger, *The Discovery of the Unconscious,* pp. 85–102, 331–409.
13. On Newcomb's role, see Albert E. Moyer, *A Scientist's Voice in American Culture: Simon Newcomb and the Rhetoric of Scientific Method* (Berkeley: University of California Press, 1992), pp. 166–182. English scientific naturalists shared the skepticism of their American counterparts. See Frank Miller Turner, *Between Science and Religion: The Reaction to Scientific Naturalism in Late Victorian England* (New Haven, Conn.: Yale University Press, 1974), pp. 6–8.
14. John J. Cerullo, *The Secularization of the Soul: Psychical Research in Modern Britain* (Philadelphia: Institute for the Study of Human Issues, 1982), pp. 44–45, 60–61, 64. Myers, *The Subliminal Consciousness,* p. 349; William James, "Report of the Committee on Mediumistic Phenomena," *Proceedings of the*

American Society for Psychical Research 1 (July 1886): 102–106. See also by James: "Circular Requesting Information on Mediumistic Phenomena," *Boston Daily Advertiser,* April 9, 1885; "Notes on Automatic Writing," *Proceedings of the American Society for Psychical Research* 1 (March 1889): 548–549, 563–564; "The Census of Hallucinations," in James, *Essays in Psychical Research,* pp. 56–57; "Telepathy," *Johnson's Universal Cyclopedia* (New York: A. J. Johnson, 1895), pp. 45–47; "Report on Mrs. Piper's Hodgson-Control," *Proceedings of the Society for Psychical Research* 23 (June 1909): 4–5.

15. "Imitation" figures in *The Soul of the Far East* (Boston: Houghton Mifflin Co., 1888), p. 219, and was the subject of Lowell's lectures to the English Law School in Tokyo, published in *Japan Weekly Mail,* March 23, April 16, and May 11, 1889. On intellectual fashions of the Japanese upper class, see *Occult Japan,* p. 18.
16. On Mori, see Lowell, "The Fate of a Japanese Reformer," *Atlantic Monthly* 66 (Nov. 1890): 680–693, and *Occult Japan,* pp. 18–19.
17. Henrika Kuklick, *The Savage Within: The Social History of British Anthropology* (Cambridge: Cambridge University Press, 1991), pp. 86, 99–100, 105–107.
18. *The Soul of the Far East,* pp. 5, 170. *Occult Japan,* p. 19, 103.
19. *Occult Japan,* pp. 20–21, 191, 25–26, 73, 198, 132.
20. Ibid., pp. 5, 148, 170–171, 183.
21. Ibid., pp. 32, 371, 378–379.
22. Ibid., p. 321.
23. On the passivity of mediums, see Moore, *In Search of White Crows,* p. 106.
24. Lowell to Putnam, June 8, 1893, Lowell family papers, HL.
25. *Occult Japan,* pp. 175–176, 225, 145. Lowell to Putnam, Aug. 29, 1891, March 7, 1893, Lowell family papers, HL.
26. *Occult Japan,* pp. 180–181, 184. Lowell to Putnam, Sept. 30, 1891, Lowell family papers, HL; W. B. Mason to Lafcadio Hearn, March 3, 1893, in Hearn, *Letters I: To and from Various Persons* (Tokyo: Tenri Central Library, 1934). See also Erwin Baelz, *Awakening Japan: The Diary of a German Doctor: Erwin Baelz,* ed. Toku Baelz (1932; Bloomington: Indiana University Press, 1974), pp. 77–78. Baelz was a member of the Asiatic Society of Japan. See *Transactions* 20 (1893): xx. It was widely believed that foxes could revenge themselves on people or be the agents of others' revenge by assuming human form and possessing their victims. See Blacker, *The Catalpa Bow,* pp. 52–53.
27. *Occult Japan,* p. 122.
28. Ibid., pp. 310, 311, 313, 312. On James's view of mental states, see *Principles of Psychology,* 1, chaps. 2–3.
29. On James's reference to Lowell, see "Human Immortality: Two Supposed Objections to the Doctrine," in *William James on Psychical Research,* ed. Gardner Murphy and Robert O. Ballou (New York: Viking, 1960), pp. 289–291. For

Spencer's views, see *First Principles* (New York: D. Appleton and Co., 1907), p. 217. On James's differences with Spencer, see James, *The Principles of Psychology,* 1, pp. 151–153, 402–403.

30. *Occult Japan,* pp. 299, 300, 303, 339; James, *The Principles of Psychology,* 1, pp. 134–141, 284, 288.
31. *Occult Japan,* pp. 321, 322.
32. Ibid., pp. 332–343.
33. James, *The Principles of Psychology,* 1, p. 399.
34. See *Occult Japan,* pp. 337–338, 377–378.
35. James, *The Principles of Psychology,* 2, p. 595; 1, pp. 227–228, 393–395.
36. Ibid., 2, pp. 363–364, 530, 538, 564, 571, 578, 596; *Occult Japan,* pp. 280, 328.
37. Faculty psychology, still taught at American universities in the 1890s, stressed the autonomy of mind from brain. On hypnotic and possession trances, see Lowell to Putnam, March 7, March 8, May 18, 1893, Lowell family papers, HL; and *Occult Japan,* pp. 343–367.
38. *Occult Japan,* p. 60. Basil Hall Chamberlain to Lafcadio Hearn, Jan. 10, 1893, in Kazuo Koizumi, ed., *Letters from Basil Hall Chamberlain* (Tokyo: Hokuseido Press, 1937), p. 3. Lafcadio Hearn found *Occult Japan* "painfully unsympathetic—Mephistophelian in a way that chills me," Hearn to Chamberlain, Jan. 1895, in Elizabeth Bisland, ed., *The Life and Letters of Lafcadio Hearn* (Boston: Houghton Mifflin Co.), vol. 2, p. 200.
39. William James, *The Varieties of Religious Experience* (New York: Longman's Green, 1958), pp. 87–108.
40. Baelz, *Awakening Japan,* p. 70. Chamberlain to Hearn, Jan. 10, 1893, in Koizumi, ed., *Letters from Basil Hall Chamberlain,* p. 2. On Hearn's intimacy with the Japanese, see Robert Rosenstone, *Mirror in the Shrine* (Cambridge, Mass.: Harvard University Press, 1988).
41. On doubts about science in the 1890s, see Henry Adams's, *The Education of Henry Adams* (Boston: Houghton Mifflin Co., 1918), pp. 331–345 and 379–390.
42. *Boston Evening Transcript.* April 28, 1908. pp. 9, 12. "Lecture before the St. Botolph Club," Jan. 2, 1895, unpublished manuscript, LOA.

8. Cosmogonist

1. Lowell's cosmogony is treated in the biographies of him by A. L. Lowell and William Graves Hoyt as an issue of marginal interest. Stephen Brush discusses Lowell's cosmogony briefly in "A Geologist among Astronomers: The Rise and Fall of the Chamberlin-Moulton Cosmogony, Part 1," *Journal for the History of Astronomy* 9 (1978): 20–21, and *Fruitful Encounters: The Origin of the Solar System and of the Moon from Chamberlin to Apollo* (Cambridge: Cambridge

University Press, 1996), pp. 48, 63, while Norriss Hetherington treats it more fully in "The Simultaneous 'Discovery' of Internal Motions in Spiral Nebulae," *Journal for the History of Astronomy* 6 (1975): 115–125.

2. On Peirce's support for the hypothesis, see Ronald Numbers, *Creation by Natural Law: Laplace's Nebular Hypothesis in American Thought* (Seattle: University of Washington Press, 1977), pp. 25, 78–87. On the nebular hypothesis in Lowell's freshman ethics course, see Stephen Bulfinch, *Manual of the Evidences of Christianity for Classes and Private Reading* (Boston: William V. Spencer, 1869), pp. 6, 22, 32.
3. See Pierre Laplace, *Exposition du système du monde* (Paris: Imprimérie du Cercle-Social, 1796), chap. 6, pp. 505–529, and note 7 and last, pp. 547–568.
4. Spencer's "Nebular Hypothesis" was originally published in *Westminster Review* (July 1858); Lowell read the same essay in *Essays: Scientific, Political, and Speculative* (London: Williams and Norgate, 1868), vol. 1, pp. 237–299. Harvard College Library Charging Records, 1875–1876, p. 369 (P. Lowell), HUA. Numbers, *Creation by Natural Law,* pp. 20, 25.
5. Lowell's graduation disquisition, "The Nebular Hypothesis," runs from pages 1–16 and then from 1–8. To distinguish the two sets of pages numbered 1–8, I have placed a (1) or a (2) after the page number. See Lowell, "The Nebular Hypothesis," pp. 2, 6–7 (2), HUA; Spencer, "The Nebular Hypothesis," pp. 276–277.
6. Laplace, *Exposition,* p. 507. See Michael Crowe, *The Extraterrestrial Life Debate, 1750–1900: The Idea of a Plurality of Worlds from Kant to Lowell* (Cambridge: Cambridge University Press, 1986), pp. 277, 289, 346. Lowell endorsed divine creation of the solar system in 1876. "The Nebular Hypothesis," pp. 7–8 (2); Numbers, *Creation by Natural Law,* pp. 78–87.
7. *Boston Commonwealth,* May 26, 1894, p. 3.
8. Lowell, *Mars* (Boston: Houghton Mifflin Co., 1895), pp. 3–5.
9. On the full draft, see "Fundamental Problems of Geology," *Carnegie Institution Yearbook No. 3 for 1904* (1905), pp. 195–258. In 1906 two summaries were published: in F. R. Moulton, *An Introduction to Astronomy* (New York: Macmillan Co., 1906), pp. 463–487, and in T. C. Chamberlin and Rollin D. Salisbury, *Geology,* vol. 2 (New York: Henry Holt and Co., 1906), pp. 1–81, which Lowell ordered. See Lowell to A. C. McClurg, Nov. 8, 1909, Lowell papers, LOA.
10. On the critics, see Stanley L. Jaki, *Planets and Planetarians* (New York: Wiley, 1977), pp. 154–185.
11. For an earlier version of the hypothesis, see T. C. Chamberlin and F. R. Moulton, "The Development of the Planetesimal Hypothesis," *Science,* xii (1900), pp. 201–208. Brush, "A Geologist among Astronomers," pp. 1–29.
12. See Brush, "A Geologist among Astronomers," pp. 1–29.
13. Moulton, *Introduction to Astronomy,* p. 464. Donald Osterbrock, *James E.*

Keeler: Pioneer American Astrophysicist and the Early Development of American Astrophysics (Cambridge: Cambridge University Press, 1984), pp. 317–318; see Brush, "A Geologist among Astronomers, pp. 17, 19. On Hale, see Helen Wright, *Explorer of the Universe: A Biography of George Ellery Hale* (New York: E. P. Dutton and Co., 1966).

14. Lowell's MIT lectures were delivered in December 1902 and published the next year as *The Solar System.* One likely source of Lowell's collision theory is James Croll, "On Geological Time and the Probable Date of the Glacial and Upper Miocene Period," *The London, Edinburgh and Dublin Philosophical Magazine and Journal* 35, series 4 (1868): 363–384.
15. Lowell, *The Evolution of* Worlds (New York: Macmillan Co., 1909), chaps. 1, 5. Lowell claimed originality for his cosmogony (Lowell to J. Ellard Gore, July 1, 1910, Lowell papers, LOA), but he had already read and annotated Moulton's *Introduction to Astronomy* (Lowell to Moulton, June 15, 1906, Lowell papers, LOA).
16. *Evolution of Worlds,* chaps. 1, 5.
17. Ibid.
18. See ibid., chap. 5, and Chamberlin and Salisbury, *Geology,* 2, pp. 62–63.
19. See *Evolution of Worlds,* pp. 145–146, and Chamberlin and Salisbury, *Geology,* 2, pp. 64–67, 70–75.
20. Moulton, *An Introduction to Astronomy,* p. 470.
21. See Chamberlin and Salisbury, *Geology,* 2, pp. 68–69, 82. On the differences between Dana and Chamberlin, see Mott T. Greene, *Geology in the Nineteenth Century: Changing Views of a Changing World* (Ithaca, N.Y.: Cornell University Press, 1982), pp. 132–142, 259–266.
22. *Evolution of Worlds,* pp. 213–216.
23. Eliot Blackwelder, "Mars as the Abode of Life," *Science* 29 (April 23, 1909): 338–340. See also Joseph Barrell, "Fair Play and Toleration in Criticism," *Science* 30 (July 2, 1909): 21–23, and T. J. J. See, "Fair Play and Toleration in Science," *Science* 30 (May 28, 1909): 859–861. On See's cosmogony, see Thomas J. Sherrill, "A Career of Controversy: The Anomaly of T. J. J. See," *Journal for the History of Astronomy* 30 (1999): 25–50.
24. For Lowell on Moulton, see "Discussion and Correspondence," *Science* 30 (Sept. 10, 1909): 338–340; Moulton's rejoinder was published in "Discussion and Correspondence," *Science* 30 (Nov. 5, 1909): 639–641. For a summary of the controversy, see William Graves Hoyt, *Lowell and Mars* (Tucson: University of Arizona Press, 1976), pp. 260–261; Hetherington, "The Simultaneous 'Discovery,'" pp. 115–125. Lowell had already pointed gently to Moulton's calculating error in "On the Velocity with Which Meteors Enter the Earth's Atmosphere," *Astronomical Journal* 26 (April 17, 1908): 1–3.
25. *Evolution of Worlds,* pp. 127–128.

26. Ibid., p. 173. *Mars as the Abode of Life,* p. viii. Hetherington identified Lowell as an exponent of the nebular hypothesis in "The Simultaneous 'Discovery,'" p. 118. On Lowell's enthusiasm for the nebular hypothesis, see Lowell, "The Revelation of Evolution: A Thought and Its Thinker," *Atlantic Monthly* 104 (Aug. 1909): 76. On Lowell's acknowledgment of weaknesses in Laplace's theory, see *The Evolution of Worlds,* p. 129.
27. On Lowell's distinction, see Lowell to John Ellard Gore, July 1, 1910; Lowell to Lester Ward, Oct. 26, 1910, Lowell papers, LOA. On Chamberlin's limited claims, see Brush, "A Geologist among Astronomers," p. 87. On Lowell's reaction to V. M. Slipher's discovery, see *Boston Advertiser,* March 24, 1913. Slipher wrote Lowell that his discovery came closer to confirming Kant than Laplace, March 26, 1913, Lowell papers, LOA. Lowell's enthusiasm for Slipher's discovery is expressed in "Our Solar System," *Popular Astronomy* 24 (Aug.–Sept. 1916): 419–427.
28. See Greene, *Geology in the Nineteenth Century,* pp. 132–142, 259–266.
29. On the rejection of Spencer's hierarchy of disciplines, see Hamilton Cravens, "The Abandonment of Evolutionary Social Theory in America: The Impact of Academic Professionalization upon American Sociological Theory, 1890–1920," *American Studies* 12 (Fall 1971): 5–19. George Ellery Hale, *The Study of Stellar Evolution: An Account of Some Recent Methods of Astrophysical Research* (Chicago: University of Chicago Press, 1909), p. 4.
30. On the changing climate of opinion, see especially T. H. Huxley's Romanes lecture of 1893 in *Evolution and Ethics* (New York: D. Appleton and Co., 1896), p. 80.

III. Lowell as Astronomer

1. See John Lankford and Ricky L. Slavings, "The Industrialization of American Astronomy, 1880–1940," *Physics Today* 49 (Jan. 1996): 34–41.
2. On Hale, see Helen Wright, *Explorer of the Universe: A Biography of George Ellery Hale* (New York: E. P. Dutton, 1966); on amateurs and professionals, see John Lankford, "Amateurs and Astrophysics: A Neglected Aspect in the Development of a Scientific Specialty," *Social Studies of Science* 11 (1981): 275–303; Marc Rothenberg, "Organization and Control: Professionals and Amateurs in American Astronomy, 1899–1918," *Social Studies of Science* 11 (1981): 303–325; John Lankford, "Amateur versus Professionals: The Controversy over Telescope Size in Late Victorian Science," *Isis* 72 (1981): 11–28; Norriss S. Hetherington, "Amateur versus Professional: The British Astronomical Association and the Controversy over Canals on Mars," *Journal of the British Astronomical Association* 86 (1976): 303–308; Allan Chapman, *The Victorian Amateur Astronomer: Independent Astronomical Research in Britain, 1820–1920*

(New York: Wiley, 1998). On Lowell's methods, see William Graves Hoyt, *Lowell and Mars* (Tucson: University of Arizona Press, 1976), pp. 48–49, 174–177.

3. See Donald Osterbrock, *Yerkes Observatory, 1892–1950: The Birth, Near Death, and Resurrection of a Scientific Research Institution* (Chicago: University of Chicago Press, 1997).
4. Lowell to Frank Very, Aug. 22, 1910, Lowell papers, LOA.
5. On the "power elite" in astronomy, see John Lankford, *American Astronomy: Community, Careers, and Power, 1859–1940* (Chicago: University of Chicago Press, 1997), chap. 7. Simon Newcomb to George Ellery Hale, Oct. 2, 1903; Hale to Newcomb, April 10, 1907, in Daniel J. Kevles, ed., *George Ellery Hale Papers, 1882–1937: Microfilm Edition,* 100 rolls (Carnegie Institution of Washington and California Institute of Technology, 1968), Mount Wilson and Palomar Observatory Library, Pasadena, Calif.; hereinafter cited as Hale papers, MWPOL.
6. See B. Z. Jones and L. G. Boyd, *The Harvard College Observatory: The First Four Directorships, 1839–1919* (Cambridge, Mass.: Harvard University Press, 1971); Donald E. Osterbrock, John R. Gustafson, and W. J. Shiloh Unruh, *Eye on the Sky: Lick Observatory's First Century* (Berkeley, Calif.: University of California Press, 1988), and Osterbrock, *Yerkes Observatory.* See also David De Vorkin, ed., *The American Astronomical Society's First Century* (New York: American Institute of Physics, 1999).
7. On V. M. Slipher's contribution, see Robert Smith, *The Expanding Universe: Astronomy's Great Debate* (Cambridge: Cambridge University Press, 1982), pp. 17–22.

9. From "Astronomical Picnic" to Observatory

1. On the founding, see William Graves Hoyt, *Lowell and Mars* (Tucson: University of Arizona Press, 1976); B. Z. Jones and L. G. Boyd, *The Harvard College Observatory: The First Four Directorships, 1839–1919* (Cambridge, Mass.: Harvard University Press, 1971); George Ernest Webb, *Tree Rings and Telescopes: The Scientific Career of A. E. Douglass* (Tucson: University of Arizona Press, 1983). These accounts are partial in both senses of the word. Boyd and Jones use the Harvard University Archives, while Hoyt draws from the Lowell Observatory Archives. Webb's focus is on Douglass.
2. On amateurs and professionals, see Part III introduction, note 2. On the Pickerings, see Howard Plotkin, "Harvard College Observatory's Boyden Station in Peru: Origin and Formative Years, 1879–1891," in A. Lafuente, A. Elena, and M. L. Ortega, eds., *Mundialization de la ciencia y cultura nacional: Actas del Congreso Internacional "Ciencia, descubrimiento y mundo colonial"* (Madrid: Doce Calles, 1993), pp. 689–705.

3. On Boston culture, see E. Digby Baltzell, *Puritan Boston and Quaker Philadelphia: Two Protestant Ethics and the Spirit of Class Authority and Leadership* (New York: Free Press, 1979); Martin Green, *The Problem of Boston: Some Readings in Cultural History* (New York: W. W. Norton and Co., 1966); Ronald Story, *The Forging of an Aristocracy: Harvard and the Boston Upper Class, 1800–1870* (Middletown, Conn.: Wesleyan University Press, 1980); and Frederic Cople Jaher, "The Proper Bostonians in the Age of Industrial Capitalism," in Jaher, ed., *The Age of Industrialism in America* (New York: Free Press, 1968), pp. 188–262.
4. *Records of the Board of Overseers of Harvard College,* vol. 13 (1891–1900), p. 77B; Samuel C. Prescott, *When MIT was "Boston Tech," 1861–1916* (Cambridge, Mass.: Technology Press, 1954), pp. 155–178.
5. See Edward Weeks, *The Lowells and Their Institute* (Boston: Little, Brown and Co., 1966). Edward Pickering to Augustus Lowell, 1879, E. C. Pickering papers, HUA. Augustus Lowell to Pickering, March 12, 1889, Lowell family papers, HL.
6. *Proceedings of the American Academy of Arts and Sciences* 35 (May 1899–May 1900): 433–436.
7. See *Appalachia: The Journal of the Appalachian Mountain Club* 1 (1876): 5, 6, 58; 5 (1887): 2, 174; and 6 (1890): 103.
8. Seth Chandler to Edward Holden, Sept. 4, 1894, Holden papers, MLSA.
9. Pickering to William Sturgis Bigelow, Oct. 4, Nov. 25, Dec. 8, 1893; George Agassiz to Pickering, Dec. 31, 1905, Jan. 29, 1906, May 2, 1910, E. C. Pickering papers, HUA.
10. See Lowell to Pickering, Sept. 9, Nov. 7, 1892; Pickering to Lowell, Sept. 10, E. C. Pickering papers; see also W. H. to E. C. Pickering, March 6, 1893, W. H. Pickering papers, HUA.
11. Lowell to Pickering, Sept. 9, Nov. 7, 1892, E. C. Pickering papers, HUA; *Records of the Board of Overseers of Harvard College,* vol. 13 (1891–1900), p. 77B.
12. John H. Conover, *The Blue Hill Meteorological Observatory: The First 100 Years, 1885–1985* (Boston: American Meteorological Society, 1990), pp. 25, 38–39.
13. Jones and Boyd, *The Harvard College Observatory,* pp. 253, 257, 313, 315, 326.
14. On Arequipa, see Plotkin, "Harvard College Observatory's Boyden Station," pp. 689–705.
15. Jones and Boyd, *The Harvard College Observatory,* p. 306; and Plotkin, "Harvard College Observatory's Boyden Station."
16. E. C. Pickering to Charles Eliot, Feb. 13, 1894, E. C. Pickering papers; W. H. Pickering to *W. R. Burrows Monthly Weather Magazine,* Feb. 6, 1894; Pickering to J. A. Brashear, Nov. 1, 1893; Pickering to the Rumford Committee, Oct. 13, 1893; Pickering to Brashear, Nov. 1, Dec. 11, 1893; Pickering to W. H. Rollins, Dec. 5, 1893; Pickering to H. A. Howe, Jan. 9, 1894; Pickering to Lowell, Jan. 17, 1894, W. H. Pickering papers, HUA.

17. W. H. to E. C. Pickering, March 6, 1893, W. H. Pickering papers, HUA; S. W. Burnham to E. E. Barnard, Nov. 16, 1894, Barnard papers, Vanderbilt University Archives. On Lowell's viewing in Japan, see Lowell to William Lowell Putnam, undated (approx. Jan. 30, 1893), Lowell family papers, HL; for Lowell's earliest remarks about the Arizona expedition, see Lowell to E. C. Pickering, Jan. 27, 1894, E. C. Pickering papers, HUA; see also Ralph Curtis to Isabella Stewart Gardner, Nov. 26, 1893, Isabella Stewart Gardner papers, GM.
18. Lowell's copy of Flammarion's book is in the Lowell Observatory Library: *La planète mars* (Paris: Gauthier-Villars et fils, 1892).
19. I am indebted to John Conover, Rotch's biographer, and Lawrence Rotch, his descendant and custodian of the Rotch diaries, who searched the entries for 1894 and later years. See also John Ritchie, Jr., "Percival Lowell," *Technology Review* 19 (1917): 36.
20. Pickering to Lowell, Jan. 17, 1894, W. H. Pickering papers, HUA. Details of the agreement between Lowell and Harvard are given in the *Boston Herald,* Feb. 13, 1894. Pickering to John Brashear, Jan. 24, 1894, W. H. Pickering papers, HUA.
21. Jones and Boyd, *The Harvard College Observatory,* pp. 326–327. Pickering to Lowell, Jan. 25, 1894, W. H. Pickering papers, HUA.
22. Lowell to E. C. Pickering, Jan. 27, 1894, E. C. Pickering papers, HUA.
23. W. H. Pickering to Lowell, Feb. 16, 1894; Lowell's telephone message to Pickering, Feb. 17, 1894, W. H. Pickering papers, HUA. Lowell to E. C. Pickering, Feb. 14, 1894, E. C. Pickering papers, HUA.
24. See Weeks, *The Lowells and Their Institute.*
25. W. H. Pickering to Lowell, Feb. 21, 1894; Pickering to Brashear, Feb. 21, 1894; Pickering to Henry Gannett, Feb. 22, 1894, W. H. Pickering papers, HUA.
26. "Arizona Astronomical Expedition," March 1, 1894, W. H. Pickering papers, HUA. On observatory finances, see E. C. Pickering to Eliot, Feb. 13, 1894, E. C. Pickering papers, HUA.
27. Chandler to Holden, Sept. 4, 1894, Holden papers, MLSA.
28. W. H. Pickering to *W. R. Burrows Monthly Weather Magazine,* Feb. 6, 1894; Pickering to Gannett, Feb. 22, 1894; Pickering to Lowell, Jan. 24, Feb. 7, 1894, W. H. Pickering papers, HUA.
29. "Instructions given to A. E. Douglass regarding the trip to Arizona," Feb. 28, 1894; Pickering to Lowell, April 20, 1894, W. H. Pickering papers, HUA.
30. Lowell telegram to Douglass, March 15, 1894, Douglass papers; Lowell to Douglass, March 22, 1894, Lowell papers, LOA.
31. W. H. Pickering to Lowell, Jan. 24, March 12, 1894, W. H. Pickering Papers, HUA. Lowell to Douglass, March 24, 28, April 16, 1894; Douglass papers, LOA.
32. W. H. to Mrs. Pickering, June 4, 1894; W. H. to E. C. Pickering, July 14, 1894, W. H. Pickering papers, HUA.
33. E. C. to W. H. Pickering, Nov. 7, 1894; W. H. to E. C. Pickering, Nov. 10, 1894, E. C. Pickering papers, HUA. Evidently there were no hard feelings between

Lowell and Pickering; see W. H. Pickering to Lowell, May 22, 1895, Lowell papers; Putnam to Douglass, Jan. 16, 1899, Douglass papers, LOA.

34. See *Astronomy and Astrophysics* 13 (1894): 538–553, 645–650; *Popular Astronomy* 2 (1894): 1–8, 52–56; *Scientific American* 37 (1894): 15588–15590, 15682–15684. Chandler to Holden, Sept. 4, 1894, Holden papers, MLSA.
35. Chandler to Holden, Sept. 4, 1894, Holden papers, MLSA.
36. Lowell to E. C. Pickering, Nov. 16, 1894, E. C. Pickering papers, HUA.
37. E. C. Pickering to Eliot, Nov. 21, Nov. 23, 1894, Eliot papers, HUA; E. C. Pickering to Augustus Lowell, 1879, Lowell family papers, HL.
38. A. L. Lowell, *The Biography of Percival Lowell* (New York: Macmillan Co., 1935), pp. 62–69.
39. *Annals of the Lowell Observatory* 1 (Boston: Houghton Mifflin Co., 1898), pp. vii–xi. See also C. O. Lampland to A. L. Lowell, Aug. 19, 1935, Lampland papers, LOA.
40. See "Arizona Astronomical Expedition," March 1, 1894, W. H. Pickering papers, HUA.
41. Lowell, *The Biography of Percival Lowell,* pp. 92–97.
42. Douglass to Lowell, Dec. 28, 1894; Lowell to Douglass, March 5, 21, 1895, Lowell papers, LOA.
43. On the Clark refractor, see "Agreement," April 25, 1895, LOA; articles based on his lectures appeared in the *Atlantic Monthly,* May–August 1895, pp. 594–603, 749–758, 106–119, 223–235. Lowell, *Mars* (Boston: Houghton Mifflin Co., 1895), p. 211.
44. *Dial* 21 (July 16, 1896): 43.
45. Douglass to Lowell, March 28, May 24, 1895, Lowell papers, LOA.
46. Lowell to Eliot Clarke, Jan. 23, Feb. 10, 1896, Clarke papers, HL; Lowell to Henry Mercer, Feb. 27, 1896, Mercer papers, Bucks County Historical Society, Doylestown, Pa. See also Lowell, *The Biography of Percival Lowell,* pp. 93–94. On Trépied, see *Catalogue de 9997 étoiles* (Paris: Gauthiers-Villars et Cie, 1924), p. 1.
47. F. R. Moulton apparently recommended Cogshall to See. In the spring of 1898, Cogshall took several postgraduate courses at the University of Chicago. Wilbur Cogshall to Douglass, July 28, 1898, Douglass papers, LOA. Drew had also studied with See at Chicago. *American Men of Science: A Biographical Dictionary* (New York: Bowker, 1906), pp. 64, 90.
48. Douglass to Lowell, April 12, May 24, 1895, Lowell papers, LOA.
49. See to William Rainey Harper, Harper papers, University of Chicago Archives. Lowell to Douglass, Sept. 18, 1895, Lowell papers, LOA. I am grateful to Professor Charles Peterson of the University of Missouri for showing me a draft of chapters 4 and 5 of his *Visionary and Dangerous Genius: The Life and Career of the Unparalleled T. J. J. See.*
50. Putnam to Douglass, Jan. 19, 1898, March 29, 1899, Putnam papers, LOA.

51. Douglass to Lowell, Aug. 1, 1995, Lowell papers; Douglass to William A. Davis, Oct. 5, 1895, Douglass papers; Lowell to Douglass, June 1, 5, 1897, Lowell papers, LOA. Webb, *Tree Rings and Telescopes,* p. 39.
52. *Boston Commonwealth,* May 26, 1894.
53. On Douglass and Pickering, see Webb, *Tree Rings and Telescopes,* pp. 28, 35, 39.
54. The microfilmed edition of the correspondence of the Lowell Observatory contains no letters on the hiring of See, Cogshall, and Drew. See Hoyt, *Lowell and Mars,* p. 119. Lowell's invitation to See is noted in T. J. J. See, "Discoveries and Measures of Double and Multiple Stars in the Southern Hemisphere Made with the Clark Refractor," *Astronomical Journal* 18 (1898): 181. On See's reputation, see John Lankford, "A Note on T. J. J. See's Observations of Craters on Mercury," *Journal for the History of Astronomy* 11 (1980): 129–132.
55. Cogshall to Lowell, July 19, 1901, Jan. 31, 1903; Lowell to Cogshall, July 7, 1901, Feb. 8, 1903, Lowell papers, LOA.
56. A detailed account of See's firing is offered by Hoyt, *Lowell and Mars,* p. 123.
57. On the Venus markings, see William Sheehan, *Planets and Perception: Telescopic Views and Interpretations, 1609–1909* (Tucson: University of Arizona Press, 1988), pp. 204–208, 218–222, 227–230.
58. On Douglass's optical experiments, see Lowell to Douglass, Sept. 5, Oct. 10, Oct. 14, Nov. 21, Nov. 22, 1900; Douglass to Lowell, Oct. 16, 1900, Lowell papers, LOA. On Lowell's recantation, see "The Markings on Venus," *Astronomische Nachrichten* 3823 (July 3, 1902): 129–131. On Lowell's doubts about double canals, see *Annals of the Lowell Observatory,* vol. 2 p. 441. On doubts about the Venus markings, see Hoyt, *Lowell and Mars,* p. 119; Douglass to Arthur Herbert Wilde, Jan. 7, 1912.
59. Putnam to Douglass, Aug. 3, 22, 1898, Putnam papers, LOA.
60. Ibid. On observatory expenses, see Putnam to Douglass, Aug. 3, 1898, Putnam papers, LOA.
61. See note 58.
62. Douglass argues in his 1948 manuscript "Lowell Observatory: Reminiscences" (p. 40) that doubts about markings on the planets led directly to the purchase of the spectrograph, Douglass papers, University of Arizona Archives.
63. Lowell to Elizabeth Lowell, April 12, 1901, Lowell family papers, HL.
64. On Pickering's Jamaica career, see Howard Plotkin, "William H. Pickering in Jamaica: The Founding of Woodlawn and Studies of Mars," *Journal for the History of Astronomy* 24 (1993): 101–122.

10. Lowell's Campaign for the Canals of Mars

1. See William Graves Hoyt, *Lowell and Mars* (Tucson: University of Arizona Press, 1976), pp. 173–175; and Lowell, *Mars and Its Canals* (New York: Macmillan Co., 1906), p. 17.

2. A. E. Douglass, "Lowell Observatory: Reminiscences," manuscript, 1948, pp. 35–40, Douglass papers, University of Arizona Archives.
3. Douglass to Arthur Herbert Wilde, Jan. 7, 1912; Lowell to Douglass, Sept. 5, Oct. 10, Oct. 14, Nov. 21, Nov. 26, 1900, Douglass papers, University of Arizona Archives.
4. On Flagstaff visual experiments, see William Sheehan, *Planets and Perception: Telescopic Views and Interpretation, 1609–1909* (Tucson: University of Arizona Press, 1988), pp. 227–229; and George Ernest Webb, *Tree Rings and Telescopes: The Scientific Career of A. E. Douglass* (Tucson: University of Arizona Press, 1983), pp. 47–48.
5. Hoyt, *Lowell and Mars,* pp. 164–166; Sheehan, *Planets and Perception,* pp. 231–232.
6. A. E. Douglass, "Atmosphere, Telescope and Observer," *Popular Astronomy* 5 (1897): 67. For Lowell's views, see "Means, Methods and Mistakes in the Study of Planetary Evolution," which was completed on April 13, 1905, submitted to the Royal Astronomical Society on Nov. 15, 1905, but never published, Lowell papers, LOA.
7. "Chart of Faint Stars for Magnitude Comparison (2nd series)," *Publications of the Astronomical Society of the Pacific* 6 (Aug. 15, 1894): 227–228.
8. On Lowell's claims, see *Monthly Notices of the Royal Astronomical Society* 66 (Dec. 1905): 56–57; *Popular Astronomy* 13 (Aug.–Sept. 1905): 391; *Proceedings of the Royal Society of London* 77 (June 1906): 132–135; *Outlook* 85 (April 13, 1907): 845; *Century* 75 (Dec. 1907): 303; *Philosophical Magazine* 15 (April 1908): 469. Lowell to Simon Newcomb, Oct. 3, 1905, Lowell papers, LOA. The rebuke of Newcomb is found in *Outlook* 85 (April 13, 1907): 845. Edward Morse wrote immediately to praise Lowell for his "dig at Newcomb," June 12, 1907, Lowell papers, LOA. For other versions of this controversy, see John Lankford, *American Astronomy: Community, Careers, and Power, 1859–1940* (Chicago: University of Chicago Press, 1997), pp. 201–203; Donald Osterbrock, "To Climb the Highest Mountain: W. W. Campbell's 1909 Mars Expedition to Mount Whitney," *Journal for the History of Astronomy* 20 (1989): 93.
9. On photographing the canals, the best account is Hoyt, *Lowell and Mars,* pp. 173–187. On the problem of visual representation in astronomy, see Jennifer Tucker, "Science Illustrated: Photographic Evidence and Social Practice in England, 1870–1920," Ph.D. diss., Johns Hopkins University, 1996, pp. 190–256, and Alex Pang, "Technology, Aesthetics, and the Development of Astrophotography at the Lick Observatory," in Timothy Lenoir, ed., *Inscribing Science: Scientific Texts and the Materiality of Communication* (Stanford: Stanford University Press, 1998), pp. 223–248.
10. Lowell to George Ritchey, March 28, May 9, 1906, Lowell papers, LOA. On this episode, see Donald Osterbrock, *Pauper and Prince: Ritchey, Hale and Big*

American Telescopes (Tucson: University of Arizona Press, 1993), pp. 89–91, 110.

11. Osterbrock, "To Climb the Highest Mountain," and David DeVorkin, "W. W. Campbell's Spectroscopic Study of the Martian Atmosphere," *Quarterly Journal of the Royal Astronomical Society* 18 (1977): 37–53.
12. V. M. Slipher to Lowell, Feb. 20, 1908; Lowell to Slipher, May 4, 1908; Slipher to Lowell, May 13, Oct. 13, Dec. 29, 1908, Lowell papers, LOA. Slipher to W. W. Campbell, April 17, 1908, V. M. Slipher papers, LOA.
13. Not all of Lowell's friends believed in the canals. See William Sturgis Bigelow to Frederick Shattuck in Akira Murakata, "Selected Letters of Dr. William Sturgis Bigelow," Ph.D. diss., George Washington University, 1971, p. 290, and Ralph Curtis to Barrett Wendell, Feb. 2, 1910, Wendell papers, HL.
14. On Morse, see Dorothy Wayman, *Edward Sylvester Morse: A Biography* (Cambridge, Mass.: Harvard University Press, 1942), especially pp. 394–397, and Robert Rosenstone, *Mirror in the Shrine: American Encounters with Meiji Japan* (Cambridge, Mass.: Harvard University Press, 1988); on Thomson, see Lowell to Elihu Thomson, May 11, 1906, April 7, 1910, Lowell papers, LOA, and David O. Woodbury, *Beloved Scientist: Elihu Thomson: A Guiding Spirit of the Electrical Age* (New York: McGraw-Hill, 1944); on Story, see Roger Cooke and V. Frederick Rickey, "W. E. Story of Hopkins and Clark," unpublished and undated manuscript; on Agassiz, see "George Russell Agassiz," *The National Cyclopedia of American Biography,* 2nd ed. (New York: J. T. White, 1937), pp. 48–49.
15. On Osborn, see Ronald Rainger, *An Agenda for Antiquity: Henry Fairfield Osborn and Vertebrate Paleontology at the American Museum of Natural History, 1890–1935* (Tuscaloosa: University of Alabama Press, 1991); on the Todds, see Elizabeth Horan, "Expedition to Esahi: The Amherst Trip to See the Eclipse of 1891," *Amherst Alumni News* (Spring 1990): 18–24; "Description of the Todd Eclipse Expeditions," typescript, Amherst College Archives.
16. Lowell to David Todd, Aug. 5, 1895, Nov. 9, 1895, April 15, 1897, Lowell papers, LOA; T. J. J. See to A. E. Douglass, Sept. 7, 1897, Douglass papers, LOA; Louise Leonard to Todd, undated (late 1899), misc. papers, LOA; Mabel to David Todd, Feb. 11, 1900, David Todd papers, YUA; Mabel Loomis Todd diary, Amherst, Dec. 26, 1899, and "Tripoli, Barbary 1900," May 24, 1900, Mabel Loomis Todd papers, YUA.
17. Ruth Owen Jones, "Those Daring Young Amherst Men (and One Intrepid Woman) in Their Flying Machines," *Amherst Alumni News* (Summer 1994): 28–31. On Lowell's participation with Todd in an "aeroplane" club, see Lowell to Todd, July 26, 1907, Lowell papers, LOA. See also "Mabel Loomis Todd Journal: South American Trip for the Opposition of Mars—for Percival Lowell," Mabel Loomis Todd papers, YUA; "David P. Todd diary, April 15, 1907–Aug. 30, 1907," David Todd papers, YUA. Regarding Lowell's presentation, see Low-

ell to Lampland, Nov. 20, 1907, Lowell papers, LOA. Regarding the Todds' talks to the National Academy and the Tech Club, see Mabel Loomis Todd diary, Nov. 20, 1907, Feb. 25, 1907, Mabel Loomis Todd papers, YUA; Leonard to C. O. Lampland, Feb. 27, 1908, Lampland papers, LOA. See also "Prof. David Todd's Hopes of Receiving Martian Messages," *Scientific American*, June 5, 1909, clipping in David Todd papers, YUA.

18. See list of public lectures, 1906–1910, Morse papers, PMS. See also Edward Morse, "My 34 Nights on Mars," *Boston Globe*, Oct. 7, 1906, and Morse, *Mars and Its Mystery* (Boston: Little, Brown and Co., 1906). See Leonard to Morse, Oct. 2, 1906; Lowell to Herbert Mills, Oct. 4, 1906; Lowell to Frederick Webster, Oct. 24, 1906; Leonard to Lowell, May 7, 1909; Lowell to Morse, Oct. 22, 1916, Lowell papers, LOA. On Agassiz's work for Lowell, see Leonard to Slipher, May 14, 1908, V. M. Slipher papers; Lowell to Agassiz, Sept. 16, 1907; Agassiz to Lowell, Sept. 27, 1907, Lowell papers, LOA. Leonard to E. C. Slipher, May 16, 1908, E. C. Slipher papers, LOA.
19. "Life on Mars," unnamed periodical, Dec. 1896, scrapbook, LOA; "The Canals of Mars," *Knowledge and Scientific News* 1, n.s. (March 1904): 37–38; Lowell to *London News*, Nov. 27, 1903, Lowell papers, LOA.
20. Henry F. Osborn to Lowell, Nov. 1, 1909, Lowell papers, LOA. See Lowell, "Oxygen and Water on Mars," and E. C. Slipher, "The Photograph in Astronomy," *American Museum Journal* 15 (May 1915): 207–209 and 210–219. Ward to Lowell, March 31, 1907, Lowell papers, LOA; Woodbury, *Beloved Scientist*, p. 256. See also "Exhibit by the Lowell Observatory, Flagstaff, Arizona," Boston Public Library, 1915. Mabel Rice Percy to Leonard, April 13, 1915; Leonard to Percy, undated, Leonard papers, LOA.
21. On Lowell's support for the Todd expedition, see Lowell to Todd, April 11, 1907, David Todd papers, YUA. On Morse's ticket, see Lowell telegram to Morse, May 4, 1905, Lowell papers, LOA.
22. On Todd's *Cosmopolitan* article, see Lowell to A. L. Everett, Dec. 19, 20, 24, 1907, Jan. 10, Feb. 8, 1908; Lowell to Todd, Dec. 31, 1907, March 5, 1908, Lowell papers, LOA.
23. Lowell to Morse, July 16, 1905, Nov. 25, 1905, no date, PMS; Ms. Beulah Hill to Lowell, June 22, 1905, Lowell to M. M. Giacobini, Nov. 25, 1905, May 25, 1906, May 7, 1907; L. F. Ward, "Mars and Its Lessons," *Brown Alumni Monthly* (March 1907): 159–165; regarding Ward's letter, see Lowell to Morse, April 5, 1907, Lowell papers, LOA. Ward to Lowell, March 31, 1907, Lowell papers, LOA.
24. See *Astronomy and Astrophysics* 13 (Oct. 1894): 647. William Story, "The Canals on Mars," *Knowledge and Scientific News* 1, n.s. (March 1904): 37–38; W. F. Denning, "The Canals of Mars," ibid. (April 1904): 67–68; E. Walter Maunder, "The 'Canals' of Mars: A Reply to Mr. Story," ibid. (May 1904): 87–89.
25. George Agassiz, "Drawings of Mars from Flagstaff," *Astronomische Nachrichten*

184 (1910): 191–192. Lowell to William Lowell Putnam, May 6, 1907, Lowell papers, LOA; on Todd's vision, see E. C. to V. M. Slipher, July 28, 1907, V. M. Slipher papers, LOA; David Todd, "Professor Todd's Own Story of the Mars Expedition: First Article Published from the Pen of the Leader of the Party of Observation," *Cosmopolitan Magazine* 44 (March 1908): 349; Mabel Loomis Todd, "Our Ruddy Neighbor Planet," *Independent* 64 (April 9, 1908): 792.

26. Thomson to Morse, July 5, 1907, Lowell papers, LOA; Thomson to Morse, undated, in Wayman, *Edward Sylvester Morse,* p. 397. On Thomson and Mars, see Woodbury, *Beloved Scientist,* pp. 257–260.
27. Mabel Loomis Todd, "Our Ruddy Neighbor Planet," p. 794; David Todd, "Professor Todd's Own Story," p. 350.
28. On Lowell's honorary degree, see Todd to Millicent Todd Bingham, undated (probably 1922), David Todd papers, YUA; see also Lowell to Todd, July 26, 1907, Lowell papers, LOA. See also "Amherst's Finale," *Boston Evening Transcript,* June 25, 1907, p. 4.
29. Lowell to Story, March 24, 1909, Lowell papers, LOA. See also "Advanced Study in Hygiene Is Greatly Desired," *Worcester Gazette,* Sept. 9, 1909, Clark University Archives; *Boston Evening Transcript,* Sept. 9, 1909, p. 2; Sept. 10, 1909, p. 4; Sept. 11, 1909, p. 2.
30. On Lowell as pioneer, see Keith Baker, *The Channels of Mars* (Austin: University of Texas Press, 1982), p. 13.
31. See "Means, Methods and Mistakes," pp. 1–5, 14–22; John Lankford, "Amateurs versus Professionals: The Controversy over Telescope Size in Late Victorian Science," *Isis* 71 (1981): 11–28.
32. Lowell to Morse, March 2, 1905, Morse papers, PMS. On Lowell's use of Dana and Merriam, see *Mars and Its Canals,* pp. 18, 19, 131, 139, 140, 379.
33. Lowell to Morse, July 16, 1905, Morse papers, PMS.
34. *Mars and Its Canals,* p. 111; Lowell's place among advocates of an extraterrestrial perspective is analyzed in Karl S. Guthke, *The Last Frontier: Imagining Other Worlds from the Copernican Revolution to Modern Science Fiction,* trans. Helen Atkins (Ithaca, N.Y.: Cornell University Press, 1990 [1983]), pp. 354–362.
35. *Mars and Its Canals,* p. 149.
36. Morse, *Mars and Its Mystery,* pp. 141, 143.
37. Ibid., pp. 50, 107–108, 112–113, 45; *Mars and Its Canals,* pp. 195, 205, 361–362.
38. *Mars,* p. 211; Ward to Lowell, March 31, 1907; Lowell to Ward, April 10, 1907, Lowell papers, LOA.
39. Morse, *Mars and Its Mystery,* pp. 11–13.
40. Ibid., pp. 5–6, 8–9; see also *Mars and Its Canals,* p. 39.
41. Morse, *Mars and Its Mystery,* pp. 71, 72.
42. Ibid., pp. 75, 77.

43. Morse, *Mars and Its Mystery*, p. 73; *Mars and Its Canals*, p. viii. In 1913 Lowell hired a journalist to explain the observatory's work in popular magazines. Lowell to Victor R. Emmanuel, Sept. 15, 1913; Leonard to Emmanuel, Sept. 26, Oct. 7, 8, 15, 31, Nov. 10, Dec. 4, 1913, Lowell papers, LOA.
44. Morse to Simon Newcomb, Dec. 3, 1905, MDLC. For critics' mixed responses to Lowell and Morse, see E. T. Brewster, "The Earth and the Heavens," *Atlantic Monthly* 100 (Aug. 1907): 262–263; Herbert A. Howe, "The Red Planet Mars," *Dial* 42 (Feb. 1, 1907): 75–77; anonymous (reported by Morse and Lowell to be Simon Newcomb), "Two Books on Mars," *Nation* 84 (April 4, 1907): 317–318; and anonymous, "New Books on Astronomy," *Independent* 61 (Dec. 1906): 1567–1568.
45. On Morse's confession, see *Mars and Its Mystery*, p. 163.

11. The Establishment Responds

1. See claimed that the astronomers were "nearly all entangled in the ring which is just like the ring in Pennsylvania." T. J. J. See to Lowell, Oct. 8, 1907, Lowell papers, LOA.
2. On the rivalry between elite astronomers, see John Lankford, *American Astronomy: Community, Careers, and Power, 1859–1940* (Chicago: Chicago University Press, 1997), pp. 198–235.
3. W. W. Campbell to George E. Hale, May 11, 1908, Hale papers, MWPOL; Hale to Campbell, Campbell papers, MLSA.
4. E. B. Frost to Hale, Nov. 27, 1907, Hale papers, MWPOL; F. R. Moulton to Campbell, Oct. 18, 1909; Campbell to Moulton, Oct. 25, 1909, Campbell papers, MLSA. Simon Newcomb to Hale, Nov. 9, 15, 1901, Hale papers, MWPOL. Hale once remarked: "Our philanthropic intent might indeed be likened to that of the good Sir Percival, when with unsparing hand he sprinkles oases upon the barren wastes of Mars," Hale to Wright, Oct. 16, 1901, Hale papers, MWPOL.
5. Newcomb to E. E. Barnard, March 9, 1896, as quoted in William Sheehan, *Planets and Perception: Telescopic Views and Interpretations, 1609–1909* (Tucson: University of Arizona Press, 1988), p. 253; Barnard to Hale, Oct. 26, 1905, Hale papers, MWPOL. Two years later, E. C. Pickering encouraged Douglass to publish his theory that "defects in the eye" explain the appearance of the canals of Mars on the telescope lens. See Pickering to A. E. Douglass, Feb. 9, 1907, E. C. Pickering papers, HUA.
6. Lowell to Newcomb, March 1, 1903; Newcomb to Lowell, March 9, 1903; Lowell to Newcomb, March 15, 1903; Newcomb to Lowell, March 23, 1903, Lowell papers, LOA. Morse to Newcomb, Nov. 5, 22, 24, Dec. 3, 4, 1905, Newcomb papers, MDLC.

7. Newcomb to E. C. Pickering, Jan. 19, 24, Feb. 11, 18, 28, March 6, 1907; Pickering to Newcomb, Jan. 21, 19, Feb. 16, 26, 1907, Newcomb papers, MDLC.
8. Newcomb to Pickering, Jan. 19, Feb. 28, 1907; Lowell disputed Harvard's failure to photograph a new star he had recently seen. Pickering to Newcomb, Jan. 27, 1907, Newcomb papers, MDLC.
9. Pickering to Newcomb, March 5, 1907; Morse to Newcomb, March 28, 1907, Newcomb papers, MDLC. Simon Newcomb, "The Optical and Psychological Principles Involved in the Interpretation of the So-Called Canals of Mars," *Astrophysical Journal* 26 (July 1907): 10–12, 14, 16.
10. Lowell, "The Canals of Mars, Optically and Psychologically Considered: A Reply to Professor Newcomb," *Astrophysical Journal* 26 (October 1907): 137. After the experiment, Newcomb wrote Campbell: "Six months ago I could not have entertained the idea of going through Flagstaff without stopping. But it was the middle of the night, and Lowell did not seem overanxious to have me look at Mars through his telescope," June 30, 1907, Campbell papers, MLSA. Lowell's supporters defended him. T. J. J. See denounced Newcomb's article as "not worth the paper it is written on . . . [it] is directly contradicted by your latest photographs," Oct. 8, 1907, Lowell papers, LOA.
11. For other accounts of the *Outlook* controversy, see Lankford, *American Astronomy,* pp. 201–203; and Donald Osterbrock, "To Climb the Highest Mountain: W. W. Campbell's 1909 Mars Expedition to Mount Whitney," *Journal for the History of Astronomy* 20 (1989): 93. Frost to Campbell, Sept. 16, 1908; see also Campbell to Frost, Sept. 21, 1900, Campbell papers, MLSA.
12. Campbell to Frost, Sept. 28, 1908, Campbell papers, MLSA.
13. Campbell to Frost, Sept. 1, 1908; Hale to Campbell, Sept. 17, 1908, Campbell papers, MLSA.
14. See draft of letter from Campbell to the editor of *Popular Astronomy,* Sept. 3, 1908, Campbell papers, MLSA.
15. V. M. Slipher to Lowell, Oct. 14, 1908; Lowell to Slipher, Oct. 19, 1908, Lowell papers, LOA; Louise Leonard (for Lowell) to Slipher, Oct. 14, 1908, V. M. Slipher papers, LOA.
16. V. M. Slipher, "Air and Large Telescopes," *Outlook* 90 (Dec. 19, 1908): 867–869.
17. Campbell to Benjamin Wheeler, May 29, June 23, 1909, Campbell papers, MLSA. Draft of Campbell letter to the editor of *Outlook,* May 28, 1909, Campbell papers, MLSA.
18. Campbell to the editor of *Outlook,* Jan. 13, 1909, Feb. 8, 1909; Wheeler to Campbell, Feb. 9, 1909, Campbell papers, MLSA.
19. Campbell to Slipher, Oct. 4, 1909, V. M. Slipher papers, LOA.
20. Slipher to Campbell, Nov. 15, 1909, ibid.
21. Campbell to Slipher, Nov. 2, 1909, ibid.
22. James Wallace of Yerkes remarked to Hale: "Lowell wants me to make a screen

[filter] for his telescope to photograph the Canals in Mars—('for a consideration' I could put them in the screen)" Jan. 30, 1904, Hale papers, MWPOL.

23. Hale to Campbell, Sept. 17, 1909, Campbell papers, MLSA.
24. Hale to Campbell, Oct. 6, 1909, Campbell papers, MLSA; Hale to Frost, Nov. 1, 1909, Hale papers, MWPOL.
25. Hale to Campbell, Nov. 1, 1909, Campbell papers, MLSA; E. E. Barnard to Hale, Oct. 26, 1905; Hale to Barnard, Nov. 9, 1905, Hale papers, MWPOL.
26. On Hale's doubts about the photographs, see Hale to Frost, Nov. 1, 1909; Campbell to Hale, Oct. 26, 1909, Hale papers, MWPOL; Hale to Campbell, Oct. 23, 1909, Campbell papers, MLSA.
27. On Douglass, see Hale to Campbell, Sept. 17, 1908; on Sykes, see Hale to Campbell, Nov. 1, 1909; on Michelson, see Hale to Campbell, Oct. 23, 1909, Campbell papers, MLSA.
28. Hale to Campbell, April 8, 26, 1910, Campbell papers, MLSA.
29. Hale wrote Campbell: "I suppose from the English descriptions of Lowell's photographs of Mars that they are probably decidedly superior to mine," May 10, 1910, Campbell papers, MLSA; Barnard to Hale, Oct. 19, 1911, Hale papers; E. E. Barnard, "An Account of Some Observations Made with the 60 Inch Reflector at Mount Wilson in November 1911," Hale papers, MWPOL. For a fuller account of this episode, see William Sheehan, *The Immortal Fire Within: The Life and Work of Edward Emerson Barnard* (Cambridge: Cambridge University Press, 1995), pp. 393–395.
30. Barnard to Hale, undated (but received by Hale on June 1, 1908), Hale papers, MWPOL. Leo Brenner, born Spiridion Gopcevic, established his own observatory in Italy and made even more extravagant claims than Lowell about visible markings on the surface of Mars. See Michael Crowe, *The Extraterrestrial Life Debate, 1750–1900: The Idea of a Plurality of Worlds from Kant to Lowell* (Cambridge: Cambridge University Press, 1986), p. 515.
31. Pickering to Hale, July 21, 1913; Hale to Pickering, Aug. 13, 1913, Hale papers, MWPOL. Lowell to Pickering, Sept. 2, 1913, Lowell papers, LOA. Campbell to Pickering, Aug. 25, 1913, Campbell papers, MLSA.
32. Campbell to Hale, May 14, 1908, Hale papers, MWPOL; Frost to Campbell, Jan. 18, 1909; Campbell to Wheeler, April 30, 1909, Campbell papers, MLSA; For more detailed accounts of the expedition, see Osterbrock, "To Climb the Highest Mountain," pp. 77–97, and David H. DeVorkin, "W. W. Campbell's Spectroscopic Study of the Martian Atmosphere," *Quarterly Journal of the Royal Astronomical Society* 18 (1977): 37–53.
33. Osterbrock, "To Climb the Highest Mountain," p. 92; Hale to Campbell, Sept. 17, 1909; F. R. Moulton to Campbell, Oct. 18, 1909; Frost to Campbell, Oct. 26, 1909; Campbell to Frost, Dec. 1, 1909, Campbell papers, MLSA.
34. Hale to Campbell, April 8, 26, May 10, 1910, Campbell papers, MLSA.

35. C. G. Abbot, "Water Vapor on Mars," (quoting Lowell's response to Campbell's Mount Whitney expedition), *Science* 31, n.s. (June 24, 1910): 987–988; Campbell to "Dear Sir" (apparently the editor of the *Astrophysical Journal,* though the letter was never published), June 1, 1910, Campbell papers, MLSA.
36. G. R. Agassiz, "The Spectrum of Mars," and W. W. Campbell, "To the Editor," *Science* 32, n.s. (Sept. 23, 1910): 407–408. Campbell to Sebastian Albrecht, Aug. 15, 1910; Albrecht to Campbell, Aug. 20, 1910, Campbell papers, MLSA. Lowell, "A New Method of Testing Spectrographically a Martian Atmosphere," and V. M. Slipher, "An Attempt to Apply Velocity-Shift to Detecting Atmospheric Lines in the Spectrum of Mars," *Lowell Observatory Bulletin* 17 (1905): 116–118.
37. E. C. Pickering to Campbell, April 8, 1905; Campbell to Pickering, Aug. 14, 1905, Campbell papers, MLSA. The early history of the telegraphic service is sketched in B. Z. Jones and L. G. Boyd, *The Harvard College Observatory: The First Four Directorships, 1839–1919* (Cambridge, Mass.: Harvard University Press, 1971), pp. 194–198.
38. Pickering to Campbell, Feb. 3, 1906, Campbell papers, MLSA. Pickering to Lowell, July 23, 1907, E. C. Pickering papers, HUA.
39. Lowell to Pickering, Aug. 1, 1907, E. C. Pickering papers, HUA. John Ritchie, Jr., to Lowell, Aug. 5, 1907; Lowell to Ritchie, Aug. 12, 1907, Lowell papers, LOA. On the Pickering-Ritchie relationship, see Lankford, *American Astronomy,* pp. 65–66.
40. Pickering to Lowell, Dec. 26, 1908; Lowell to Pickering, Jan. 21, 1909, Pickering papers, HUA; Lowell to H. Kobold, Jan. 21, 1909; "International Arrangement for the Distribution of Planetary News," Feb. 1, 1909, Lowell papers, LOA.
41. Lowell to A. L. Lowell, Sept. 20, 1909; A. L. Lowell to Lowell, Nov. 12, 1909; "Lowell Slights Harvard," *Boston Post,* undated, Lowell papers, LOA; for Pickering's explanation of his problem with Lowell, see Pickering to Campbell, Dec. 2, 1909, Campbell papers, MLSA. Plans of the International Solar Conference to visit Lowell Observatory were almost cancelled during the dispute over the telegraph service. See Campbell to Pickering, May 26, 1910, Campbell papers, MLSA; Lowell to Pickering, June 2, 1910, Lowell papers, LOA.
42. On honorary organizations, see Lankford, *American Astronomy,* pp. 248ff. For other accounts of this event, see Lankford, *American Astronomy,* pp. 265–266, and William Graves Hoyt, *Lowell and Mars* (Tucson: University of Arizona Press, 1976), pp. 275–276.
43. Lowell to Edwin Hall, March 15, 1909; Lowell to John Trowbridge, Nov. 12, 1909; Petition to Council of American Academy, Feb. 28, 1910; "Quadric" (Story) to "Mars" (Lowell), March 7, 1910; Lowell papers, LOA. E. C. Pickering to Campbell, Jan. 30, 1910, Campbell papers, MLSA. Barrett Wendell wrote to Frederic Stimson: "Percy is a bit stirred up by astronomic disputes in the Am

Acad touching on candidates and a medal. I am going to the meeting on Wed. to vote on his side," Jan. 8, 1915, Wendell papers, HL.

44. Leonard to Slipher, March 11, 1910, V. M. Slipher papers, LOA; Slipher to Lowell, March 16, 1910; Lowell to A. L. Lowell, Nov. 12, 1909; Elihu Thomson to Lowell, March 10, 1910, Lowell papers, LOA.
45. Lowell to George Davidson, May 18, 1910; July 7, 1910; S. W. Burnham to Lowell, Oct. 8, 1910; Lowell to council of the academy, Oct. 31, 1910; Hall report, Nov. 11, 1910; Lowell to Davidson, April 29, 1911; Lowell to C. O. Lampland, Dec. 12, 1910; Lowell to Charles Cross, Oct. 7, 1913; Lowell to H. W. Tyler, Dec. 15, 1914, Lowell papers, LOA.
46. Lowell to Trowbridge, Dec. 9, 1914; Lowell to Burnham, undated; "Mars" to "Quadric," Jan. 4, 1915; Lowell to Trowbridge, Jan. 11, 1915; Lowell to Lampland, undated (context indicates Feb. 1915); Lowell to P. B. Wilson, Feb. 4, 1915, Lowell papers, LOA.
47. Lankford, *American Astronomy,* p. 266.
48. On Antoniadi, see William Sheehan, *Planets and Perception: Telescopic Views and Interpretations, 1609–1909* (Tucson: University of Arizona Press, 1988), pp. 236–242.

12. The Search for Recognition

1. Carl Lampland told Clyde Tombaugh that the Pluto project was undertaken by Lowell "to improve his credibility among other astronomers." Tombaugh, "The Discovery of Pluto: Some Generally Unknown Aspects of the Story," *Mercury: The Journal of the Astronomical Society of the Pacific* 15 (1986): 66.
2. On Slipher's relationship with Lowell, see Robert Smith: "Gold Medals and Red Shifts," in William Lowell Putnam, et al., *The Explorers of Mars Hill: A Centennial History of Lowell Observatory, 1894–1994* (West Kennebunk, Me.: Phoenix Publishing, 1994), pp. 43–65; and Smith, "New Uses for Old Instruments: V. M. Slipher and the Move from the Expansion of the Universe to the Light of the Night Sky," paper delivered at the third British–North American History of Science meeting, July 23–26, 1996, Edinburgh. See also David DeVorkin, "A Summer Place: Interludes at Lowell Observatory," a chapter in his forthcoming biography *Henry Norris Russell: Dean of American Astronomers.*
3. On Miller's status with the National Academy of Sciences, see John Lankford, *American Astronomy: Community, Careers, and Power* (Chicago: Chicago University Press, 1997), p. 253.
4. V. M. Slipher to John Miller, Oct. 18, Nov. 23, 1908, V. M. Slipher papers, LOA.
5. Miller to Slipher, Nov. 17, 1908, ibid.
6. Miller to Slipher, Jan. 2, 1909, ibid.

7. Miller to Slipher, Nov. 17, 1908, ibid.
8. Slipher to Campbell, Nov. 15, 1909, ibid.
9. Slipher to Lowell, May 7, 1913; Lowell to Slipher, Aug. 10, 1914, Lowell papers, LOA.
10. Slipher to Lowell, Aug. 5, 1914; Lowell to C. O. Lampland, April 10, 1910, Lowell papers; Slipher to Miller, Dec. 10, 1915, V. M. Slipher papers, LOA.
11. Miller to Slipher, June 9, 1913, April 8, 1916; see also Jan. 4, 1917, V. M. Slipher papers, LOA.
12. "On Martian Longitudes," *Astrophysical Journal* 1 (May 1895): 393–399; A. E. Douglass to William Lowell Putnam, March 12, 1911, Douglass papers, LOA. Slipher to Lowell, Dec. 29, 1908, Lowell papers, LOA. On Lowell's difficulties with the *Astrophysical Journal,* see Donald Osterbrock, *James Keeler: Pioneer American Astrophysicist and the Early Development of American Astrophysics* (Cambridge: Cambridge University Press, 1984), pp. 208–209. On the journal as an exercise in power, see Lankford, *American Astronomy,* pp. 227–235. See also V. M. Slipher, "Observations of Standard Velocity Stars," *Astrophysical Journal* 22 (Dec. 1905): 318–340. Louise Leonard to V. M. Slipher, Jan. 17, 1911, V. M. Slipher papers, LOA.
13. "The International Union for Co-operation in Solar Research," *Astrophysical Journal* 32 (Oct. 1910): 258–262. W. W. Campbell to E. C. Pickering, May 26, 1910, Campbell papers, MLSA; Lowell to Pickering, June 2, 1910, Lowell papers, LOA.
14. E. E. Barnard to G. E. Hale, Oct. 19, 1911, Hale papers, MWPOL. On the various problems that beset the comet committee, see John Lankford, "How Not to Run an International Halley Watch," *Sky and Telescope* 68 (Dec. 1985): 530–533.
15. Lowell to Slipher, Jan. 16, 1911; Slipher to Lampland, Jan. 27, 1911, V. M. Slipher papers; Slipher to Lowell, Feb. 2, 1911, Lowell papers, LOA.
16. Lowell to chairman of the comet committee, Jan. 18, 1911, Lowell papers; Slipher to Leonard, Feb. 3, 1911, V. M. Slipher papers, LOA.
17. Lowell's Clark University lecture was "cordially welcomed" (*Boston Evening Transcript,* Sept. 9, 1909, p. 2) and was published as "The Planet Venus," *Popular Science Monthly* 75 (Dec. 1909): 521–537. William Sturgis Bigelow to Henry Cabot Lodge, Feb. 7, 1916, in Akira Murakata, ed., "Selected Letters of Dr. William Sturgis Bigelow," Ph.D. diss., George Washington University, 1971, p. 406.
18. Slipher to Lowell, April 7, 1904, Lowell papers, LOA; Campbell to Slipher, June 6, 1911, March 30, 1914; Campbell to Slipher and Lampland, Dec. 6, 1916, V. M. Slipher papers, LOA; Hale to Frost, Feb. 15, 1915, Hale papers, MWPOL.
19. See William Graves Hoyt, *Planets X and Pluto* (Tucson: University of Arizona Press, 1980), pp. 75–81.
20. On Humason, see Gibson Reaves, "The Mount Wilson Search for a Trans-Nep-

tunian Planet, 1919–1920," paper delivered at the Astronomical Society of the Pacific meeting, Berkeley, Calif., June 23, 1989.

21. *The Solar System: Six Lectures Delivered at Massachusetts Institute of Technology in December, 1902* (Boston: Houghton Mifflin Co., 1903), pp. 17, 115.
22. See Lowell, *The Evolution of Worlds* (New York: Macmillan Co., 1909), pp. 122–126.
23. On Vulcan, see Richard Baum and William Sheehan, *In Search of Planet Vulcan: The Ghost in Newton's Clockwise Universe* (New York and London: Plenum Trade, 1997.)
24. Hoyt, *Planets X*, pp. 75–77, 85, 109.
25. Ibid., pp. 115, 130. See also C. O. Lampland, untitled, undated typewritten text, LOA (internal evidence indicates the text was written after 1930).
26. Lowell, "Memoir on a Trans-Neptunian Planet," *Memoirs of the Lowell Observatory* 1, no. 1 (1915): 103–104; see also Lowell, "Preliminary Investigation into the Position of a Supposed Trans-Neptunian Planet," typewritten text with handwritten corrections and signature, dated Jan. 12, 1909; untitled handwritten drafts and sixteen-page typewritten text with handwritten corrections and signature, ca. June 1909, LOA.
27. "Memoir on a Trans-Neptunian Planet," p. 104.
28. Not only were their orbits substantially different, but also Pluto's mass was approximately one-sixtieth of the mass predicted by Lowell for Planet X. A. L. Lowell, *The Biography of Percival Lowell* (New York: Macmillan Co., 1935), p. 192.
29. Hoyt, *Planets X*, pp. 98–99.
30. Lowell to Slipher, Jan. 23, 1907; Lowell to Lampland, April 10, 1910; Lowell to Slipher, March 2, 1913; Lowell to trustees of the Boston Public Library, Nov. 24, 1914; Lowell to H. W. Tyler, Dec. 15, 1914; Lowell to George Ritchey, June 13, 1911, Lowell papers, LOA. On Lampland's photographs of spiral nebulae, see Norriss S. Hetherington, "The Simultaneous 'Discovery' of Internal Motions in Spiral Nebulae," *Journal for the History of Astronomy* 6 (1975): 120–121.
31. Lowell to E. C. Slipher, July 13, 1915, Lowell papers, LOA. E. C. Slipher's article was entitled, "The Photograph in Astronomy," *American Museum Journal* 15 (1915): 211–219.

Conclusion

1. On Lowell and Flagstaff, see William Graves Hoyt, *Lowell and Mars* (Tucson: University of Arizona Press, 1976), and William Lowell Putnam et al., *The Explorers of Mars Hill: A Centennial History of Lowell Observatory, 1894–1994* (West Kennebunk, Me.: Phoenix Publishing, 1994). See also *An Address: Com-*

memoration Day, May 30, 1903, Flagstaff, Arizona (Flagstaff: C. M. Funston, 1903), p. 11; Platt Cline, *They Came to the Mountain: The Story of Flagstaff's Beginnings* (Flagstaff: Northern Arizona University, 1976); *Who's Who in Arizona,* vol. 1: *1913* (I. Connors, Press of the Arizona Daily Star, 1913); George F. Sparks, ed., *A Many Colored Toga: The Diary of Henry Fountain Ashurst* (Tucson: University of Arizona Press, 1962).

2. Lowell to Frederic Stimson, Jan. 25, 1916, A. L. Lowell papers, HUA. On the house, see Henry Giclas, "Lowell Observatory," typescript from the author, p. 35; Lowell to V. M. Slipher, Oct. 17, 1902, Nov. 17, 1902; on the tennis court, see Lowell to Slipher, Aug. 24, 1903, Lowell papers, LOA. On food and drink, see Lowell to H. Jevne, April 8, 1903, Oct. 12, 1909, Nov. 5, 1909, Lowell papers, LOA; A. E. Douglass to Charles Deuel, April 9, 1900, Douglass papers, LOA. On servants, see Leonard to Slipher, March 5, 1905, Jan. 24, 1907, March 5, 1907, March 27, 1909, Nov. 21, 1913, V. M. Slipher papers, LOA.
3. "Lowell and Doe to Moqui," *Coconino Sun,* Aug. 24, 1901, p. 1; "Doe, Wife, Lowell, Leonard, Morse to Canyon," ibid., June 3, 1905, p. 1; Lowell to Edward Doe, April 29, May 13, 1902; C. O. Lampland to Slipher, Sept. 26, 1907; Lowell to Louise Leonard, Sept. 28, 1914; Lowell to Doe, Feb. 25, 1913, Jan. 7, 1908, Lowell papers, LOA.
4. Lowell to Stimson, Nov. 15, 1915, A. L. Lowell papers, HUA. For Sargent's appraisal of Lowell's botanical prowess, see A. L. Lowell, *The Biography of Percival Lowell* (New York: Macmillan Co., 1935), pp. 105–106; Charles Sprague Sargent to Lowell, Oct. 31, 1910; Lowell to Sargent, Jan. 18, 1914, Sargent papers, Arnold Arboretum Archives, Boston. Lowell to Lampland, Nov. 10, 1915, Lowell papers; Lampland to Sargent, Dec. 21, 1916, Lampland papers, LOA. Louise Leonard, *Percival Lowell: An Afterglow* (Boston: Richard G. Badger, 1921), p. 159. Lowell's botanical research merits further study by a specialist in the field.
5. On Lowell's Boston-centered view, see Lowell to Lampland, Nov. 10, 1915, Lowell papers, LOA; Leonard, *Percival Lowell,* pp. 126–127; Lowell to Stimson, Nov. 15, 1915, A. L. Lowell papers, HUA; "The Boston Athenaeum," *Boston Evening Transcript,* March 7, 1903. Lowell to Amy Lowell, March 1, 1903, Amy Lowell papers, HL.
6. Lowell, "Two Stars," typescript, copy in LOA, p. 3; Lowell, "Untitled," manuscript, copy in LOA.
7. On Lowell's contributions, see Lowell to Leo Verkamp, May 23, 1905, June 26, 1906; Verkamp to Lowell, May 10, 24, 1907; Lowell to Leonard, April 5, 1916; T. A. Riordan to Lowell, April 27, 1907, Lowell papers, LOA. *Coconino Sun,* July 6, 1901, p. 1; May 30, 1903, p. 1; July 4, 1903, p. 1; May 23, 1907, p. 1; May 30, 1907, p. 1; Aug. 22, 1907, p. 1; July 20, 1905, p. 1; Dec. 22, 1911, p. 1. On Lowell and the *Coconino Sun,* see Lowell to Lampland, June 16, 1904; Lowell to Slipher, Nov. 3, 1905; Lowell to Doe, Jan. 2, 16, 30, 1906, Lowell papers, LOA.

8. On the opera house, see *Coconino Sun,* Feb. 14, 1903, p. 1. On the hotel project, see Lowell to Doe, Oct. 6, 1905; Lowell to A. T. Cornish, Oct. 6, 1905; Lowell to E. P. Ripley, Dec. 28, 1905, Lowell papers, LOA; *Coconino Sun,* June 10, 1906, p. 8; June 17, 1906, p. 1.
9. For Lowell's anti-Progressive views, see "The Portent of Socialism," typescript copy of speech delivered to the Victorian Club, Dec. 2, 1910, Lowell papers, LOA. On Prohibition, see Lowell to Doe, Dec. 18, 1914; Lowell to M. E. Fellows, Nov. 19, 1914, Lowell papers, LOA. Slipher to Leonard, Dec. 14, 1914, V. M. Slipher papers, LOA.
10. See Lowell, "Immigration versus the United States" (Lynn, Mass.: Thomas P. Nichols and Son, 1916), pp. 2, 12; *Coconino Sun,* Oct. 15, 1915, pp. 1–2. Lowell to Henry Cabot Lodge, Oct. 22, 1901; Lowell to Boies Penrose, Dec. 30, 1902; Lowell to Doe, Nov. 28, 1905; Lowell to Joseph Lee, March 24, 1909; Lowell to C. E. Hughes, June 30, 1916, Lowell papers, LOA. *Coconino Sun,* Feb. 10, 1916, p. 2; Feb. 18, 1916, pp. 1–2, Feb. 25, pp. 1–2; April 14, 1916, p. 6; April 28, 1916, p. 1; May 5, 1916, p. 6. *Boston Herald,* May 6, 1916. See also William C. Widenor, *Henry Cabot Lodge and the Search for an American Foreign Policy* (Berkeley: University of California Press, 1980), pp. 57–61, and Barbara Miller Solomon, *Ancestors and Immigrants: A Changing New England Tradition* (Cambridge, Mass.: Harvard University Press, 1956). On unity, see "Two Stars," pp. 11–12. On Lowell's campaign for the Senate, see Lowell to Francis Crable, May 2, 9, 10, 26, June 5, 16, 1916; Lowell to Murray Crane, June 1, 12, 16, 1916; Crable to Lowell, June 14, 1916. Lowell deposited $1,000 in Crable's bank for campaign expenses, April 22, 1916, Lowell papers, LOA.
11. Ralph Curtis to Stimson, Sept. 24, 1915, Stimson papers, MHS; Lowell to Hall Caine, Nov. 19, 1914; Lowell to American Society of Aeronautic Engineers, Aug. 20, 1915; Lowell to Ellsworth Huntington, Nov. 8, 1915, Lowell papers, LOA. Curtis to Barrett Wendell, Jan. 13, 1915, Wendell papers, HL. Lowell to Basil Hall Chamberlain, Jan. 26, 1916, in the author's possession.
12. On trust arrangements, see Putnam et al., *Explorers of Mars Hill,* pp. 90–91.
13. *Coconino Sun,* Nov. 17, 1916, pp. 1, 4; Nov. 24, 1916, p. 1.
14. Slipher to Wilbur Cogshall, Dec. 27, 1916, V. M. Slipher papers, LOA.
15. Stimson to Wendell, undated (context indicates early Jan. 1917), Wendell papers, HL.
16. Lowell, *The Soul of the Far East* (Boston: Houghton Mifflin Co., 1888), pp. 21–22.

Index